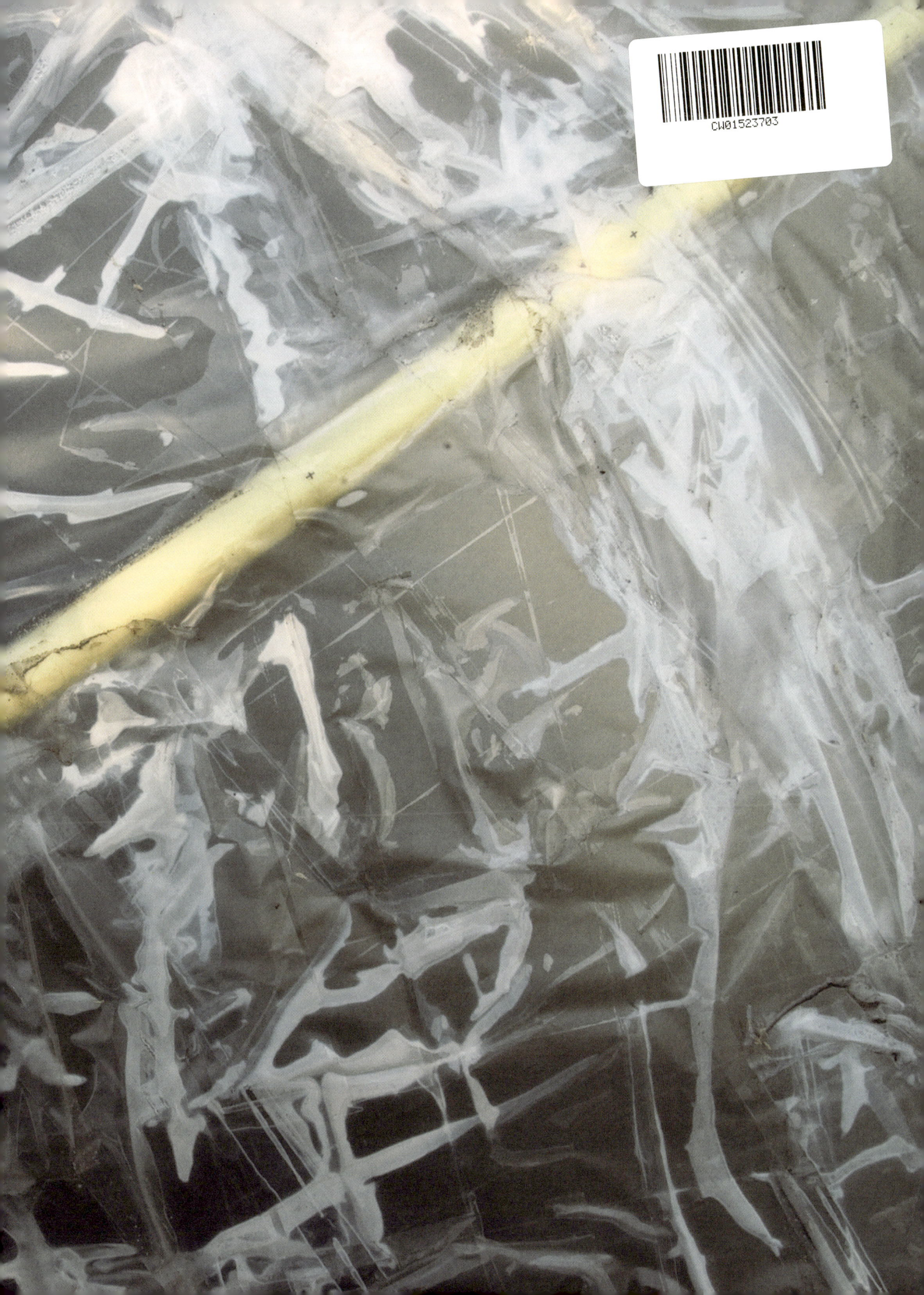

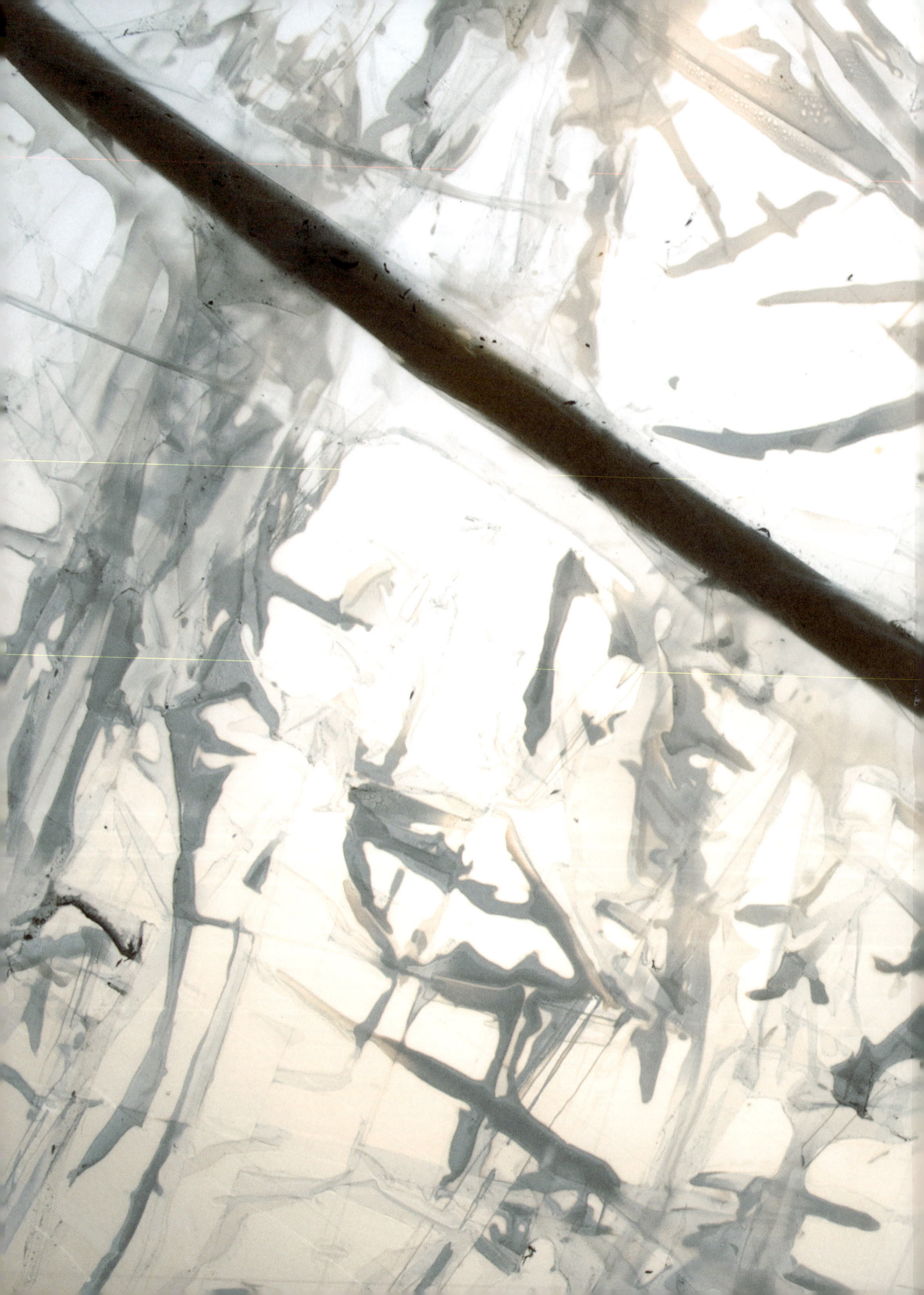

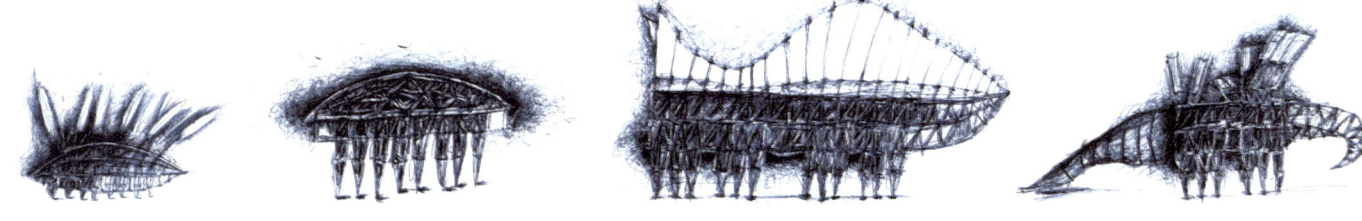

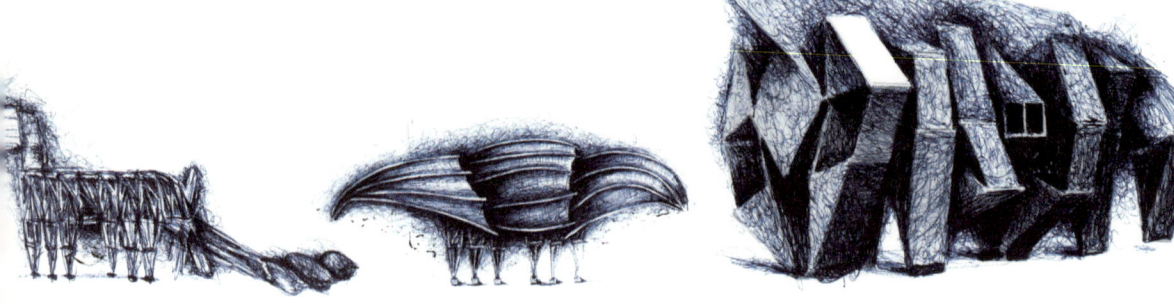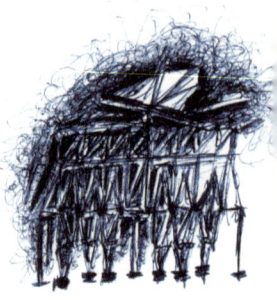

Theo Jansen

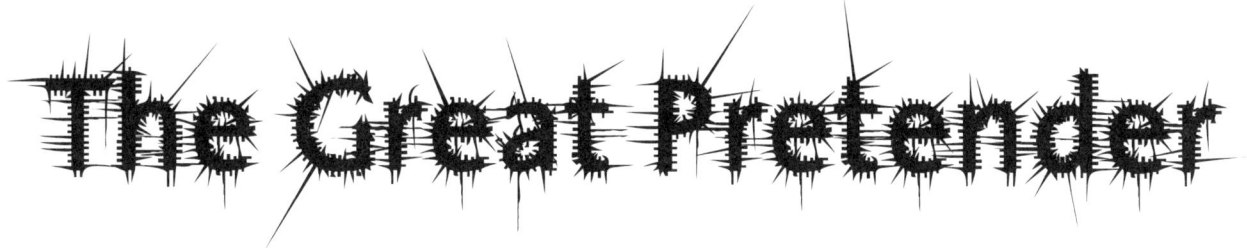

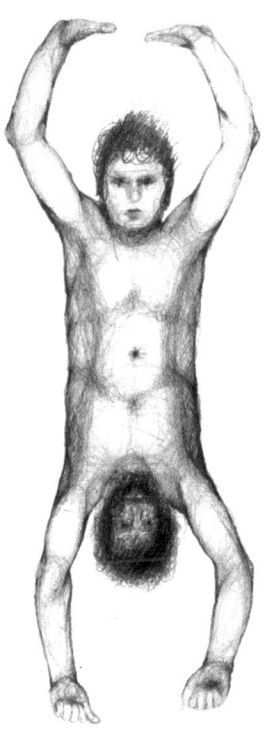

Second edition

010 Publishers, Rotterdam 2009

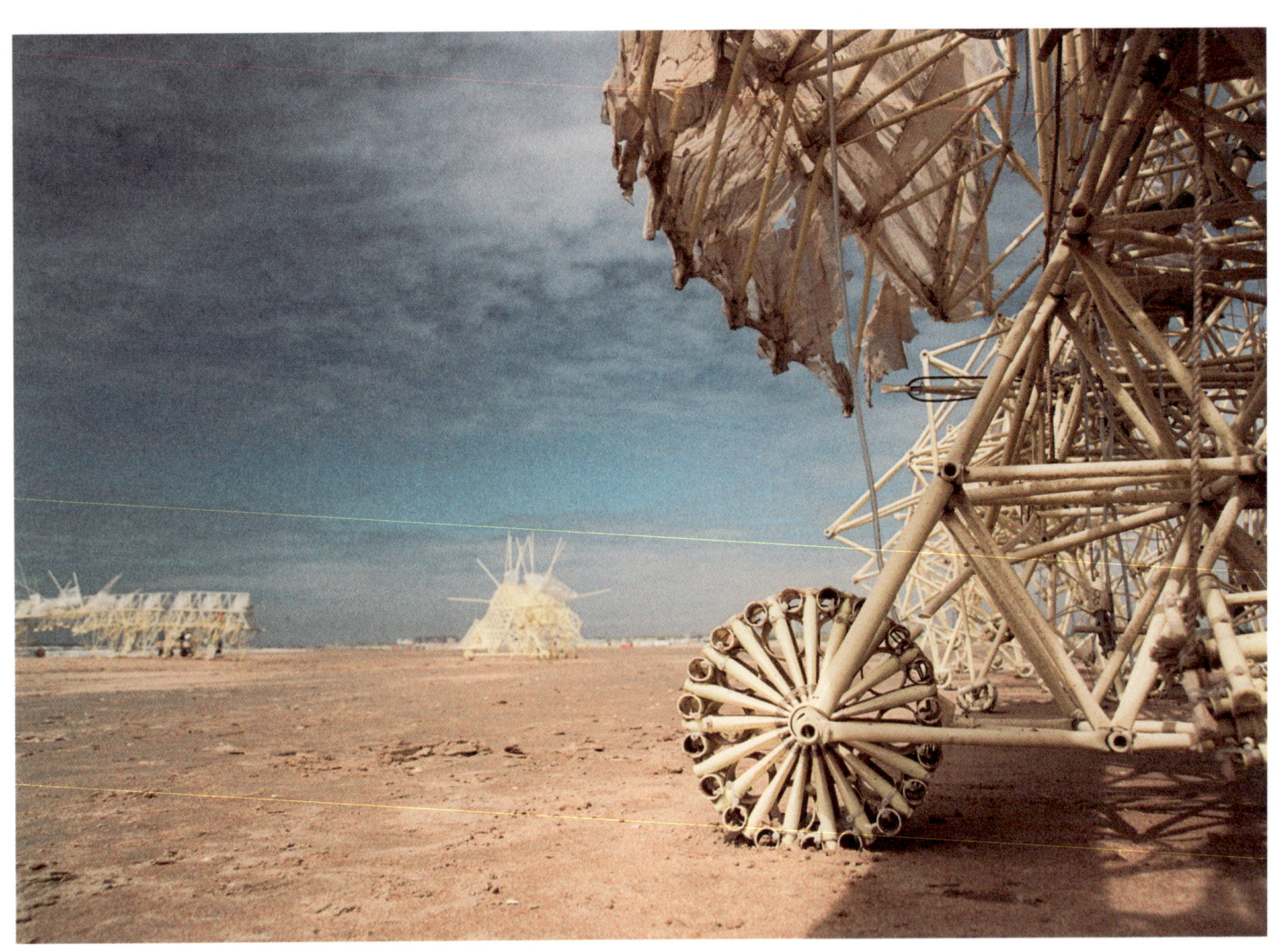

Foreword

The right-hand pages of this book present a continuous narrative. In seven chapters I describe the seven periods comprising the development of the beach animals to date. I also look beyond beach animal history in an attempt to draw some universal conclusions.

The left-hand pages bear no relation to each other and only sporadically to those on the right. They are there to add a lighter touch, a moment of respite from the rigours of the narrative. Consisting of photos and short passages of text, they give a look-in at my work from outside, unlike the pages opposite which describe the fantasy world in my mind. The majority of texts on the left consist of pieces published earlier in the Dutch daily paper *de Volkskrant* and translated for this book.

All compilation work was carried out by Johannes Niemeijer, to whom I owe a special debt of gratitude. In fact he was instrumental in getting me to write the book in the first place. Thanks go also to all those who assisted me in assembling the beach animals and generally lugging them around.

DVD with beach animal videos at the back of the book

Left half

Beach animals
10, 16-17, 20, 26-27, 38-39, 48-49, 60-61, 66, 70, 72, 86-88, 90, 98-99, 106-107, 110-111, 116, 120, 126-127, 138-139, 146, 150-152, 164-165, 174-175, 186-187, 200, 206, 214-215, 216, 220, 226-228

Columns
Cloud no. 57 22
Dream-gazing 30
Sculpting 44
Putting the alarm clock forward 23 million years 52
Machete 62
Floating 80
Coupe de grace 102
Let go of your ego 114
Football 128
Seam 142
Out among the clouds, what a treat that was 156
Urrggghh! 170
Pitch 184
Time map 198
Impossible! 210
A tinful of screws from heaven 224
Shafts of air 234

Fossils
8, 12, 14, 18, 24, 34, 40, 50, 54, 58, 68, 78, 84, 92, 104, 108, 124, 132, 136, 144, 154, 158, 162, 166, 172, 176, 182, 188, 192, 194, 196, 202, 212, 218, 222, 234

Moulds
32, 36, 64, 74, 96, 100, 112, 118, 130, 134, 140, 148, 168, 178

Miscellaneous
28, 42, 56, 76 (Painting-machine) 82, 95, 122, 160, 180, 190 (Lissajous), 204, 208, 230

Characteristics of the species
236

Family tree
238

Credits
240

Right half

I Pregluton – the period before the Gluton (before 1990) 9
Memory in reverse. Animaris Lineamentum
[the pleasures of reproduction, stick insects, computer worms, quadrupeds]

II Gluton – the tape period (1990-1991) 29
Plastic tubing. Looking back. Animaris Vulgaris
[Three-part leg, adhesive tape, cable ties, the capacity of being able, restrictions, engineers, forgetting existing nature]

III Chorda – the strap period (1991-1993) 45
Evolutionary method. Twelve holy numbers. Legs. The benefits of speed
[Joints, action of the two-part leg, twelve holy numbers, the Bambi principle, leg thicknesses, triangles, the benefits of locomotion, rerunning old programs]

IV Calidum – the hot period (1993-1994) 65
Heat gun. Sand. Animaris Speculator
[Animaris Currens Ventosa, Glutonia sand mutation, regular sand mutation, digging sand, reversing direction, Animaris Speculator]

V Tepideem – the less-hot period (1994-1997) 89
Animaris Ancora. Animari Geneticae. Genes of electrical tubing. Pregnancy and death. The form of things. Symmetry in the human body. Cells. Symmetry in Animaris Geneticus. The code of things
[propellers, Animari Geneticae, reproduction, information flow, plugs, Animaris Geneticus, pregnancy, corpses, plastic genes, coded world]

VI Lignatum – the wood period (1997-2001) 135
Pallets. Transport. Walking
[sandwich, reinventing the wheel, Animaris Rhinoceros Transport]

VII Vaporum – the pneumatic period (2001-2006) 149
Muscles. Universal anatomy. Self-locomotion. Vermiculi. Sealant and O rings. New nerve cells. Automatic reproduction. Self-regulating walking races. Self-operating plugs

VIII Cerebrum – the brains period (2006-present) 195
Pedometer. Soap, DNA and SM. In and out. Between one shell and another. Invention of the 'I'. The Great Pretender

Biography of Theo Jansen 239

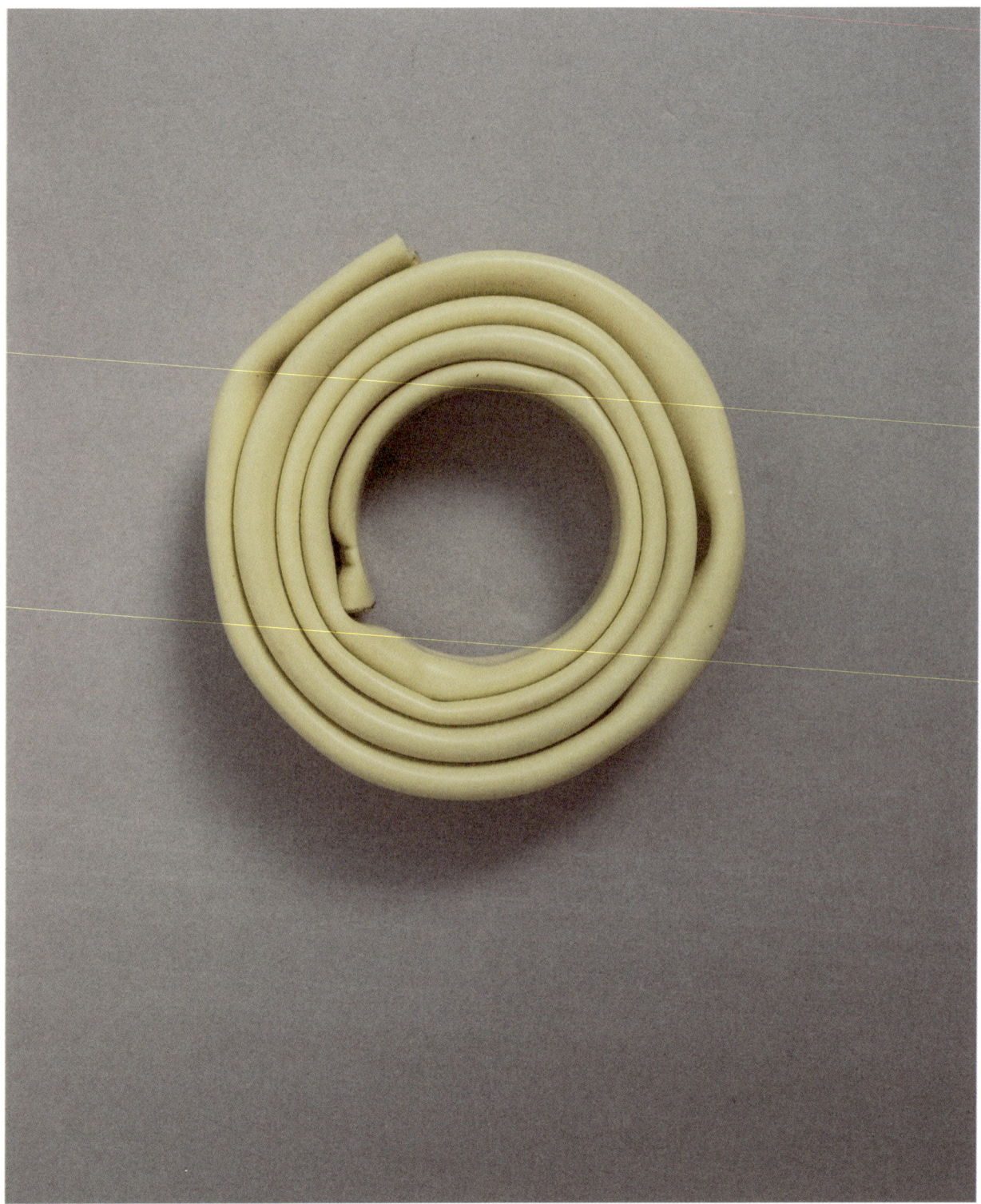

coil wheel

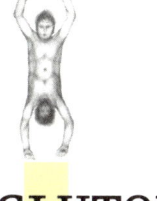

PREGLUTON
the period before the Gluton (before 1990)

Memory in reverse

In the Pregluton, there were no animals of plastic tubing as yet. This initial period was marked by an absence of matter. Life then consisted exclusively of dreaming about life. I made virtual animals in the computer, life-forms subject to an artificial evolution. I also went walking a lot on the beach and in the streets of Scheveningen for inspiration for my columns. Since 1986 I have been writing pieces for *de Volkskrant* national daily on matters that interest me. Often these are technical things, fantasies or musings. That one of these columns would ultimately define the rest of my life was something I had no idea of at the time. The beach animals began as a newspaper item.

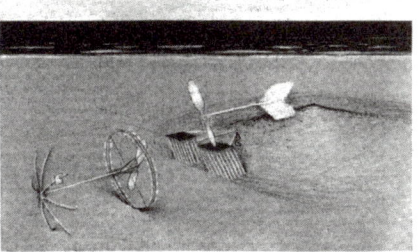

**column in *de Volkskrant*
24-02-1990
(translation on p. 240)**

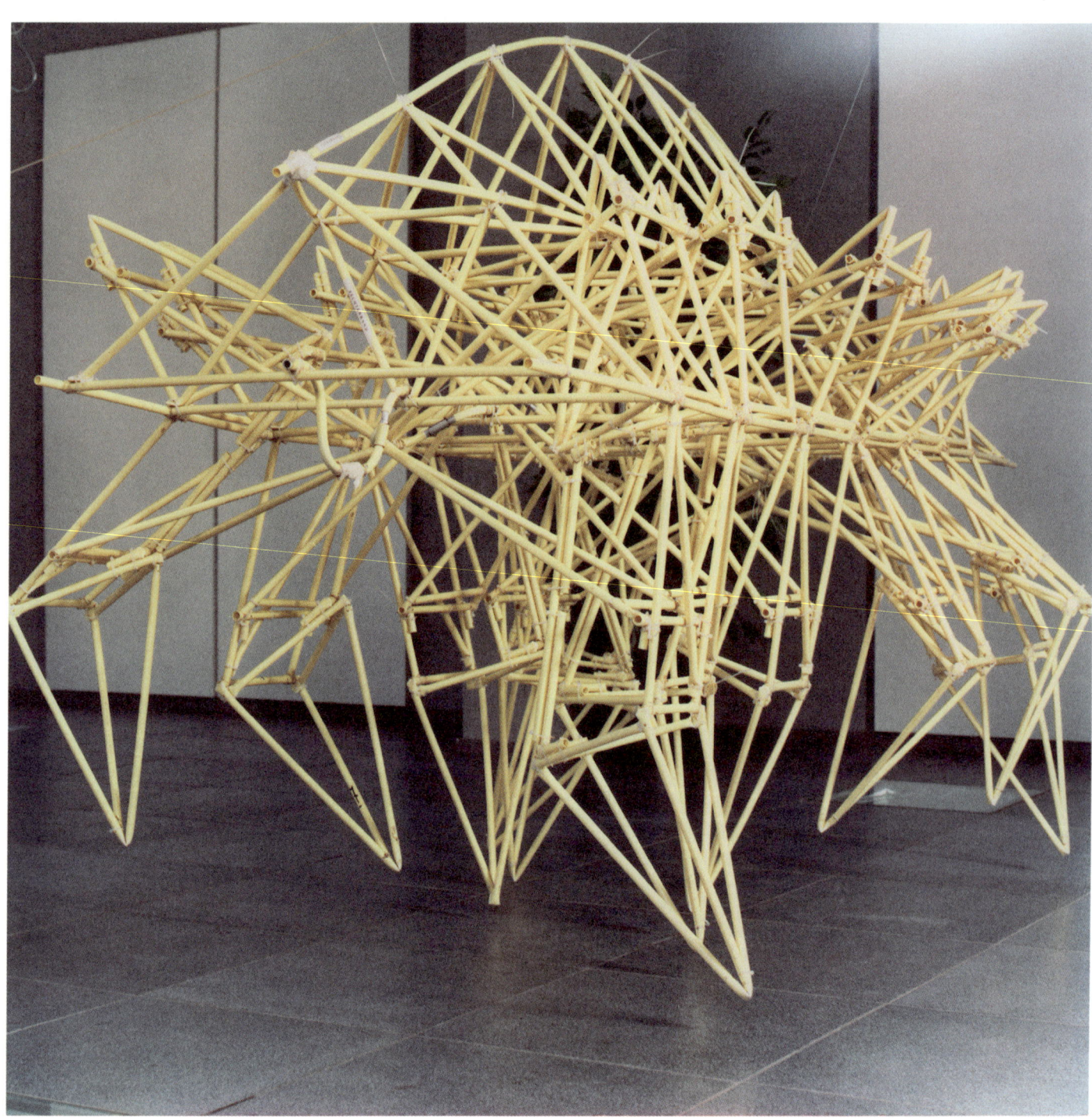
Animaris Currens. The first beach animal to walk

Nothing happened for a full six months following publication. Then, one warm September day, I went to the Gamma DIY store to buy a few of those tubes. I played around with them for a bit. It seemed you could do all sorts of things with them. That same day I decided to devote a year of my life to the tubes. That was seventeen years ago.

Scheveningen is a small seaside resort that comes under The Hague. It's where I spent my youth. It seems as though reliving my earliest years gave me an open mind. I looked at the world as if I were a child again; as though I were observing the wonders of creation for the first time. 'Resetting' was my word for it. Just as you reset a computer to start from scratch, so I 'reset' my brains by indulging in long walks taking in places from my childhood.

During my early years I must have conducted myself like an extraterrestrial visitor disguised as an earthling, as inconspicuous as a tree in a forest. I've discovered this since re-encountering people I used to know. Schoolfellows, local boys – they simply fail to recognize me. I can describe myself until I'm blue in the face, but still I draw a blank with them. This happens to me a little too often. Such encounters take place in the tram, in Scheveningen, in Amsterdam, everywhere. Whereupon I slink off in shame. I'm ashamed at my incapacity for forgetting. I am fated with an excellent memory. Details of earlier times are imprinted in my head, the stair rods at our house, the wrinkles on the forehead of our physics teacher, Mr Koesmans, the nicotine-stained teeth of my Dutch language teacher, Mr Bulsing, the faces of classmates and more than anything else the taste of school milk, which is different to that of ordinary milk. A small bottle with a silver top. I just can't forget things. Worse than that, my memories keep waylaying me. They flit across my mind at odd moments.

One's childhood years should provide a stock of memories, memories on which you can draw to create new memories. That stock is complete at the end of your life. And then you die. It would all seem to be for nothing. But of course this isn't so. There is such a thing as a memory in reverse. Something that enters the world from the mind instead of the other way round. Memories in reverse are born in the brain as an idea. Well-formed memories in reverse have the capacity to be remembered, even if the body has died. The ultimate example of a memory in reverse is Einstein's Theory of Relativity, which came into being in Einstein's grey matter in 1905 to then be disseminated all over the world. You might also call it a departing memory. It has settled in books and in all the heads on all

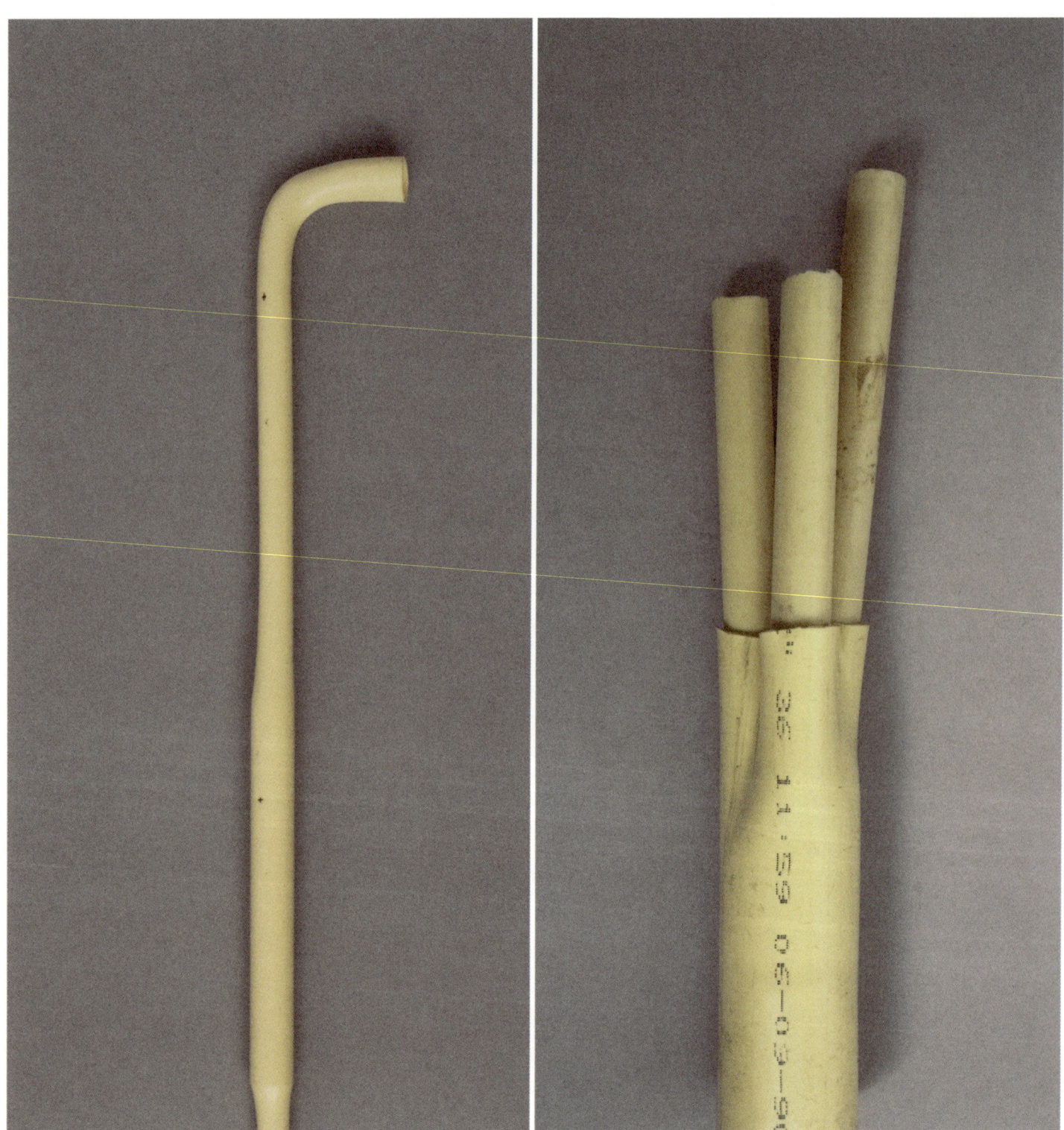

Animaris Excelsus, thumb joint of drive muscle, tri-split joint

those bodies. Despite the fact that Einstein is dead, his departed memory will live on.

Besides the stock of memories, there is also the stock of relics. Objects you lug along with you every time you move house: important letters, keepsakes, embroidery. This stock too is complete at the end of your life. As a rule it ends up on the rubbish heap. But sometimes these objects have the capacity to reproduce. Aeroplanes, for example. Whenever I drive past Amsterdam Airport and see the big planes roaring into the sky, my thoughts inevitably turn to the Wright brothers. I pretend they are on the back seat of my car and that I, representative of the present day, am showing them what their work has led to, in the hope they would be impressed. Granted, their tinkering on the beach at Kitty Hawk produced a rinky-dink affair but that plane has certainly has its fair share of offspring.

Memories in reverse have the ability to be remembered far and wide after death. Nor is it just famous and successful scientists and inventors who try to attain immortality; we all make our own attempts to achieve just that. Most people reproduce. Many of these will be unaware of why they do it. So to them I say: you're doing it to be remembered after death. They're laying on a stock of children. Children not only inherit genes, they are pre-programmed by their parents. Parents don't just leave behind a genetic trail but also a way of thinking and a way of acting.

The beach animals will be my brainchildren, my memories in reverse. Just like real children, they will be patronized, mollycoddled, cared for and trained to withstand the perils of the beach. There comes a time when they get shown the door. Off to the beach with you! Then they must fend for themselves. Once that happens, I can breathe my last with a light heart, knowing for certain they will get by.

In the Pregluton I fell under the spell of a book. This was *The Blind Watchmaker*, written by Richard Dawkins, a British zoologist and a compelling author to boot. Dawkins cites a number of striking examples from evolution. One of these was the stick insect.

Millions of years ago, the stick insect was just an insect that lived in trees. It didn't look at all like a stick. And like all other insects, this pre-stick was eaten by birds. Yet at that stage of evolution, birds still had fairly poor eyesight. They had difficulty observing the creepy-crawlies among the branches. These days,

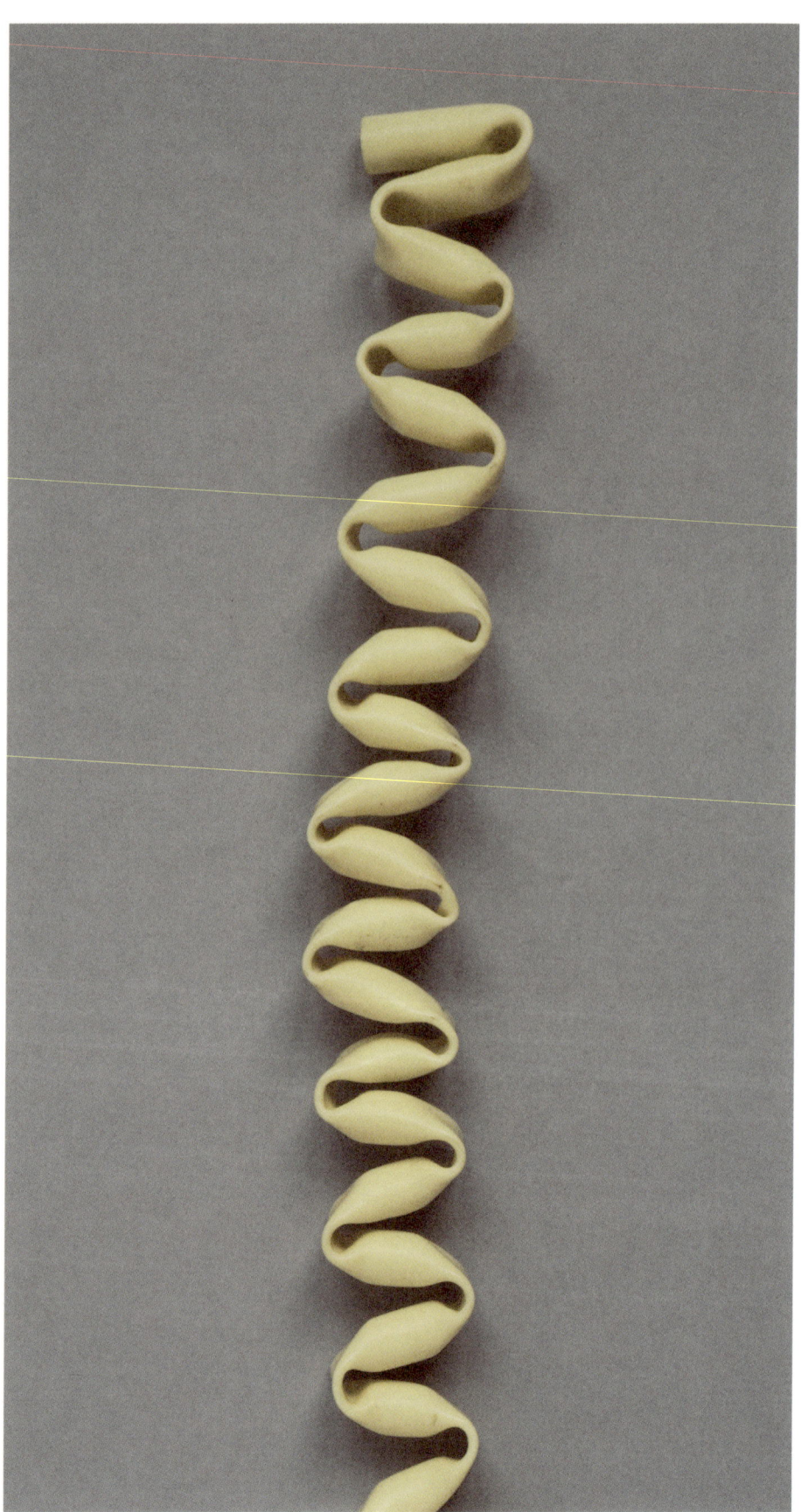

plastic spring

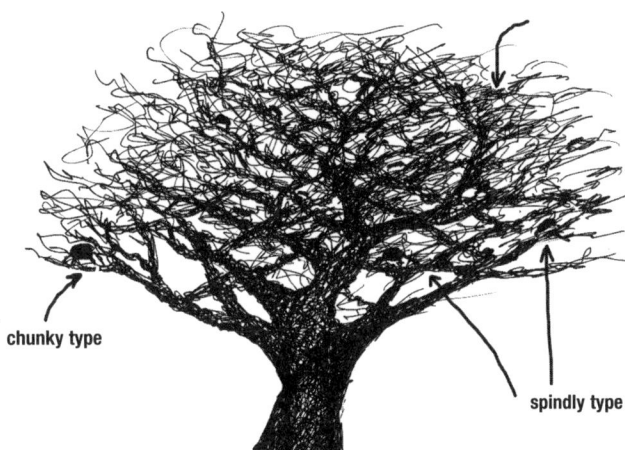

insects of a single species are never entirely identical. The same holds for us people. You have fat people and thin people and every nose is different. To us, insects of the same species all look the same yet they are individuals. The whim of nature guarantees variation. Spindly and chunky types of stick insect must have populated the trees too.

Imagine being a bird with poor eyesight, and it's immediately clear which examples of the insect world of yore would be the prime candidates for a meal: the chunky ones. The more spindly variety probably shared the same fate, only it took longer to find them. They had more time to reproduce. The upshot was that more spindly types were born. There was even a likelihood that in time an entire tree was commandeered by the spindly dynasty. Yet even in that situation variety reared its head. There were *chunky spindly* types and others you could describe as *spindly spindly*. The outcome is clear enough; the spindly spindly types produced more offspring. Over millions of years the insect bodies become more and more elongated to became the stick insect we know today. That's the way evolution works. Today's stick insect even has patterns resembling the leaf wounds found on real twigs. It may well be that scars such as hereditary pimples and knobbles gradually took on the appearance of real leaf scars as a result of selective feeding behaviour among birds. On the other hand, as time went by, the birds' eyesight improved. Each species had created the other, in a manner of speaking. The birds moulded the insects into a wondrous stick on legs and the insects are responsible for the birds' perfect vision. An absolute miracle and, at the same time, perfectly logical.

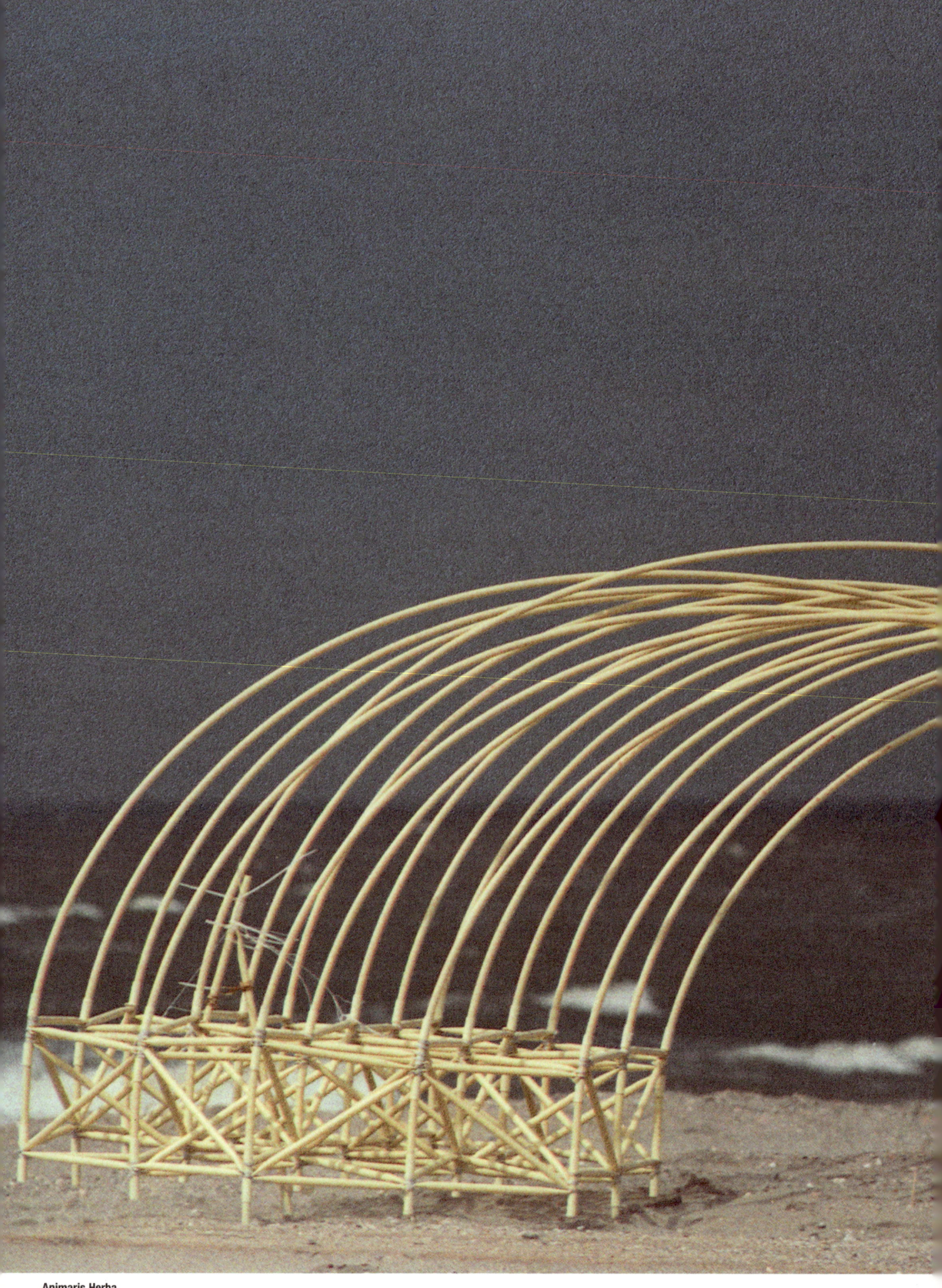
Animaris Herba

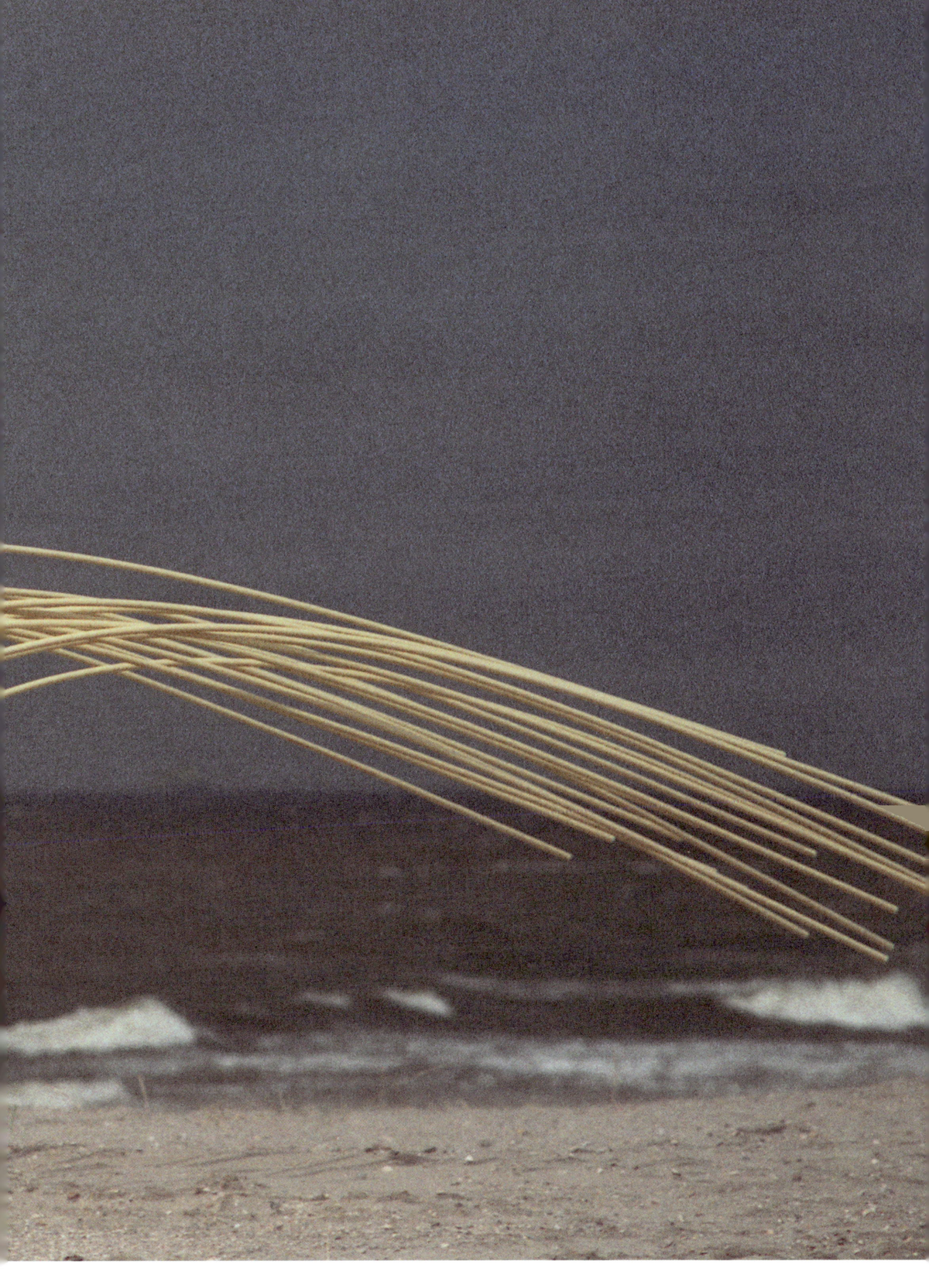

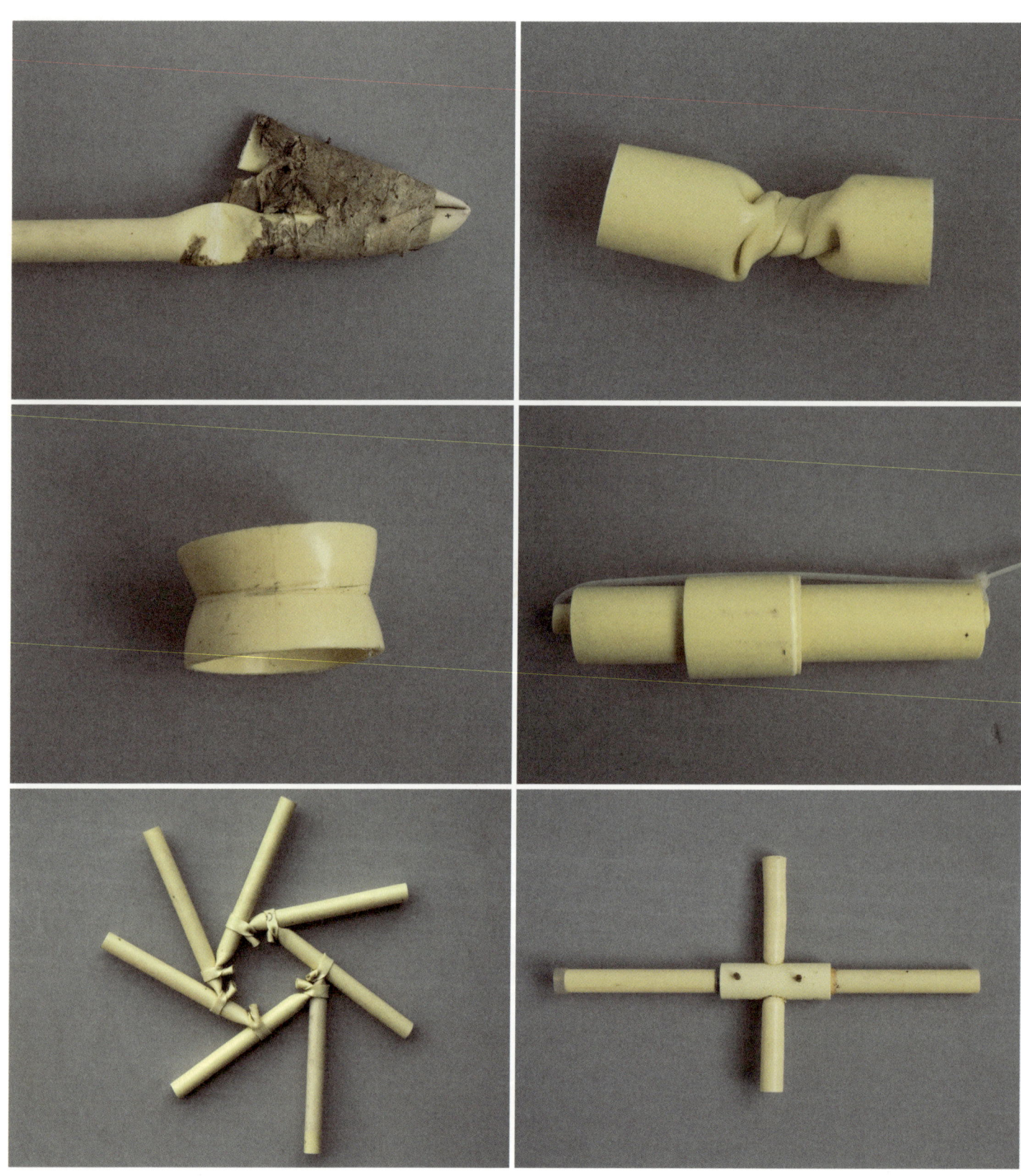

barbed hook leg of Animaris Sabulosa, rope end
rotating vertebra, precursor of foot of Animaris Rigide
example of a tube-gathering joint, on/off stick

Animaris Lineamentum

I so wanted to observe the phenomenon of evolution with my own eyes. Which is why I constructed a computer creature. It was just a line (Lineamentum) that shifted across the computer screen. Whenever it left the screen at the top it appeared again at the bottom edge. The same went for left and right. If it moved off screen at the left it immediately re-entered stage right. So the creature was imprisoned in the universe of the computer screen. Selection took place here too, just as with the stick insect. Only this time not birds but the creatures themselves did the selecting. At the tip of the travelling line was a sharp (virtual) point. If this came into contact with the flank of another such line-creature it gave it a fatal sting. As a result the population on screen gradually dwindled. Once the lines had shifted 60 times, spring arrived in line country and it was time to reproduce.

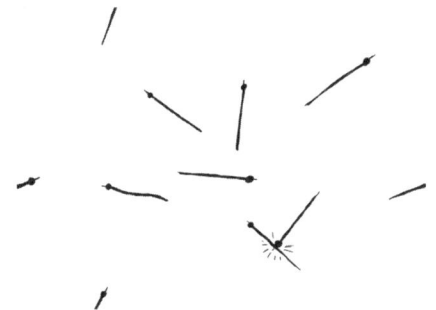

stung in the flank

I must say something about the anatomy of this creature. It consisted of just four parts, four segments. In the drawing below all the segments are straight.

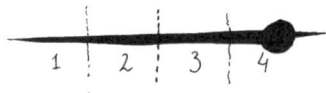

anatomy of Animaris Lineamentum

But they can also be bent: either one way or the other.

straight and bent

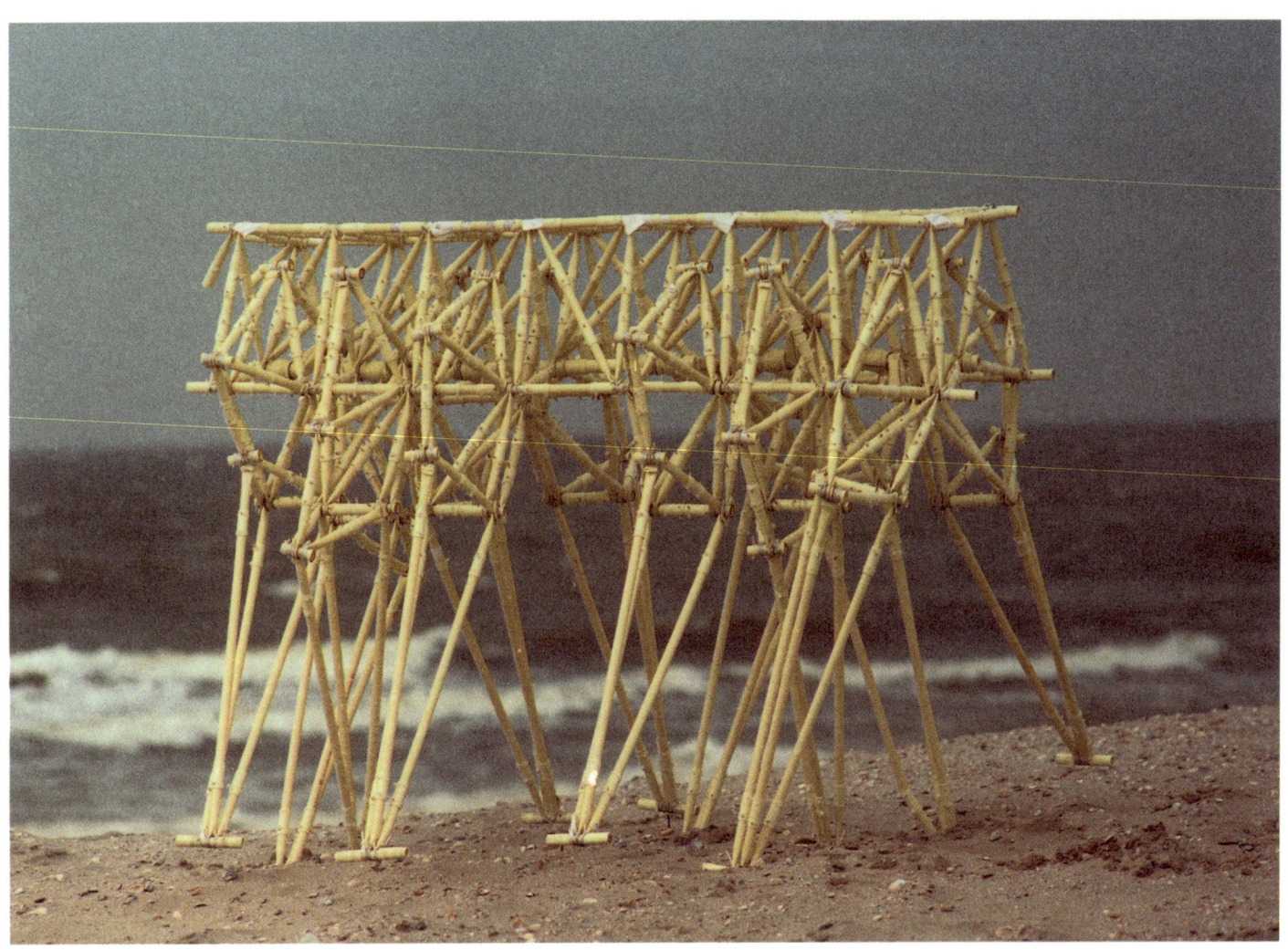

mutant of Animaris Geneticus: Animaris Propagare

Using the bent and straight body parts you can create quite a variety of forms.

different combinations

Reproduction was achieved by copying the body parts (segments) of the remaining creatures and reassembling them until the screen was chock-full of lines on the move. Then the business of selecting began all over again.
Now there were times when body parts were copied incorrectly. There was a likelihood of a deviant type, a mutant, turning up in a particular generation. I had made provision for this by including an element of chance in the computer program. A deviation such as this could be favourable or unfavourable.

Suppose that a population begins as straight lines, then a favourable deviation would be, say, a curl at the tail end. This curl would slightly shorten the creature and make it marginally less likely to get stung. Its descendants would increase in the fullness of time. The screen might even end up jam-packed with this mutant.

a curl in the tail

Were this to be the case, a mutant with a double curl might well occur at some point. A double curl is slightly more favourable again. What it all boils down to is that our line is undergoing an evolution. It begins as a straight line and rolls itself up as protection against the sharp points round about. This evolutionary process lasts roughly one night. You turn your computer on in the evening and in the morning there they all are rolled up on the screen. If you study a print-out of the different generations, you can see sudden swings. All at once the image changes, as if the evolution suddenly surged – something that really did happen in biological evolution.

Cloud no. 57

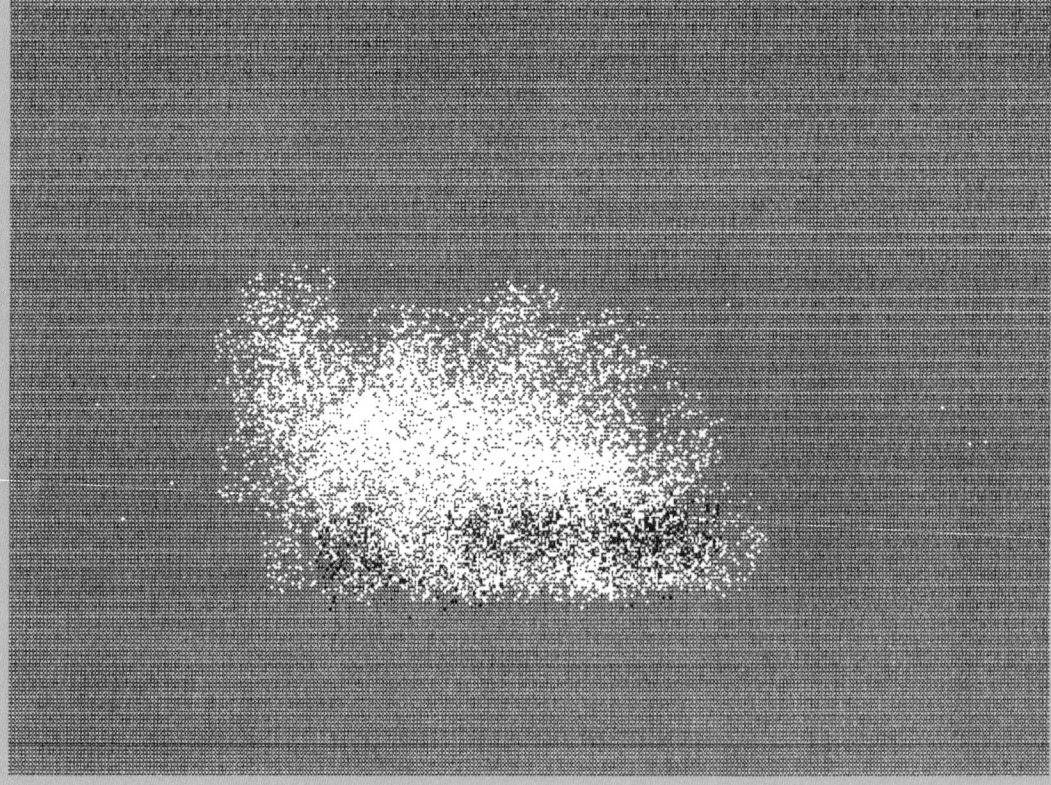

For dancing lessons we had a dancing teacher. Our dancing teacher taught us the foxtrot. He had an adam's apple so large that the word foxtrot took on an entirely different meaning for me. '1,...1, 2, 3, ...1, ...1, 2, 3, ...1, ...1, 2, 3,...' he used to shout. Merely to master this rhythm is asking a lot of a Dutchman. The ladies had their toes trampled mercilessly at times. The dancing teacher found a solution to this. He sang 'See the mooooooooon shine through the treeeeees ...', in Dutch, through a microphone. This most familiar of Santa Claus songs proved to fit the rhythm of the foxtrot exactly. From that time on, I could dance the foxtrot and didn't hesitate to ask the ladies to dance whenever the occasion presented itself. Once you can dance the foxtrot, you can dance 'em all.

For drawing lessons we had a drawing teacher. We never drew, however; we painted with powder paints.

I liked painting clouds best. I used to look outside and try to paint a particular cloud exactly the way it was. It never worked, try as I might; they looked too stiff, too *painted*. The drawing teacher insisted that it was only possible to paint nature if the painter could feel its essence. This struck me as a trifle exaggerated; a camera can paint clouds and there's no camera on earth that can feel the essence of nature. So what's the difference between a camera and a painter?

I finally found out when I exchanged powder paints for a brain-machine. 'Draw a cloud', I asked the brain-machine. 'OK', it replied, but left it at that. It didn't know what a cloud was. But before I could pass that information on to it, I had to find out myself.

Clouds are formed when rising water vapour reaches a certain height where it cools off and condenses. As there is no bathroom mirror up there, the condensed droplets drift around aimlessly. If you replace the word aimlessly with the word random and the word droplets with the word pixels (dots on a computer monitor), the brain-machine *will* know what to do. It draws the cloud the way it comes into being: warm pixels rise up and become visible above the stratum of condensed droplets.

Once my brain-machine had grasped the principle of the cloud, there was no stopping it. One cloud after another saw the light of day. One night I just let the brain-machine get on with it. I lay down on my bed with the blissful feeling that something was at work for me in the living room. That night I dreamed about the gurgling adam's apple of my dancing teacher. Not a Santa Claus song this time, but the screech of a matrix printer rose from his throat.

In the morning I found a hundred clouds on the living room floor. All of them different. My favourite is cloud no. 57 and it's reproduced here. Once you can draw one cloud, you can draw 'em all, it seems. Who drew the cloud? Who should sign the drawing, the brain-machine, the printer or I?

I didn't draw the cloud. I might have dreamt it though.

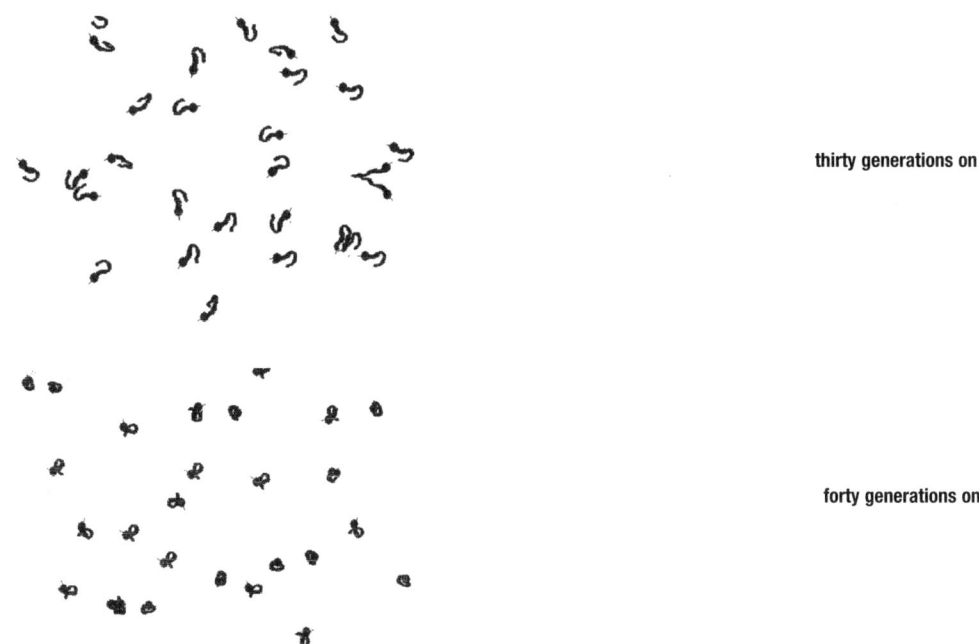

thirty generations on

forty generations on

The strength of the line creature lay in its simplicity. It had only four body parts (genes) and proved capable during the course of one night to adopt a smart form and more or less avoid the spines around it. It isn't difficult to see rolling up as an intelligent act. Just imagine it was you living in an environment bristling with spines; like the line creature you'd make yourself as small as possible and curl up (out of fear).

I also came up with the idea for a walking creature in the computer. This square beast (Quadrupes) could stretch and bend its legs and move them back and forth. Each leg was able to thrash in a repeating combination of four movements. Two hundred such creatures whose legs made this thrashing movement were created in the computer. A subroutine had been built into the program, one that could assess the effect of this thrashing on the creatures' bodies. The routine calculated the speed at which the creatures moved.

And now for the selection: the more rapid creatures earned the privilege to reproduce. The inherited thrashing movements were copied and, mixed with mutations, distributed at random to a subsequent generation once more of 200 creatures. As time went by, the random thrashing developed into walking and, five generations on, Quadrupes even broke into a gallop on occasion.

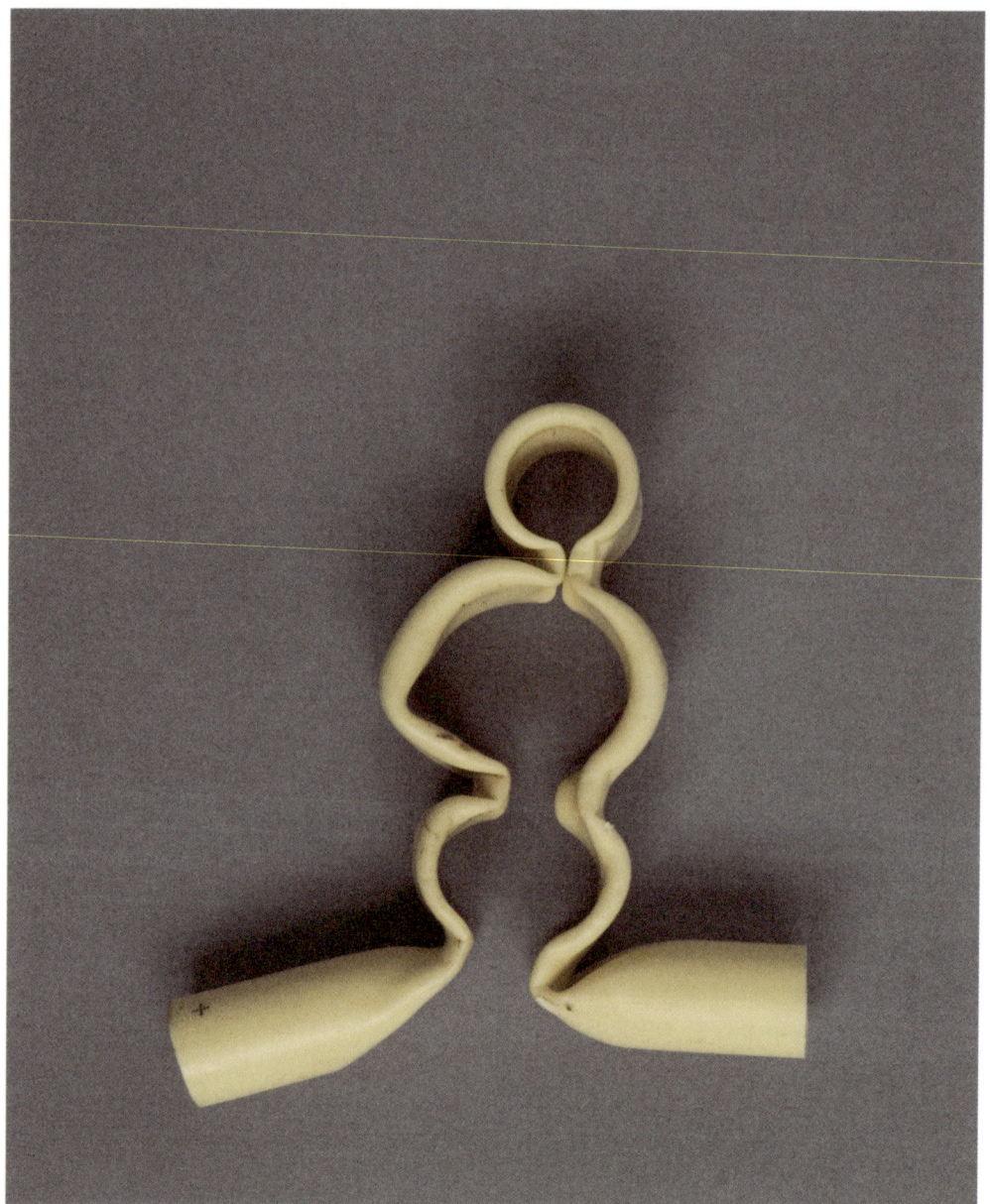
Animaris Currens Vaporis, muscle bundler

Walking was not only the passion of Quadrupes. It became my passion too. During one of my wanderings along the beach at Scheveningen, it dawned on me that I should pay a visit to the Gamma store. It was time to buy a length of plastic tube. This was in 1990.

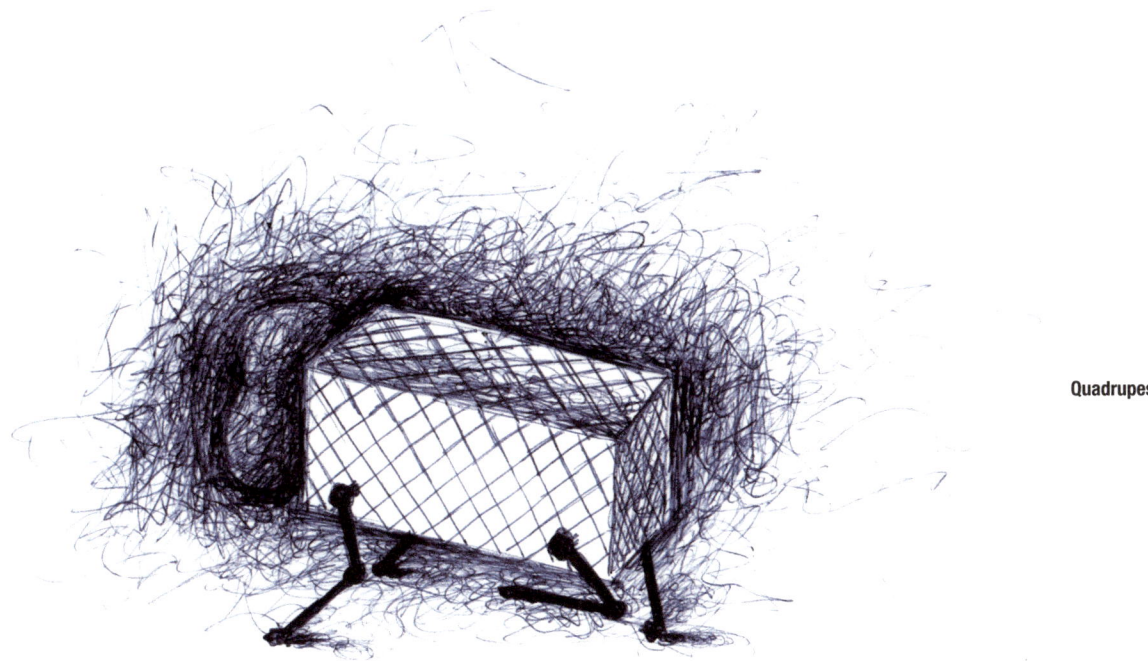

Quadrupes

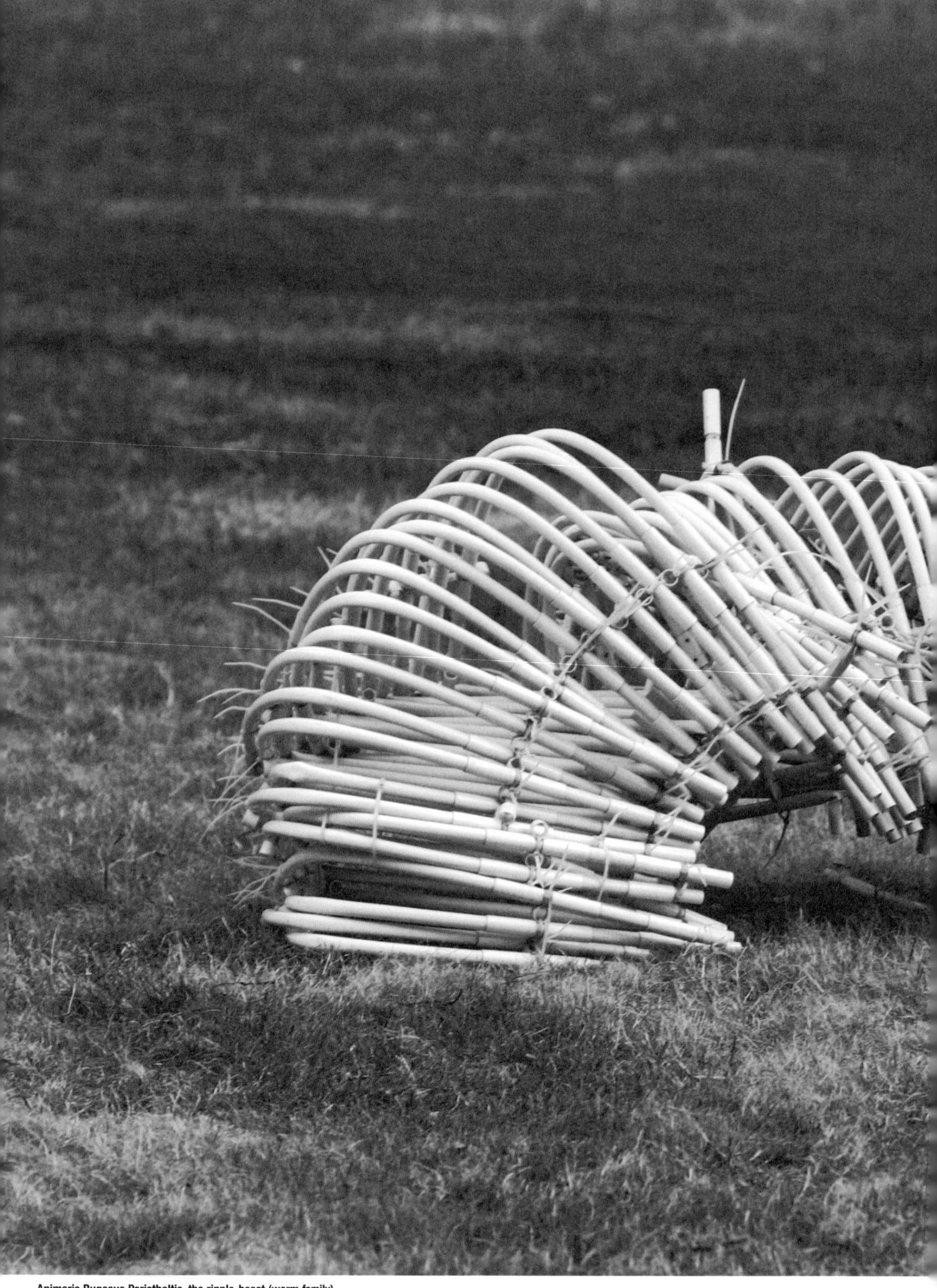
Animaris Rugosus Peristhaltis, the ripple-beast (worm family)

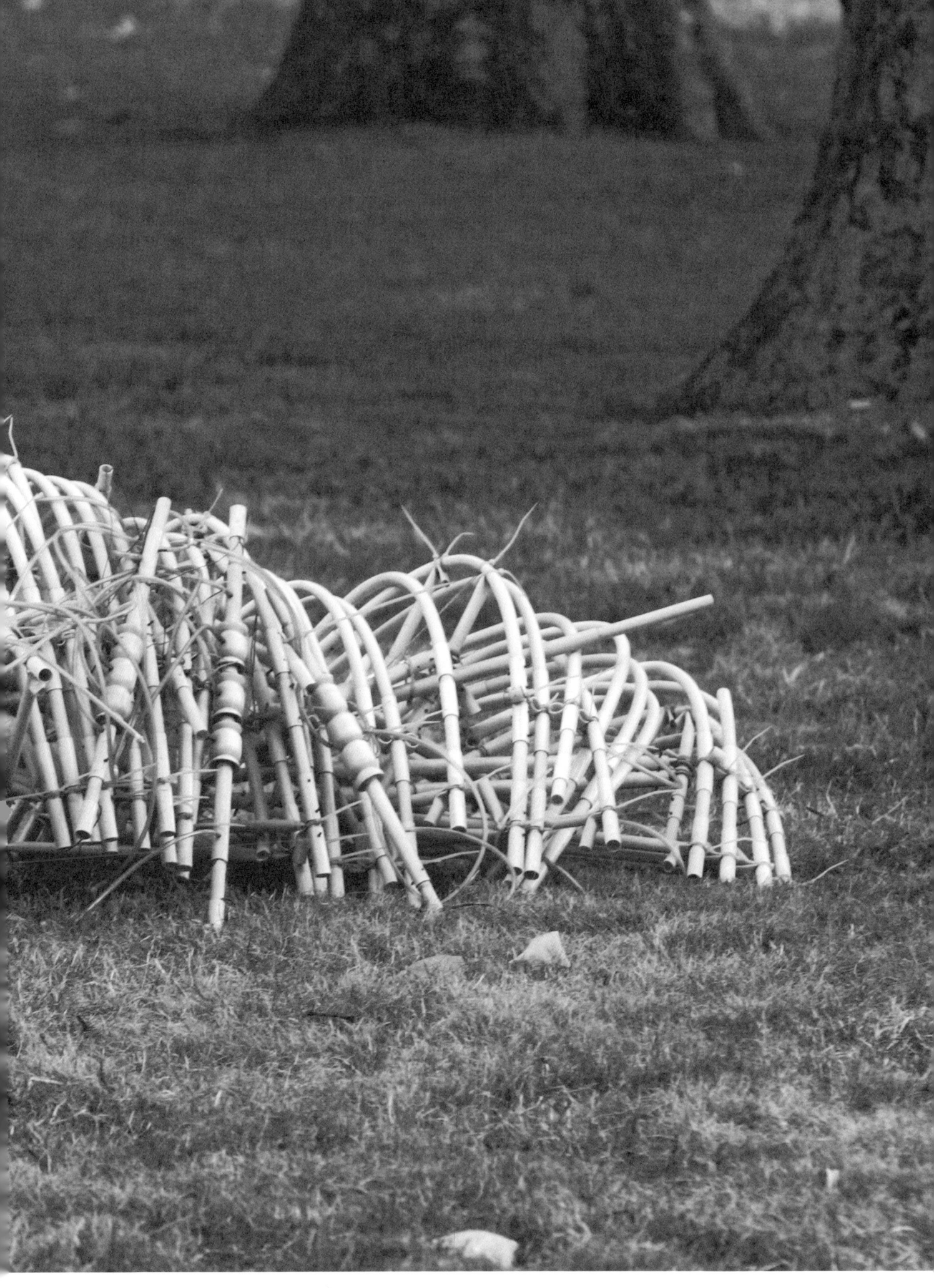

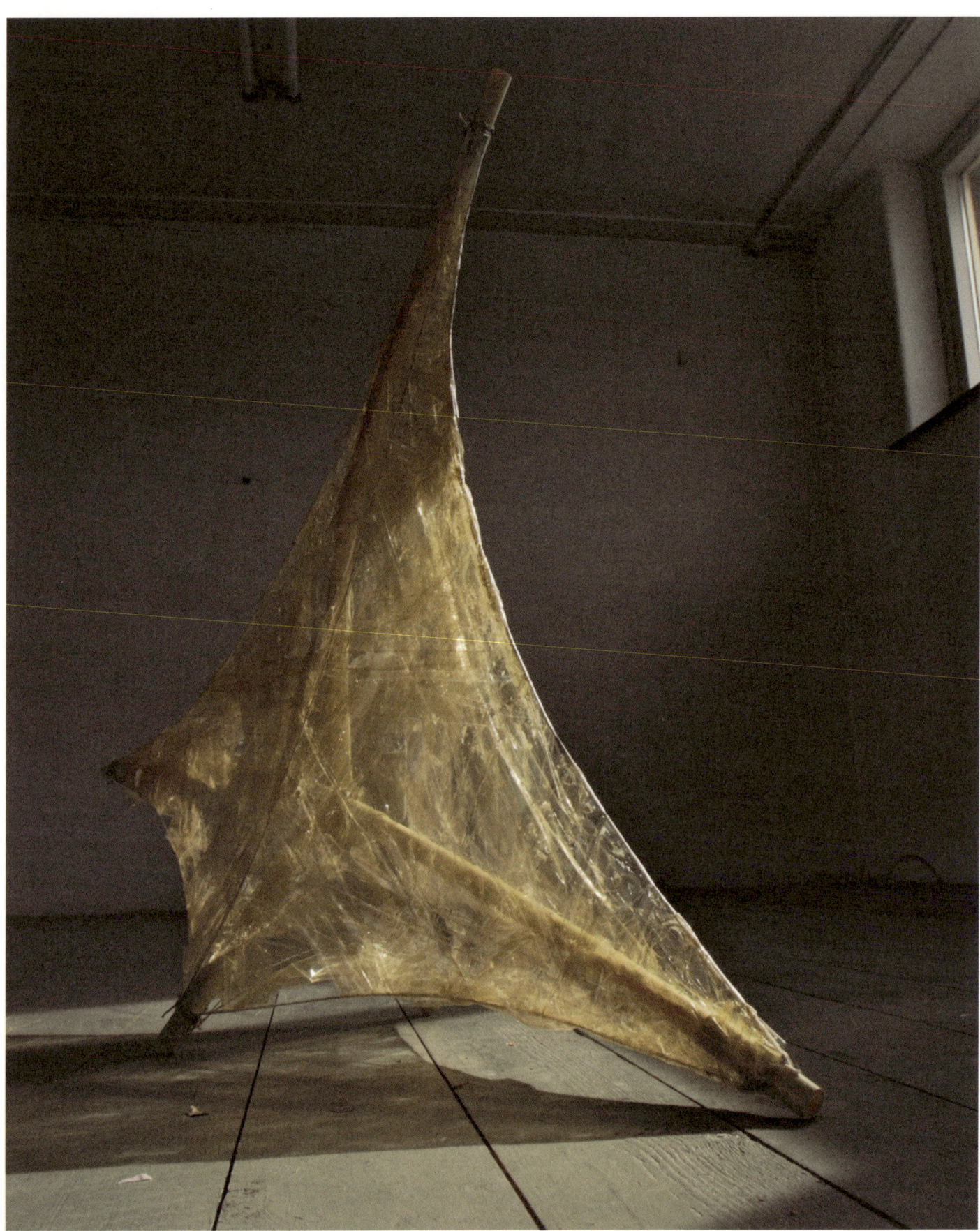

experiment with cellotape to create more rigidity

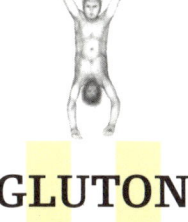

GLUTON
the tape period (1990-1991)

Plastic tubing

The adhesive tape period or Gluton was brief, just a year. In that year I bought a large consignment of plastic tubing. This yellowish tube is typically Dutch. In Germany they have grey plastic tubes, in America they make them of metal. Every country has its own kind of tube. Since 1947 Dutch law has decreed that this yellow tube be used to conduct electricity cables in houses. In the late 1940s the plastic insulation used for electrical cables was nowhere near as satisfactory as it is today. So it was decided that the wiring in a house should be replaceable. And plastic tubing was a way of doing it.

Something like six million kilometres of tubing has been produced over the years – and this is a conservative estimate. It defines the look of streets in the Netherlands. It can be found among the rubble in skips. And tied to the roof racks of the delivery vans of service engineers. In the 1980s manufacturers changed the colour of the tube to the yellowish hue it still has today. I used this tubing as early as 1979 to make a flying saucer which flew over Delft and caused a near-riot.

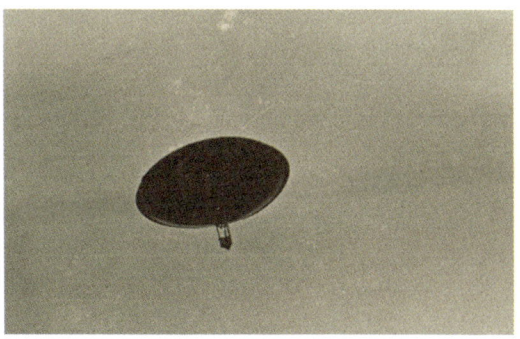

UFO

Originally the tubes had been white. The celebrated hula hoops were made from them back in the 1960s. They were also popular among kids as blowpipes for firing paper darts. That was my first encounter with these tubes. I was an expert in

Dream-gazing

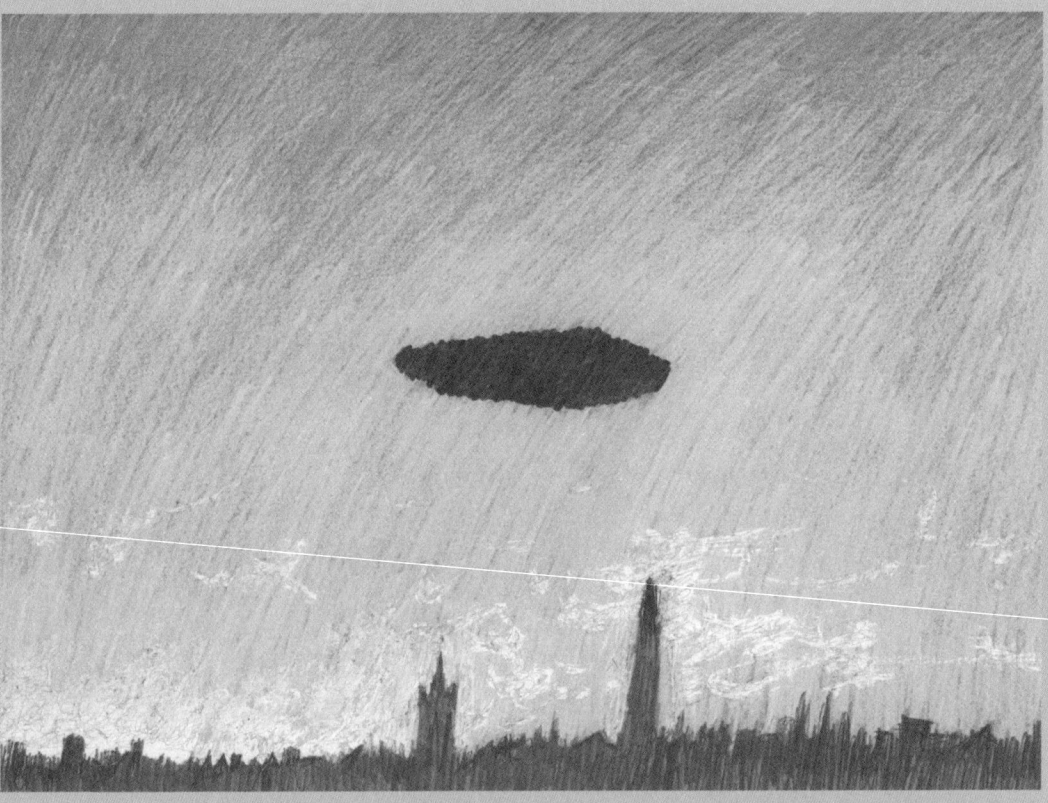

Real, authentic Dutch honey comes from the island of Tiengemeten in Zuid-Holland province. It's not yellow but white. I've noticed it tastes of cows. The bees there prefer *turd blossoms*; there's no other possible explanation. You can taste the countryside. OK, it's nice stuff, but .... Real, authentic salami should be bought at the market in Padua. This sausage is slightly brown inside instead of red. It's delicious, but at the same time I have to confess it tastes of the sties where the pigs are fattened. A trifle dungy .... The goats and kids in the children's farm at Blijdorp Zoo in Rotterdam smell of goat cheese. Maybe they aren't milked as often as they should be, so that the smell can't escape. An extract smelling strongly of urine is exuded from the animals' pores. I love goat cheese, but whenever I slide a piece into my mouth, I must confess I can taste the goat's wee-wee. So there's nothing dirty about cowpats and urine. In fact they're downright tasty in the right concentration. Evidently, the *setting* in which you taste or smell a thing is important. The honey, salami and cheese are there in the shops. You can be confident that the Commodities Act doesn't allow inedible food on the shelves. For this reason you assume it's edible and all at once urine tastes like a delicacy, particularly if it's expensive goat cheese. Our senses are the slaves of knowledge. They don't taste what they *should* taste. As soon as you *think* you're tasting something, you taste it. As soon as you think you see a thing, you see it. In 1980 a large piece of agricultural plastic film floated across the town of Delft. I saw it take off. It was circular with something resembling a hoop along its circumference. A gust of wind carried it high into the sky. Shortly after, I heard the wail of police sirens through the streets. The town was in uproar. I came across people who swore blind they had seen a flying saucer. I tried to convince them it had been a plastic bag, but it didn't work. They didn't want to believe me. On further consideration they managed to convince *me*. Now I believe them. I know for certain they saw a flying saucer. Their brains projected the thing onto their retinas. Though in reality it was a plastic bag that flew overhead, they saw a flying saucer. The thing even radiated heat, and there was a smell of ozone once it had passed. It shot into the clouds at great speed and was as big as Delft University's nuclear reactor. Just as we dream, so we observe. Everything around us is merely an invitation to observe. Observations begin with light pulses, odour pulses, sound pulses. These pulses switch on the internal film projector, open the doors of the internal aroma cupboard, tune the internal radio to the internal frequency. We dream our lives. We see what we think.

firing darts into open windows. You could write messages on them. Not that I ever did, but you could. You could write 'I love you' and then fire the dart into a girl's bedroom. The mere possibility of this form of airmail delivery fascinated me.

Exposure to sun and rain causes the tubes to fade to the same white they had been in the '60s. They also become more bone-like with time. In the beach animal bone-yard at my lab in Ypenburg near The Hague can be seen the fossils of

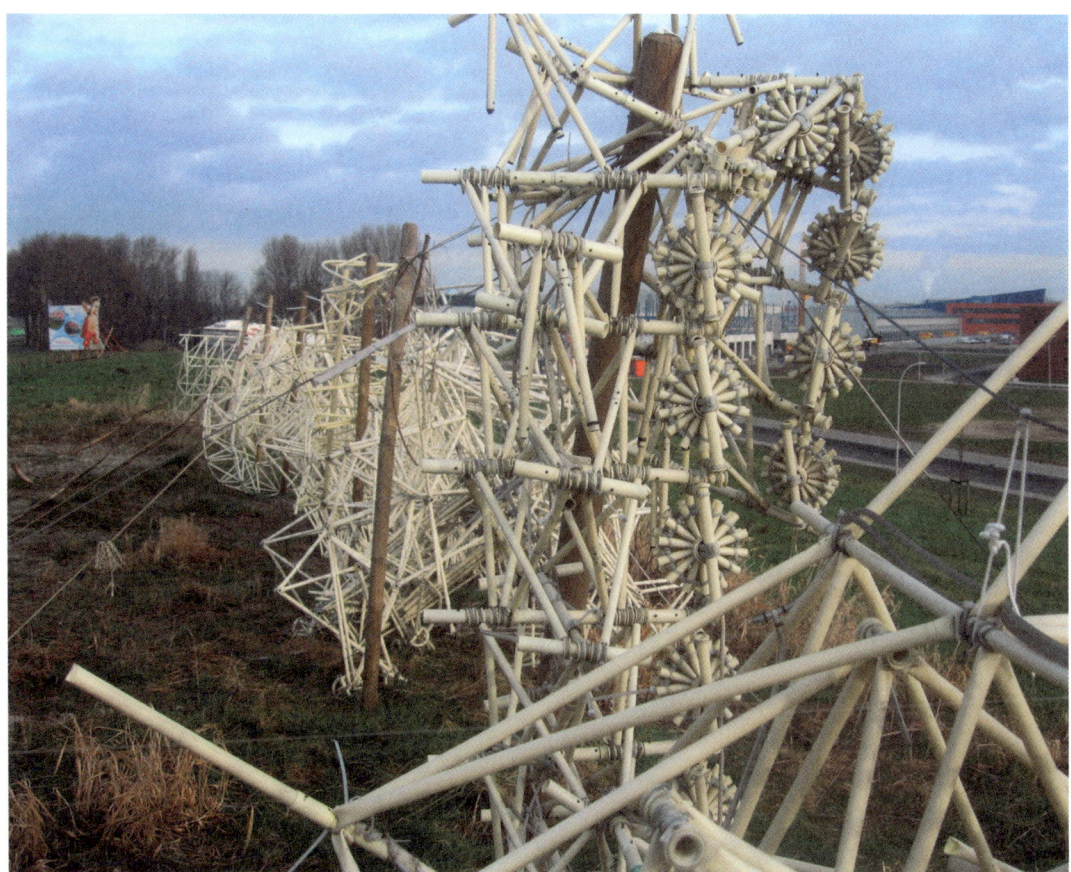

bone-yard

extinct species bleaching in the sun. Their age can be estimated from their colour. And fairly accurately too, using the chart given below.

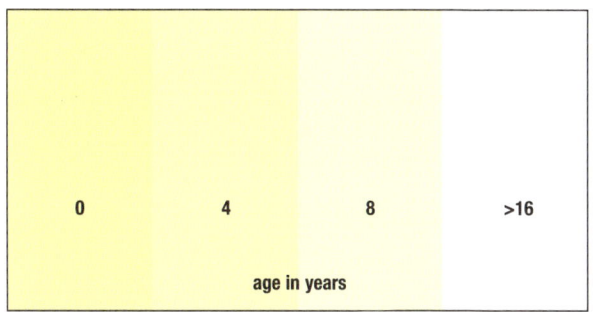

chart

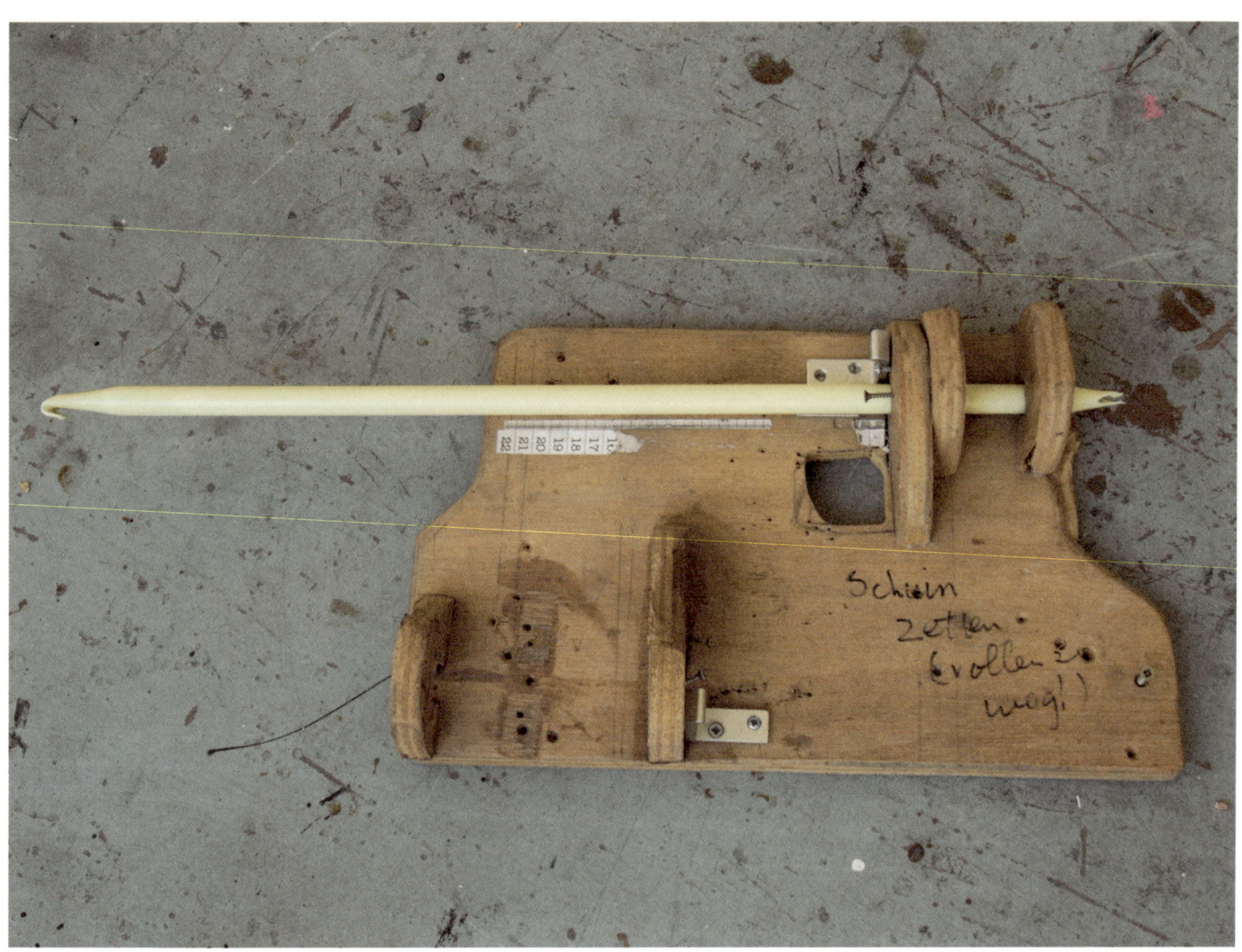

tube-cutting mould

Plastic tubing costs ten eurocents a metre, which means that a large beach animal (10 metres long, 4 metres wide and 4 metres high) costs about 100 euros in tubing (roughly 130 US dollars). The first obstacle on the path to artificial life was the problem of connecting the tubes: how were they to be fastened together? I started by sawing pieces of tube and winding adhesive tape round their extremities. Out of this first means of fastening came the first beach animal: Animaris Vulgaris.

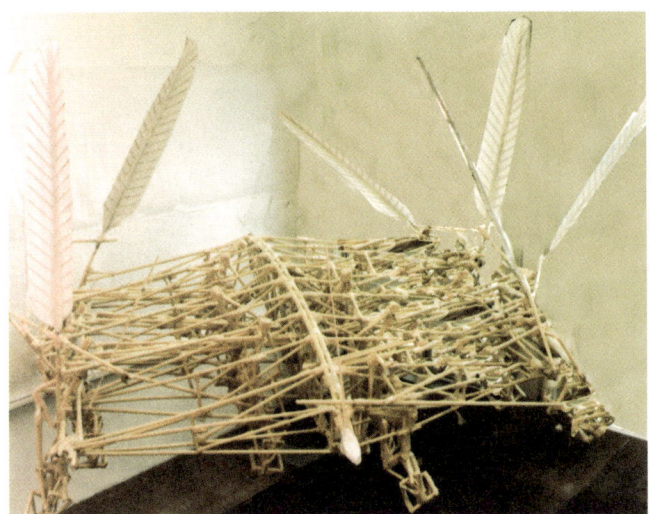

Animaris Vulgaris

Looking back
I look back at Animaris Vulgaris with a twinge of sadness. What a sorry sight it is too. Whatever made me think I could get it to walk? Some kind of irrational optimism, no doubt. Irrational optimism is something only we humans possess. Ten thousand years ago we weren't capable of too much. There were few tools in those days, maybe some early axes and spearheads. You need tools to be able to make other tools. But there were no tools to speak of. It all began with our own nails and teeth. This is where today's endless sea of equipment and skills originally came from. And there's us thinking we can do it all. We think we owe it to our own capacities, but believe me, it's the beginning stages that cost the most effort. It's a bit like removing wallpaper. This is no easy task at first. At times it's fixed so securely that you have to resort to a stripping knife. Once there's a hole, though, you can easily tear off the rest.
What you need is the *capacity of being able*. This is the art of escaping, of breaking out of the cramped conditions of being unable. Suppose I can't find my glasses

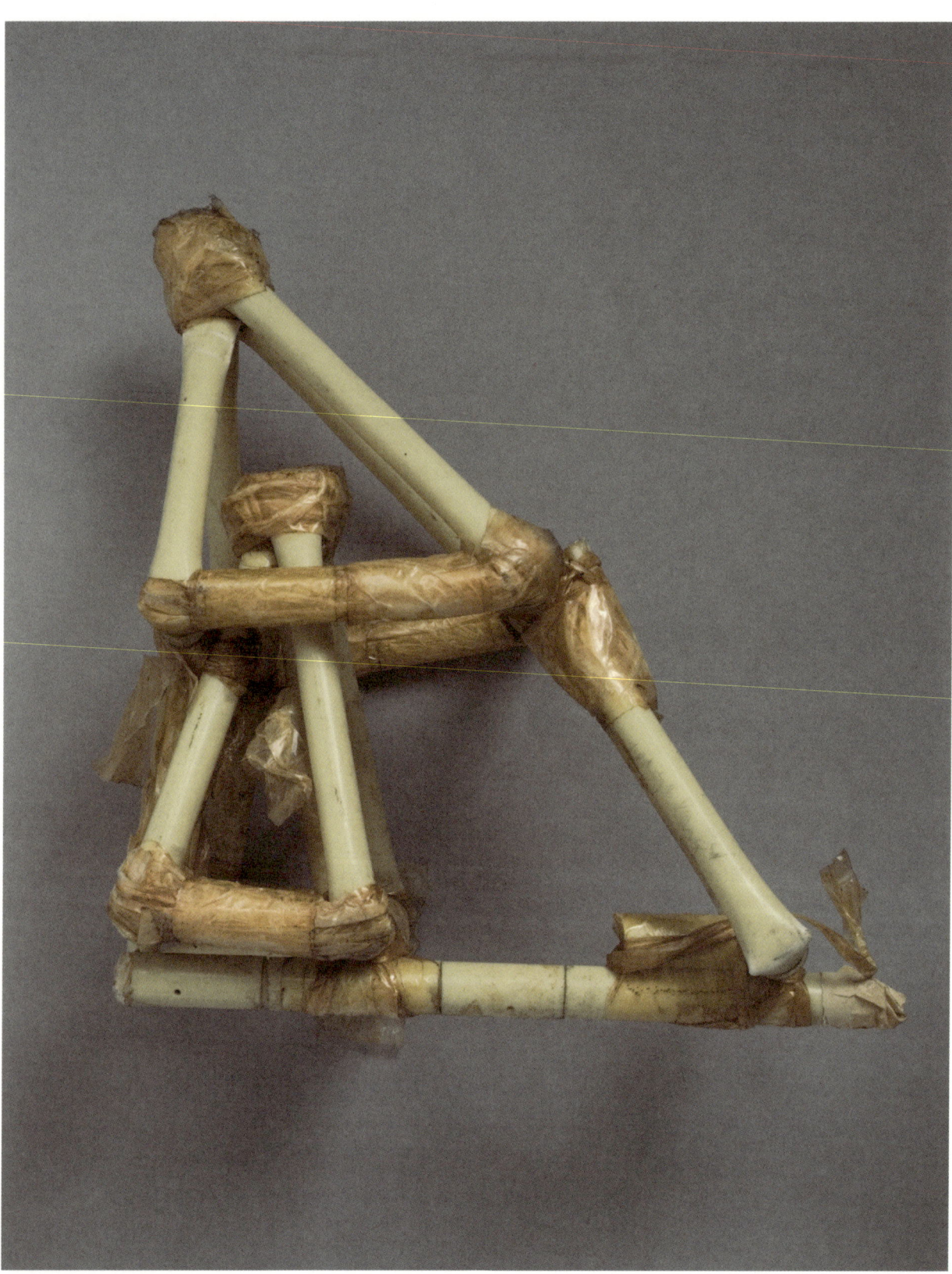

leg of Animaris Vulgaris

because without my glasses I can't find anything. Or you want to get into your house but the key's indoors. This puts you, so to speak, in the cramped conditions of being unable. The art of being able consists of escaping from this restricted state. Irrational optimism is a way out. Obviously it's possible to find your glasses without glasses. And get into your home without the key. Being able is something you have to master, and then all doors will open for you. I must admit to having had my share of desperate moments. There are even two aluminium tubes in Animaris Vulgaris. It was despair that drove me to this blemish, this youthful sin against the theory of limiting one's materials.

I want to make everything out of plastic tubing. Just as nature as we know it consists largely of protein, I want to make my own life-forms from a single material. You can use protein to make skin, eyes, lungs. Protein is multi-purpose stuff. So is tubing. It's flexible, but exceedingly rigid when used in a triangular construction. You can run pistons through it, store air in it, all sorts. I only discovered the wide range of its uses after many peregrinations through being-able country. Given the restrictions of this material I was forced to seek out escape routes that were neither logical nor obvious. The strategy I followed to assemble the animals is in fact the complete opposite of that taken by an engineer.
Suppose that engineers at a university of technology were to be commissioned to make something that could move of its own volition along the beach. What would you expect them to do? You can bet your life they would be ready in three months and also that they would have assembled stainless steel robot-like devices armed with sensors, cameras and light cells. Devices that are first thought out and then assembled. That's how engineers work. They have ideas and then they make these ideas happen. First they pore over books, then they open all the drawers in their workplace and take out what they need. It's a working method that gives rapid and reliable results, no two ways about it.

Countermanding that is the fact that any such devices engineers at these universities would develop would all be much alike. This is because our brains are much alike. We think we have exceptional brains (and of course we do) but they are embarrassingly alike in many ways. Everything we think up can in principle be thought up by someone else. Now real ideas, as evolution shows us, occur by sheer chance. The idea for the beach animals was one such accident. It

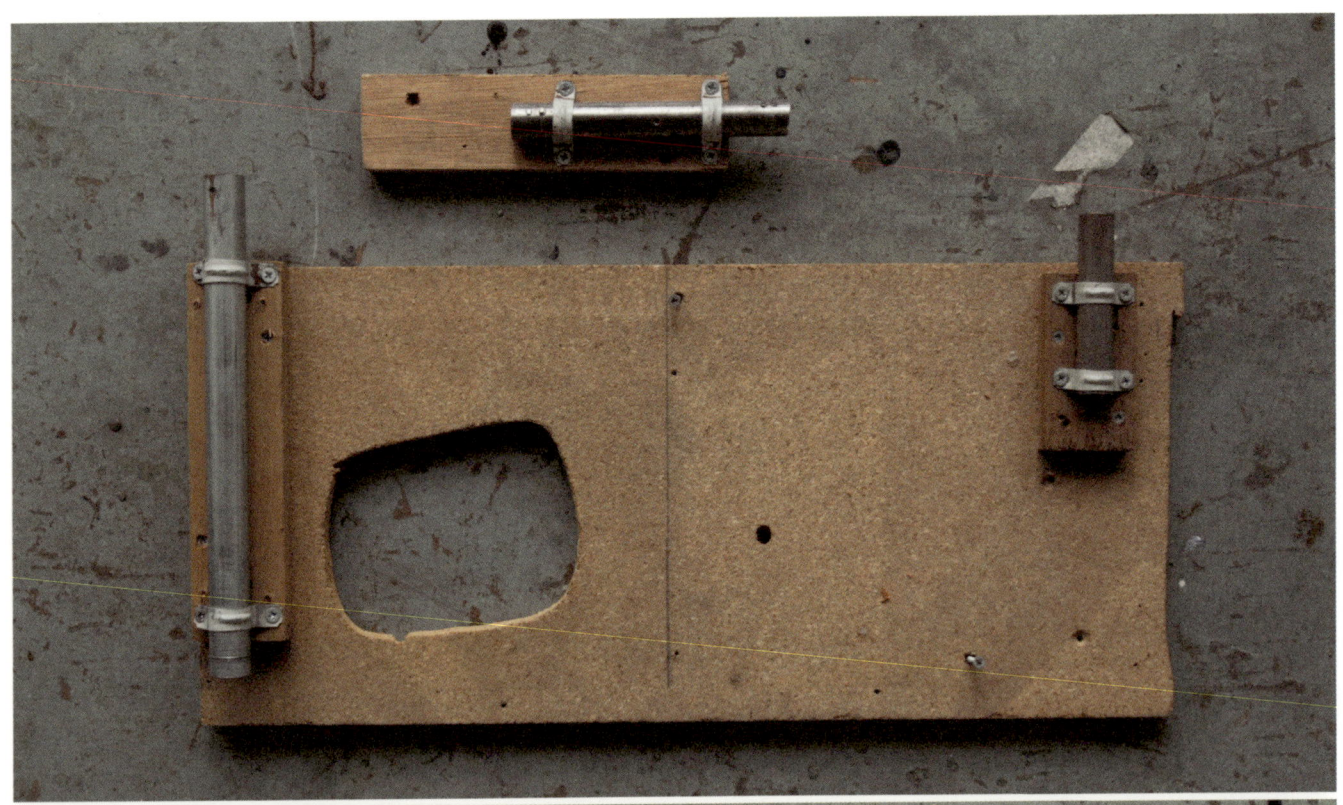

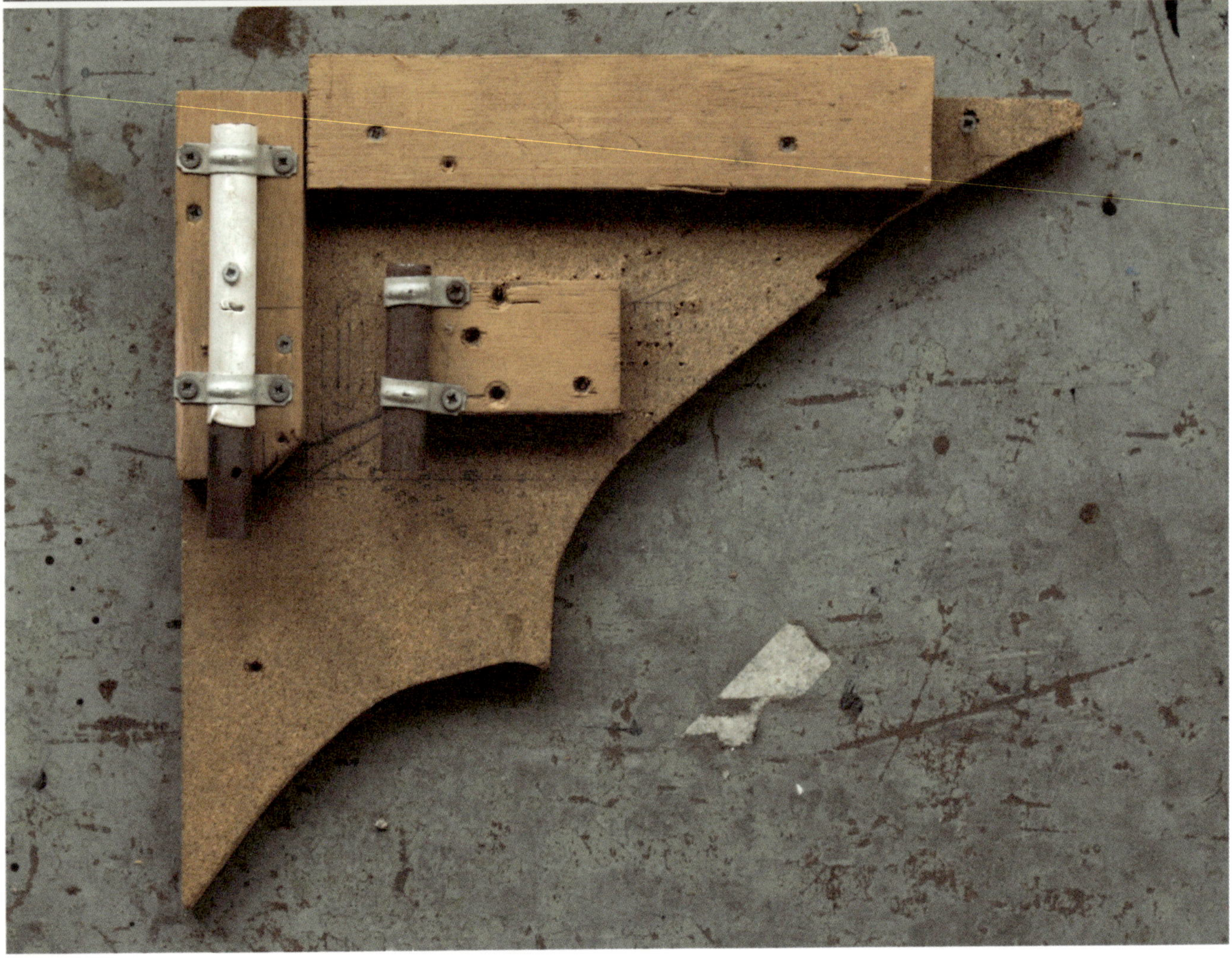

**tension mould,
crankshaft mould**

came about after I had been fooling around with plastic tubes for quite a time. It was the beach animals themselves that let me make them. And the plastic tubing showed me how.

Remarkably, chance is more likely to play a role when there are restrictions. Financial restrictions, for example, may mean that drawers in the workplace stay closed. This necessitates looking for other possibilities elsewhere. During this search new ideas automatically emerge, ideas that are often better than the ones you first had. Again, the restrictions of the plastic tubing oblige you to look for technical solutions that are less than obvious. All that searching and fooling around takes longer than the engineer's way of going about things. You might compare the engineer's method with a motorway. It takes you where you want to go, fast. However, everyone is travelling in the same direction. In the other approach, which I shall call the artist's method, your destination has yet to be decided. You park your car along the hard shoulder and scramble down the bank, machete in hand, hacking a path through the undergrowth. You'll probably never arrive at a destination in the accepted sense of the word, but you are very likely to call in at places where no-one has ever been before.

I've described the situation pretty much in black-and-white terms. I know from practice that there are plenty of engineers who scramble down the bank at times and artists who join the onslaught of vehicles. What is handy about the artist's method is that you yourself don't have to devise or invent anything. The material does that for you. So it was the plastic tubes that put the idea of a new nature into my head. Not a nature of protein like the one we know, but a nature of yellow tubing.

All this time, I tried to put the 'real' nature out of my mind. I really did try to start all over again, with a clean slate. It then transpires that animals don't always have to eat. Beach animals live on wind instead of food. They get their camouflage from sand clinging to the adhesive tape (see Animaris Sabulosa in the chapter on the Calidum period). Cannibalistic reproduction is another way of iconoclastically railing against existing nature (Animaris Geneticus in the Tepideem period).

Though I did my best to forget 'real' nature, I couldn't avoid resorting to its principles at times. One occasion was when I was developing the beach animal's leg. I could find no better, energy-efficient device for perambulating across sandy surfaces than the one already existing in old nature. I don't think there is any-

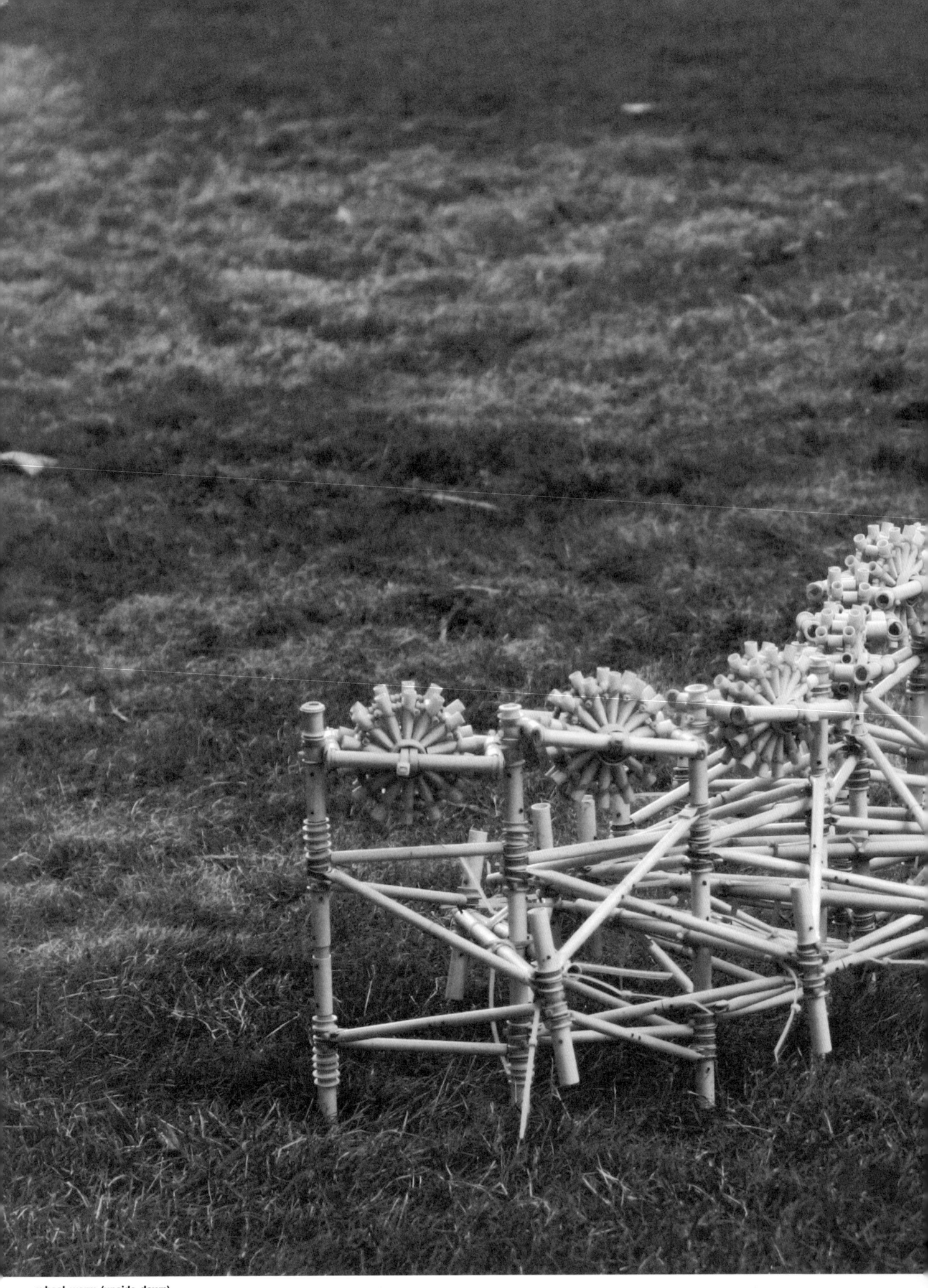

wheel-worm (upside-down)

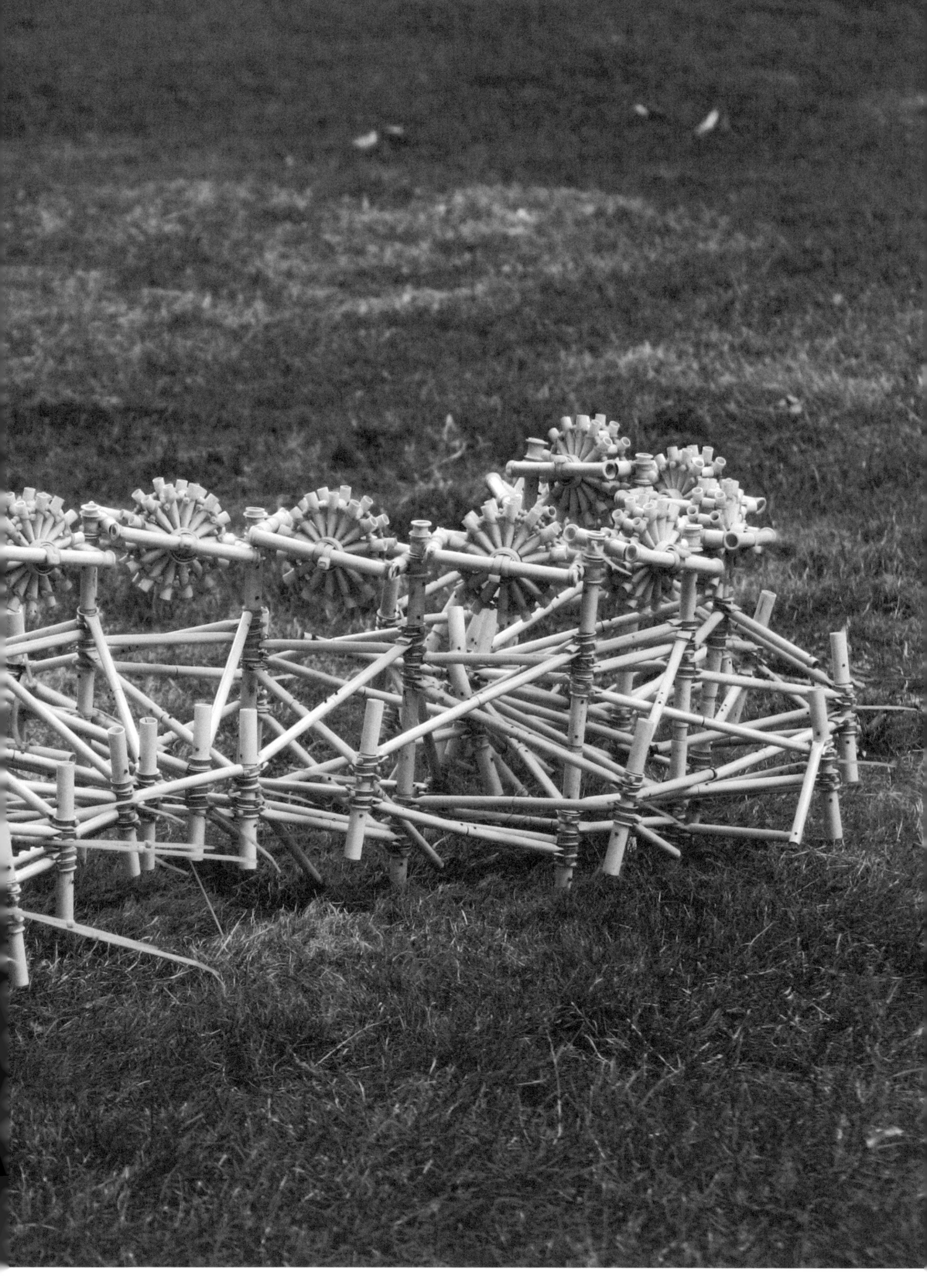

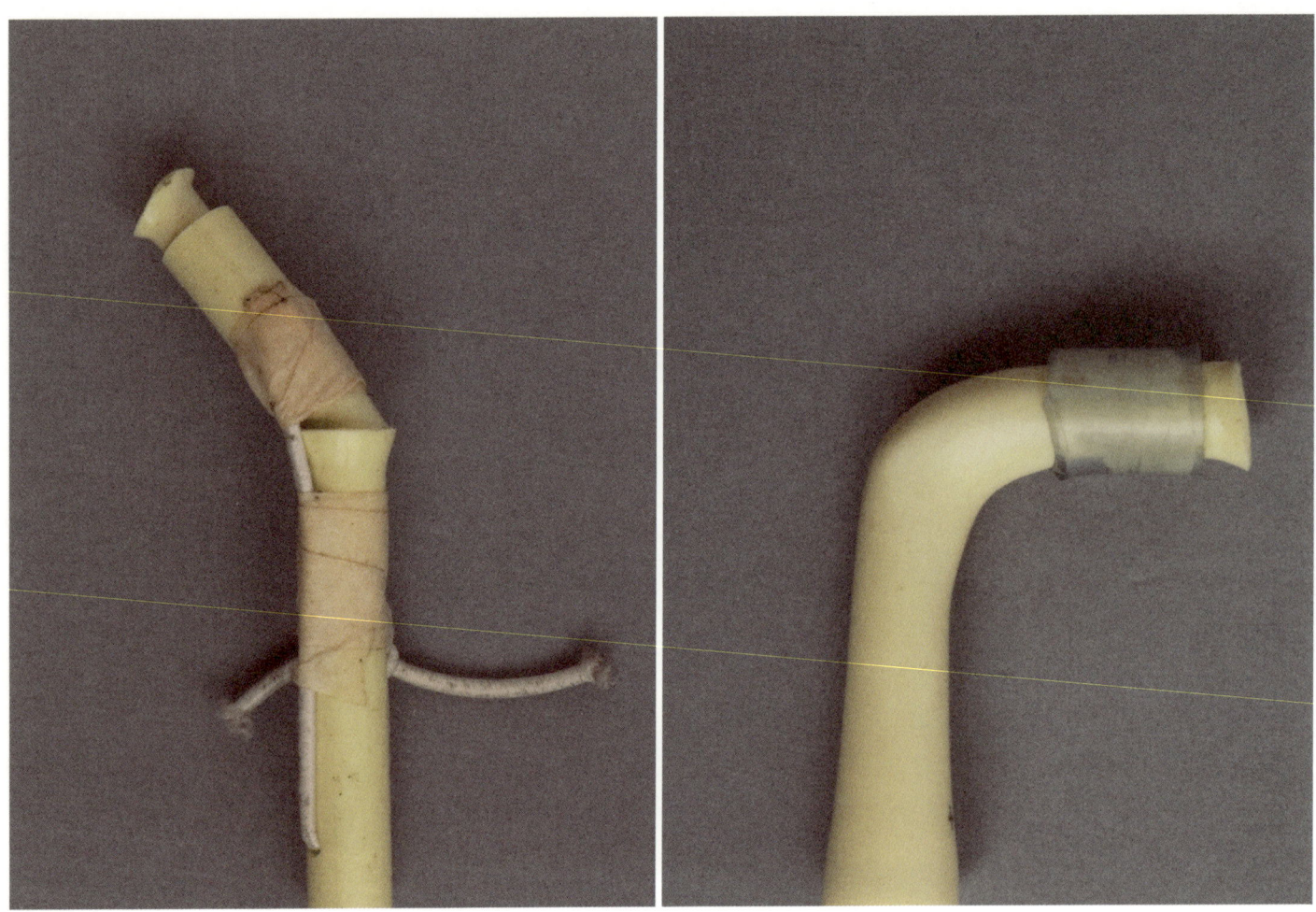

feeler, thumb joint of drive muscle (Animaris Excelsus)

thing that can beat good old legs. Now I'm working on muscles, nerves, brains. I wasn't looking for them but they happen to come in handy if you wish to survive on the beach.

I take comfort in the thought that these parallels have occurred in biological evolution. Consider the fish and the dolphin. They are unrelated. As you know, the dolphin is a mammal; the fish is a fish. And yet they still have more or less the same shape. Evidently nature couldn't come up with an aquadynamic form other than that of the fish: fattish at the front and gradually narrowing to a point at the tail.

I have come to empathize with the Creator. Not just in the tussle with stuff but also in the sheer pleasure of creating. You can't imagine the excitement that possesses me when something works, even though it may be a mere detail.

Animaris Vulgaris

This animal, now deceased, had 28 legs, attached by small pieces of tube to a rotating spine, or crankshaft, in the middle. Each crank was set differently so that the positions of all legs were different too. If the one leg was on the ground, the other was lifted. Like a human leg, each leg of Animaris Vulgaris consisted of three distinct parts. We have an upper leg, a lower leg and a foot. These pivot at their joints. With a little imagination the toes too can be regarded as independent members but this isn't our concern here. The principle of the leg of Animaris Vulgaris basically was that the upper and lower leg moved in such a way that the toe (at the end of the foot) moved back and forth describing a horizontal straight line.

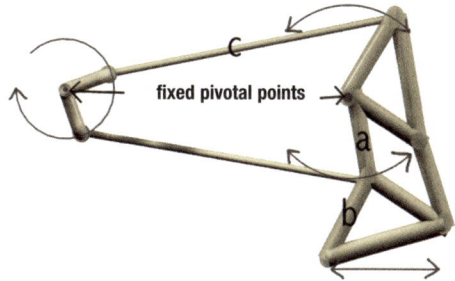

leg of Animaris Vulgaris

Upper leg and lower leg were connected by a rotating crankshaft. The crank of the upper leg was rotated 90 degrees relative to that of the lower. Providing the

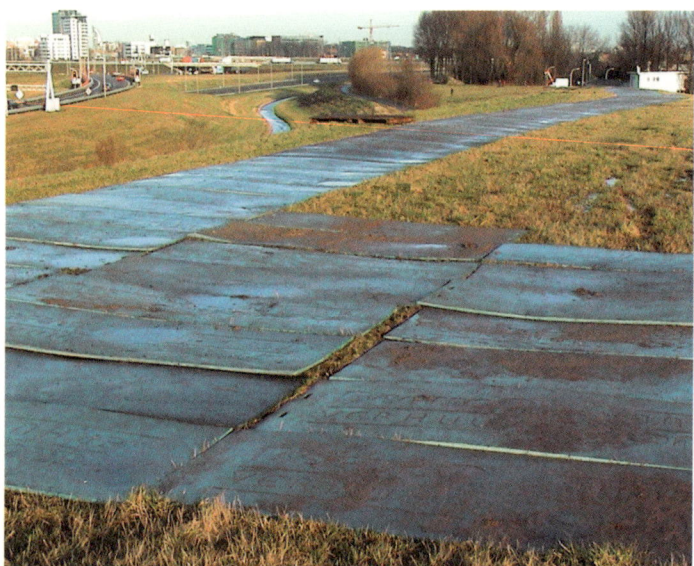
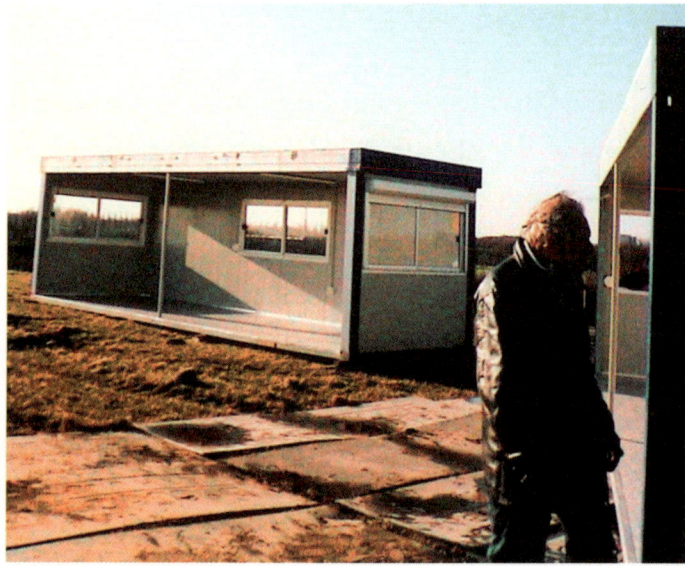
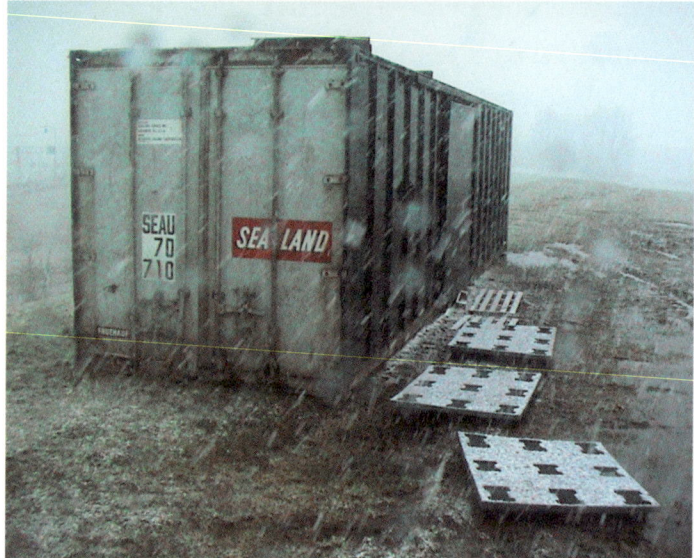
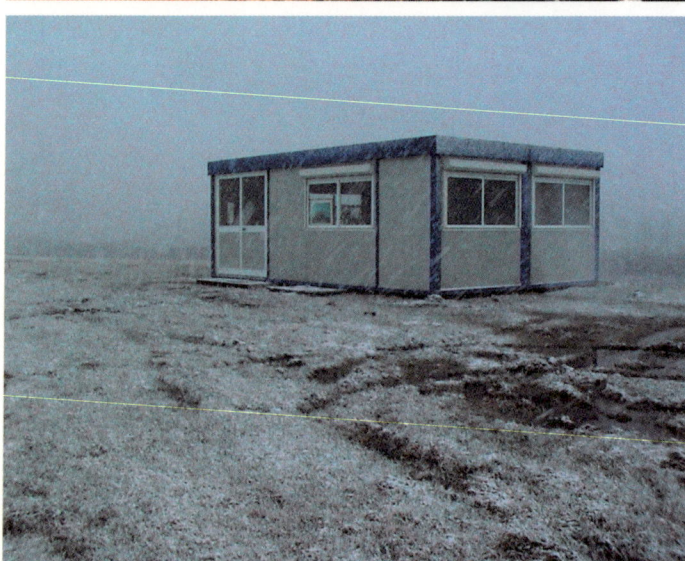
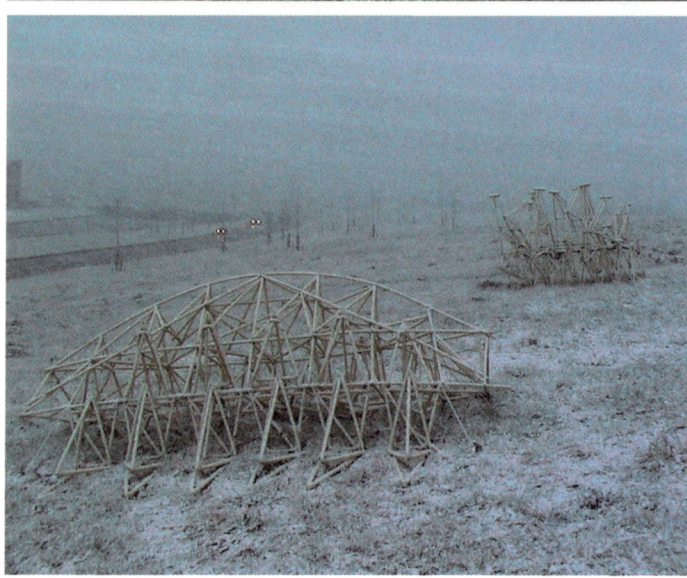
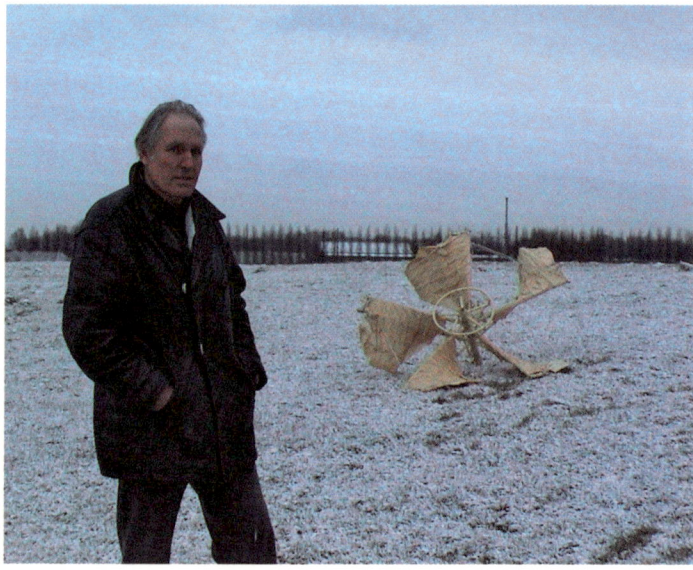

In the winter of 2003 a portocabin was placed above the windy noise barrier along the motorway between The Hague and Rotterdam.
It became the new workplace for beach animals. A beach was laid out on which to experiment with the animals; a 'limbo' before the real beach.
It was an art commission by Projectbureau Ypenburg to accompany the construction of the residential district beyond the noise barrier on the former airfield.
The commission ran its course in three years but the workplace has remained.

rods are long enough, the sum of the upper and lower leg describes a more or less straight line (it's the sum of a sine and a cosine). If the foot had been fixed rigidly to the lower leg, the animal would merely have described a horizontal back-and-forth action. It would not have moved forwards as it wouldn't have lifted its foot from the ground. And that was precisely the purpose of the foot, the third and lowermost part of the leg. The ankle joint allowed the foot to tilt sideways. In taking a step, first the toe touched the ground and described a straight line, so that the hip joint described a straight line also. On its way back, the foot was tilted sideways so that there was no further contact with the ground. In a manner of speaking, the animal was raising its leg. The cranks of the 28 legs were rotated relative to one another so that at a given moment each leg was at a different stage of the process. This complex of movements caused the animal to move sideways and in such a way that the hip joint described a straight line. The animal seemed to be moving on wheels. At least that was the idea. But it didn't work that way in practice. Animaris Vulgaris has never been able to stand up. It could only move its legs when lying on its back. I had spent a year working on it. Although it was hardly a success in technical terms, I had learnt a lot along the way.

It was the adhesive tape that caused Animaris Vulgaris to fail. The joints and connections were not rigid and strong enough to carry the body. This would all change in the Chorda with the arrival of the cable tie, a nylon strap mainly used for organizing wires and cables. I also discovered the computer (an Atari) as a tool for making beach animals' feet.

Sculpting

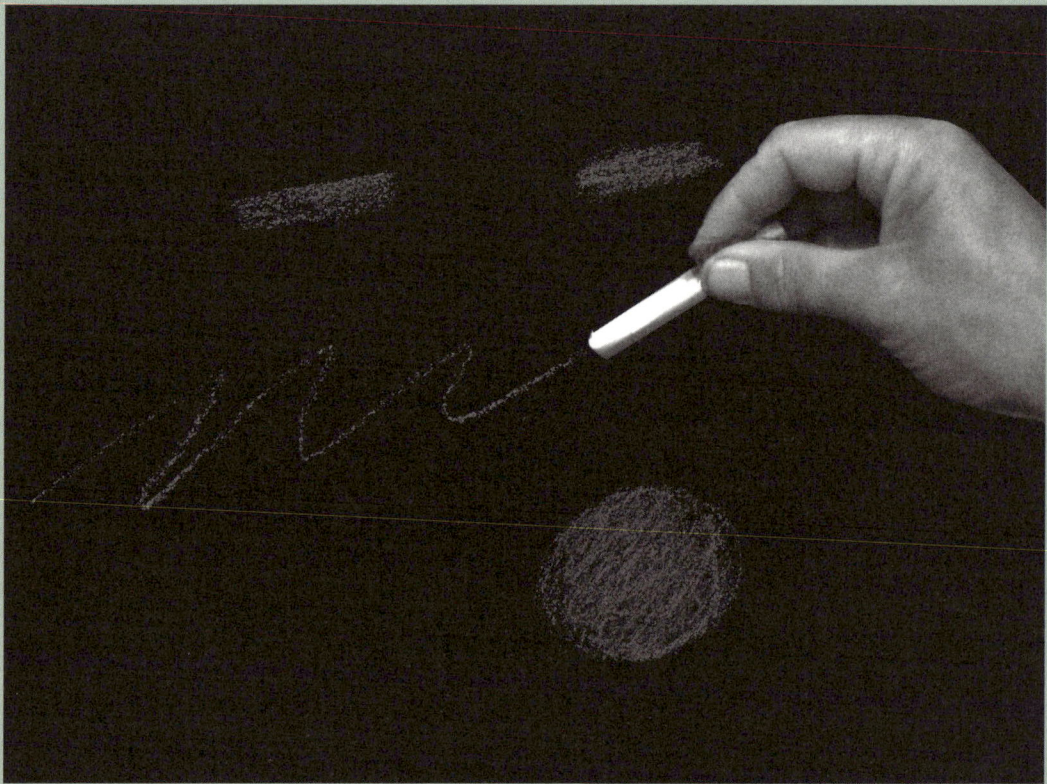

Students are young, uninhibited, impetuous, often innocent, unimportant and uncivilized. Your dealings with them make you yourself uncivilized, uninhibited, impetuous. And that feels good. It gives you a glimpse of a time before it became mandatory to do useful things. Your days were spent loafing around. Oceans of time stretched out before you. One of my students spent his time doing research into a stick of blackboard chalk. He measured the length of the line you could draw with one stick on a school board: 578 metres. I was impressed. As much by the length as by the courage it took to shamelessly occupy yourself with seemingly nonsensical matters in this age of economizing. My mind ran riot.

So a stick of chalk writes 578 metres. How many megabytes would that be? The line of a single letter on a school board measures ten centimetres on average. With one stick of chalk you can set up a school board document of 1300 words. That's a cool 65 kilobytes! It means that 20 sticks of chalk have the same capacity as a diskette.

There is one difference. The information in the chalk needs to be extracted by hand, the way a sculptor carves a sculpture from marble. The image is already in there; it just needs extracting. Carving, to quote Michelangelo, is easy; you just go down to the skin and stop.

Writing on the board changes the shape of the chalk. Writing is sculpting. A writer is a sculptor of words. A draughtsman is a sculptor of lines. Both word documents and image files fit into the chalk, just as on a diskette.

Draw a line on the board whose length is equal to that of the chalk itself. You now have a self-portrait of a stick of chalk. The self-portrait is a copy of the piece of chalk, only very thin and flat. A line 578 metres long is the same as an extremely long flat piece of chalk.

A large circle drawn on the board and shaded in with chalk can be seen as an extremely thick and extremely short piece of chalk. It's easy to change the shape of a piece of chalk, but a piece of chalk it will remain.

The chalk is itself a line, a thick three-dimensional chalk line. A line so thick that you can hold it but also throw it away. It is a true-to-nature self-portrait. So true to nature in fact that no distinction can be made between the drawn image and the original. Image and original are one and the same. It is a thing, like every other thing. A tree is a three-dimensional drawing done in wood. A lifting crane is a three-dimensional drawing done in iron. Or rather a crane is a three-dimensional drawing of a crane done in crane. The skyline is a line done in buildings. The equator is a thin line done in chalk. This is a chalk that doesn't give off powder when you draw with it. It can be used to draw an infinitely long line.

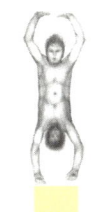

CHORDA
the strap period (1991-1993)

This period owes its name to the strategy of strapping the animals together with cable ties. Cable ties are used in orthopaedic surgery to hold together two parts of a fractured bone, by electricians for bundling cable, and for attaching price tags to articles of clothing. They are nylon tapes with a ribbed surface, a pointed tip at one end and a mouth-like casing at the other. To secure the tape, the tip of the tail is placed in the casing and pulled through. It's one-way traffic only; a flexible ratchet pressing against the ribs prevents the tail from slipping out. The mouth can ingest but not disgorge. The trade calls it a cable tie. It does bear some resemblance to a tie, as in collar and tie; more importantly, it allows you to work fast and efficiently.

I saw on TV that the police use cable ties as handcuffs to restrain captured criminals. That must pinch like hell. But that's what the cable tie is all about. Give the free end a tug with a pairs of pliers and the thing is literally as tight as a noose. Besides being tight-fitting, cable ties appeal to the imagination. To me they resemble an animal of sorts, a snake biting its own tail. The tail fits in the mouth the way a male plug fits into a female connector. Push the tail of one cable tie into the mouth of another and from two cable ties you get a new one that's almost twice as long. With a packet of cable ties (500 items) you could make a cable tie long enough to restrain a sizeable office building.

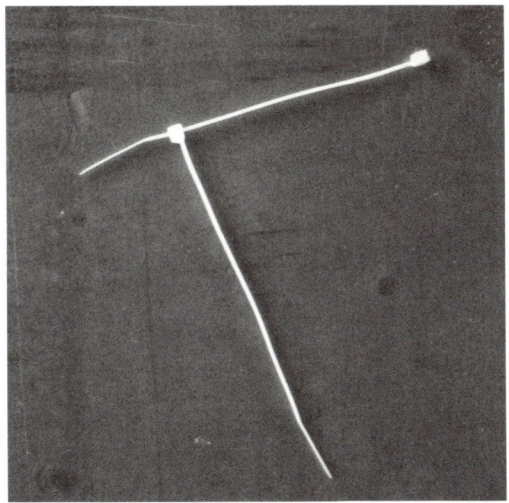

two cable ties

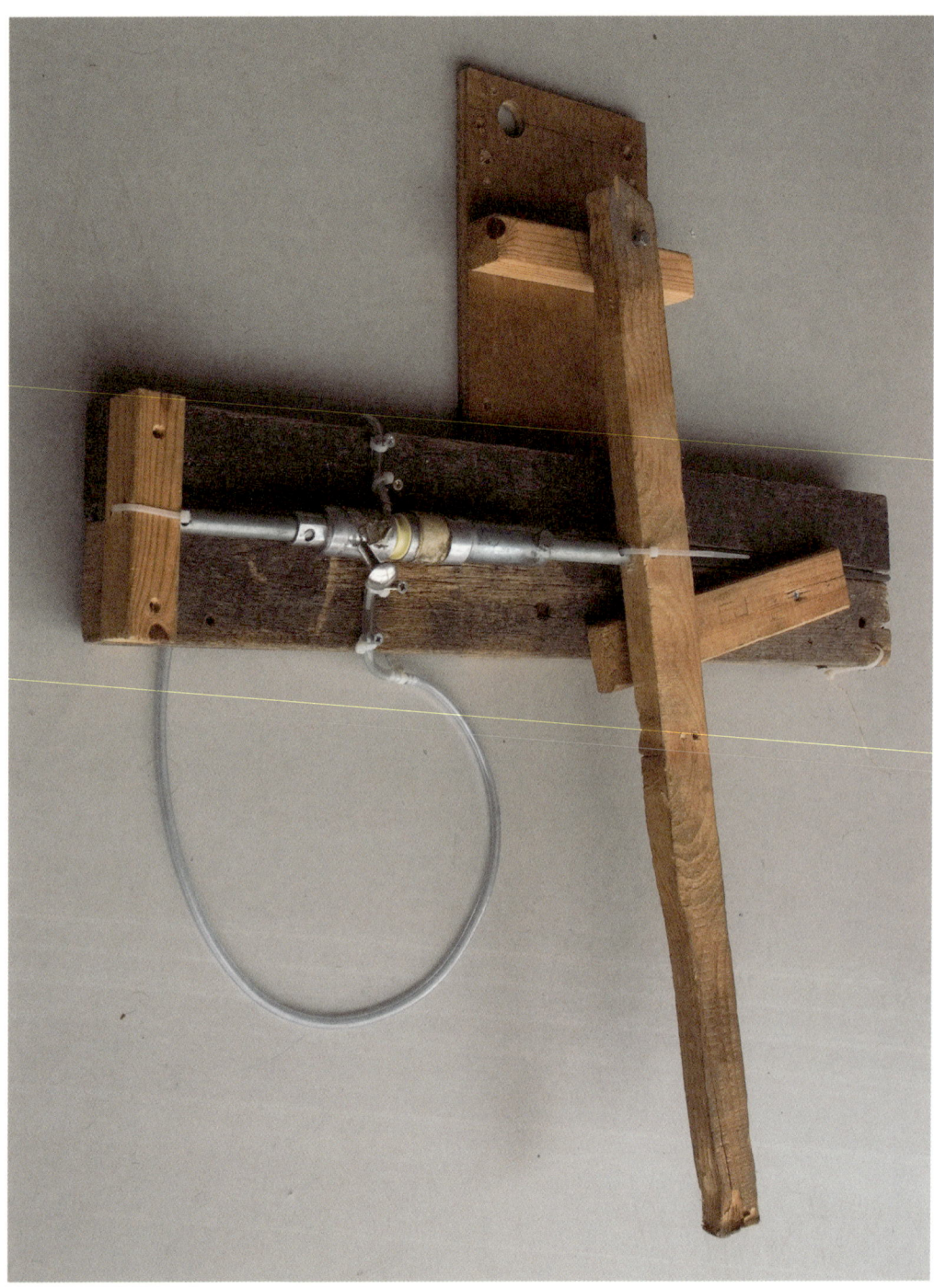

ring-maker

These tight-fitting nylon tapes proved far more effective in binding the tubes than the adhesive variety. Indeed, Animaris Currens Vulgaris was the first animal of mine that could both stand *and* walk. The tubes were bent into shape and then strapped together with a cable tie. A ring 19mm in section wrapped round a thin piece of tube 16mm in section served as a joint. At that time, joints often seized up due to sand caught between the tubes. This problem was later solved by flaring the ring on both sides. This gave a shape much like the rim of a bicycle wheel. Because of this shape, the ring simply ejected any sand that had penetrated the joint. Flaring is a technique that would be often resorted to later on in the beach animals' evolution. It entails heating the edge of a tube and making it bulge by temporarily inserting a thicker metal flare to produce a trumpet-like extremity.

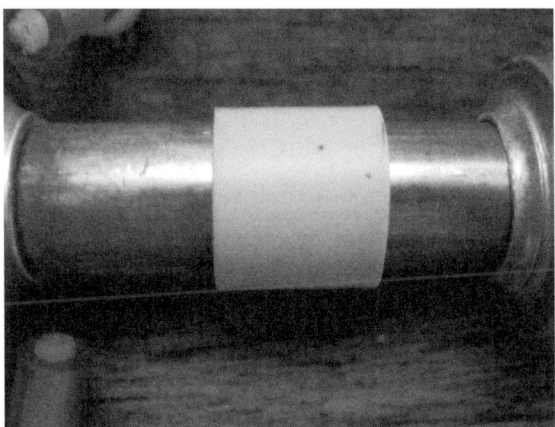
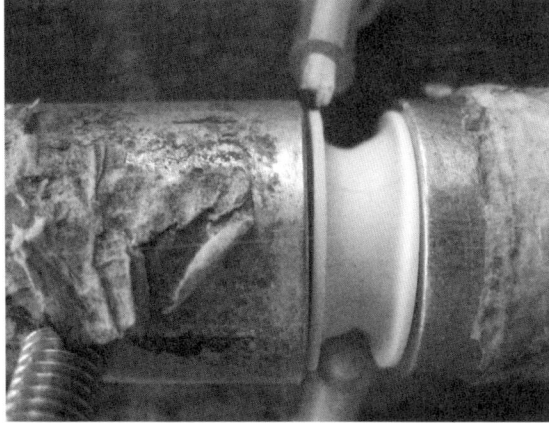
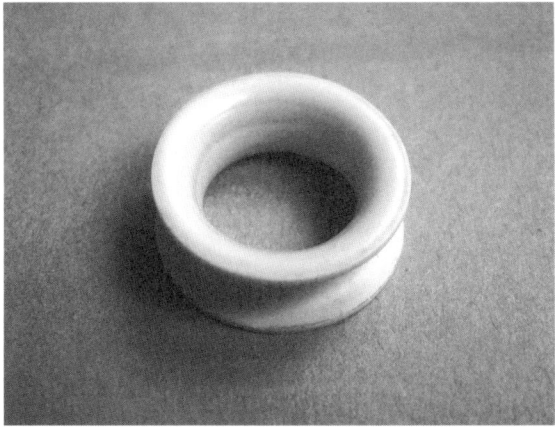
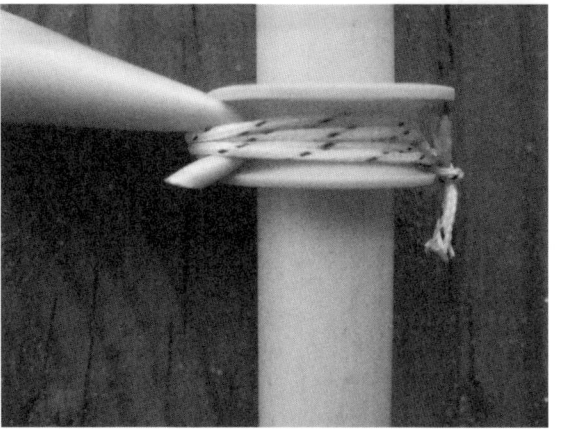

construction and use of the ring

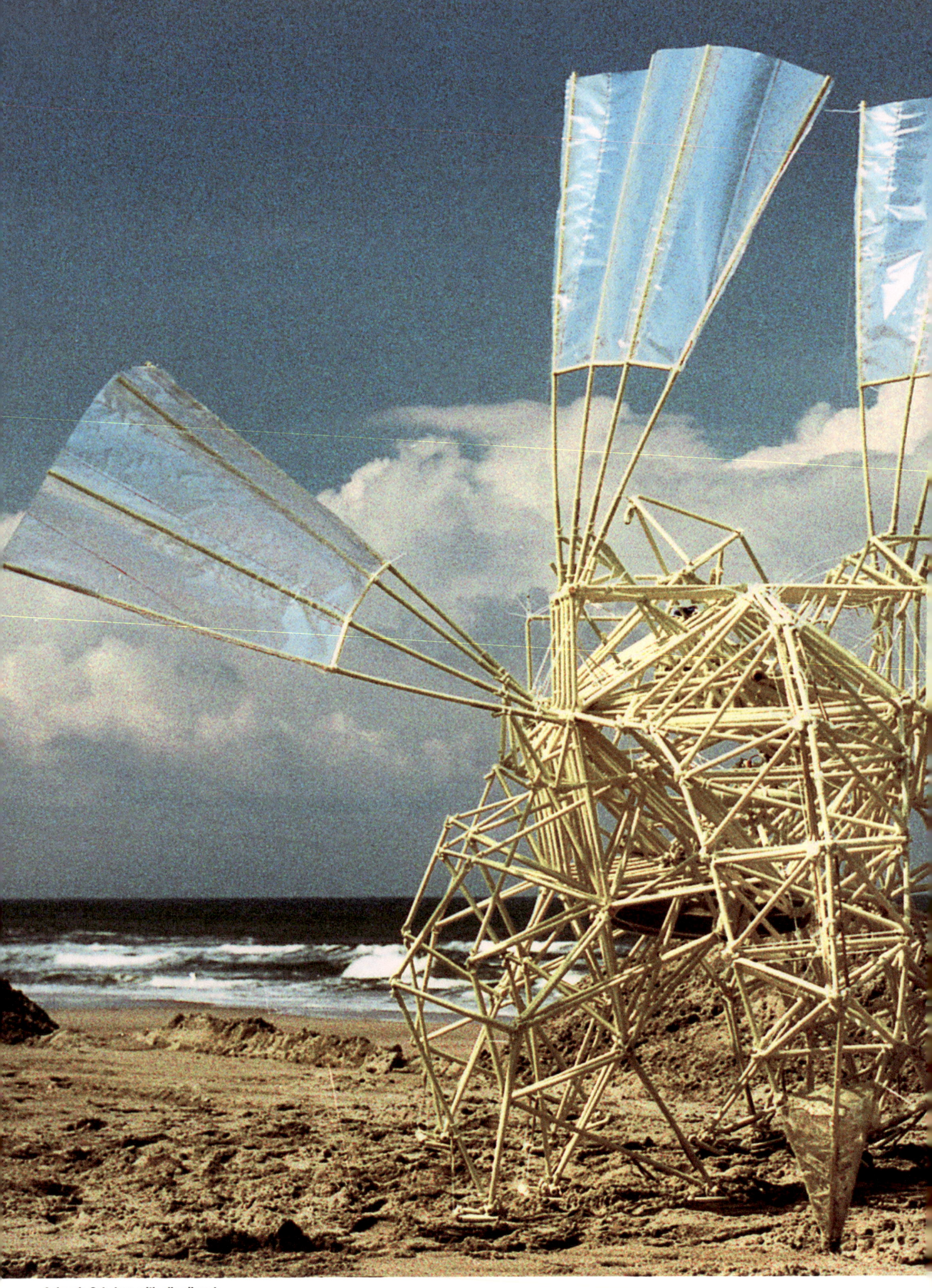
Animaris Sabulosa with all sails set

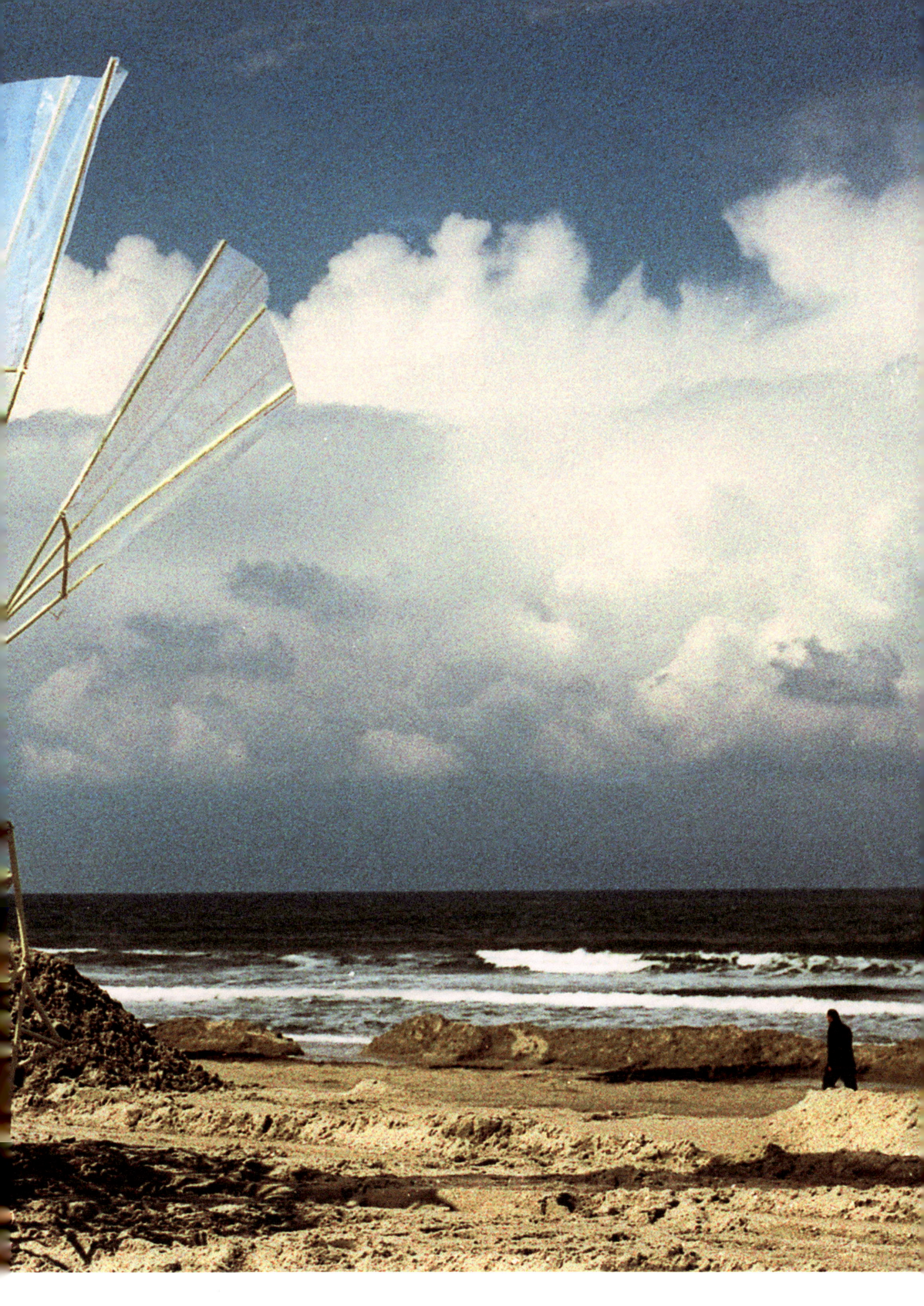

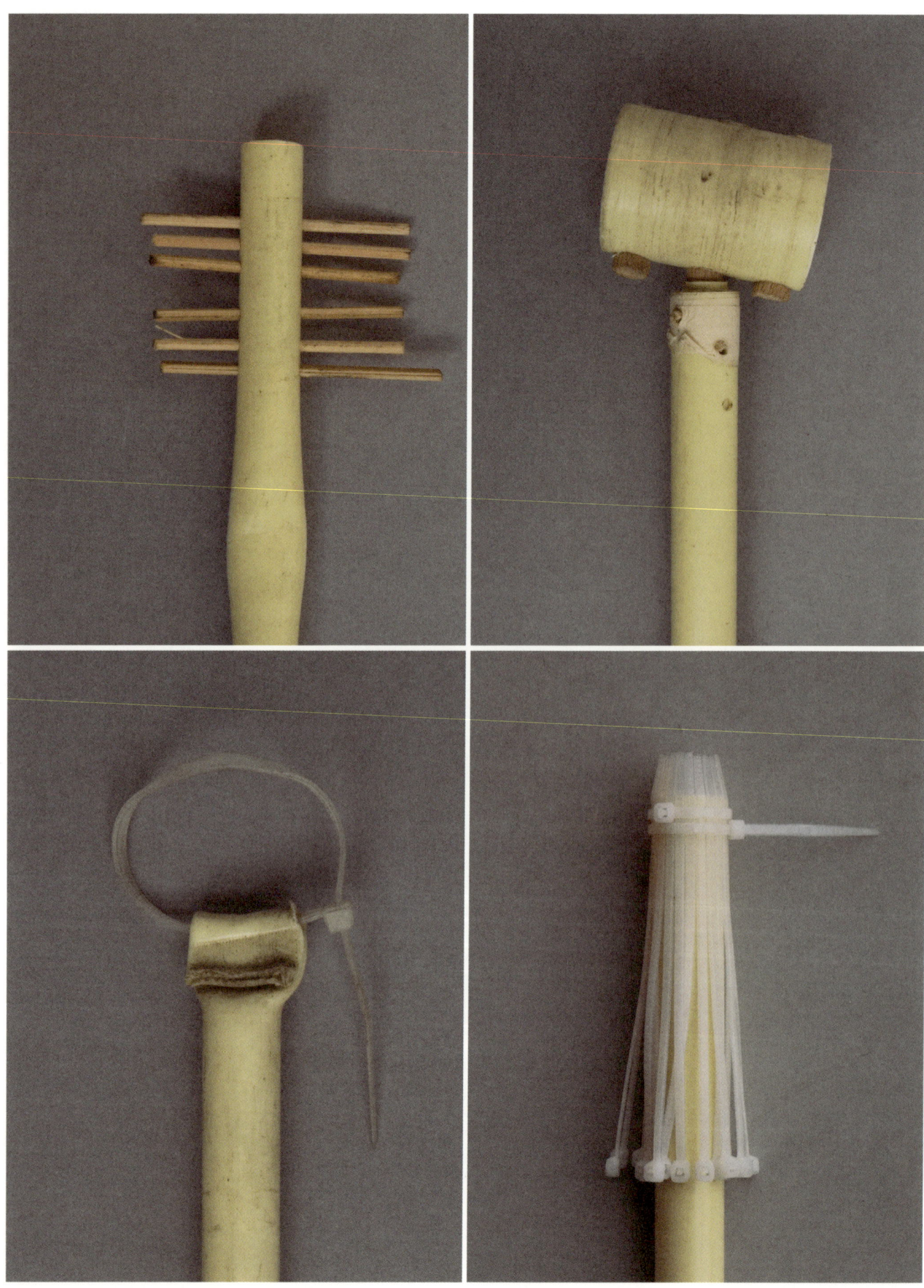

feeler (Animaris Sabulosa), end of reverse muscle,
head of F cell (Animaris Sabulosa), dust excluder (Animaris Excelsus)

Back in the Gluton the leg had been divided into three – upper leg, lower leg, foot – just like its human counterpart. With the Chorda came the two-part leg, consisting of upper leg and lower leg only. From that period on, the foot would be a fixed component of the lower leg. At the end of the Gluton, despair drove me to consider wheels as a means of locomotion. But legs prove to be more efficient on sand than wheels. Wheels have to work their way through the sand and shift relatively more of it as a result. Try pulling a cart through loose sand and it's hard work. The advantage of wheels, however, is that they don't lurch; their axle is at a constant height, which saves energy. But the legs made during the Chorda have this same advantage; they don't lurch either. The upper and lower leg parts move relative to one another in such a way that the hip joint (at the juncture with the upper leg) remains at a constant height, just as with the axle of a wheel. But they don't have the wheel's disadvantages; they don't need to touch every inch of the ground along the way, as a wheel has to. Legs can leave out patches of ground by stepping over them. Which is why you can better have legs than wheels on sandy ground. The idea came to me during the course of one night at the end of 1991. It was one or two in the morning when it struck me that the leg could have a far simpler structure. Until then, the upper leg and lower leg had had two separate cranks that differed in phase by 90 degrees (sine and cosine). It was now apparent they could share the same crank. The calculations made during the Gluton were based on the assumption that the crankshaft's centre of rotation is at infinity. If the crankshaft is placed nearer the hip joint, the push rods automatically form a right angle or thereabouts at the crank.

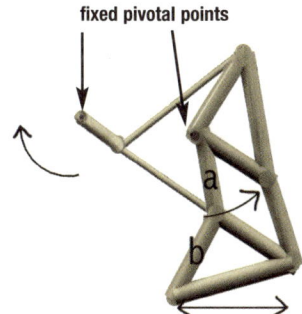

a new way of walking

A few hours later that same night I hit upon the idea of having a computer program to calculate the ratio between the lengths of the rods. This program would

Putting the alarm clock forward 23 million years

He didn't belong to my regular circle of friends. I spoke to him now and then in a café, or bumped into him in the street. He was a friend of friends. All the same, I was moved on hearing that he didn't have long to live, so I sent him a letter. Not long after, he invited me over.

Once inside his home, he assured me that the deceased was still alive; not just that, it was his birthday. A few guests were still there. One thing was certain: this wasn't going to be a fun occasion. Nor had that ever been the intention.

He wanted to show us his artistic works before he should die. These were two books he had had published with the help of friends. He also wanted to show us some eight-millimetre films he had made.

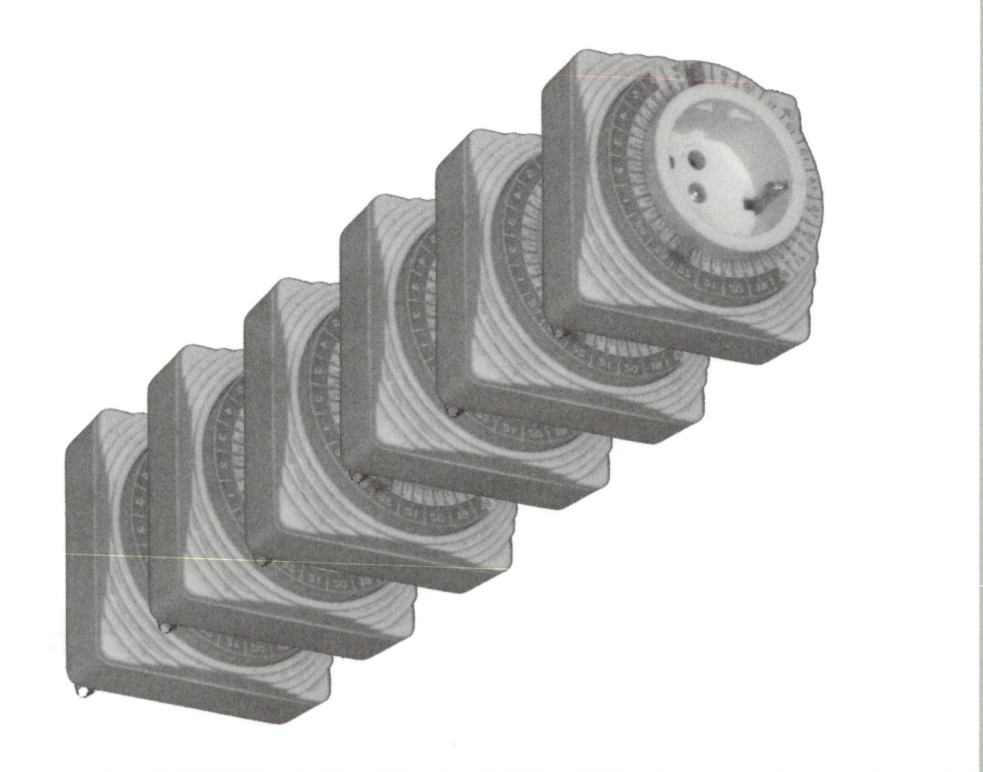

Accompanied by the rattle of the projector, we were treated to images of the city coming awake: a drawbridge rising, a boat passing through, the bridge descending, a tram crossing it, cyclists. And images of the beach in which I recognized many people I knew. The films were entertaining and full of humour.

After that, the atmosphere became more relaxed, convivial even. After a while, he asked me if I would invent something for him. He wanted his films to be automatically reshown in twenty-five years' time.

We racked our brains for a bit: a long candle that would take twenty-five years to burn down and let loose a cord that would throw a switch and start the projector. There were one or two variations on this theme.

Finally we arrived at the idea of a long-term mechanical plug-in timer. The timers you can buy are in fact wall sockets. The only difference is that they are not always electrically charged. You can programme the timer switch to be turned on between, say, nine o'clock and quarter past nine. It is impossible to programme periods longer than twenty-four hours, let alone periods of twenty-five years.

But let's say we were to insert a second timer into the socket of the first. Then this second timer would only be activated whenever the first was turned on, say between nine and quarter past. In short, the timer would move forward only fifteen minutes each day instead of twenty-four hours.

So it would only have gone full circle (24 x 4 quarters of an hour) in 96 days. Let's say that the second timer is also programmed to be activated between nine and quarter past; then it receives electric current only once in those 96 days, between nine o'clock and quarter past nine.

Let's now add a third timer to the process. This will be activated only once in every 7236 (= 96 x 96) days, more than twenty years in other words. If you add a fourth, a fifth and even a sixth timer, the last-named will only receive electricity once in 23 million years.

It's difficult setting the timers for twenty-five years from now. All the same, I did buy six plug-in timers and they are here in the wall socket, as a memorial to Hans Berger.

work on the principle of evolution. These two discoveries form the bedrock of all further developments in beach animal history.

Evolutionary method

I can best explain how the leg works using a model I once made from a sheet of plywood. I often take this model to lectures to demonstrate the leg's action.

In the middle of each beach animal is a kind of spine, more specifically a crankshaft. The remarkable thing about this spine is that it can rotate. In the model, my hand turns the crank of the crankshaft. This rotation is converted by 11 small rods into a walking movement drawn by a small pencil at the end of the leg. Let's call this pencil the toe.

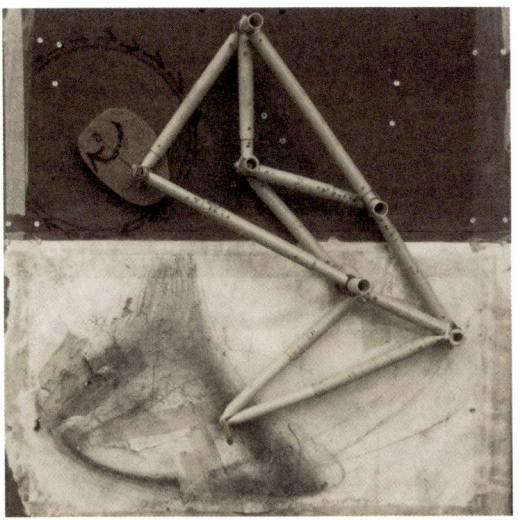

leg demonstration model

Ideally, the pencil describes a kind of triangle with rounded corners and a horizontal base. Whenever the toe is on this base, it touches the ground and carries the animal. It describes a horizontal line, or rather the entire animal does, since the toe is carrying the animal. The same holds for a wheel; the axle also describes a horizontal straight line. The beach animal doesn't lurch. When the toe reaches the end of the base (at right), the leg is lifted whereupon it rapidly

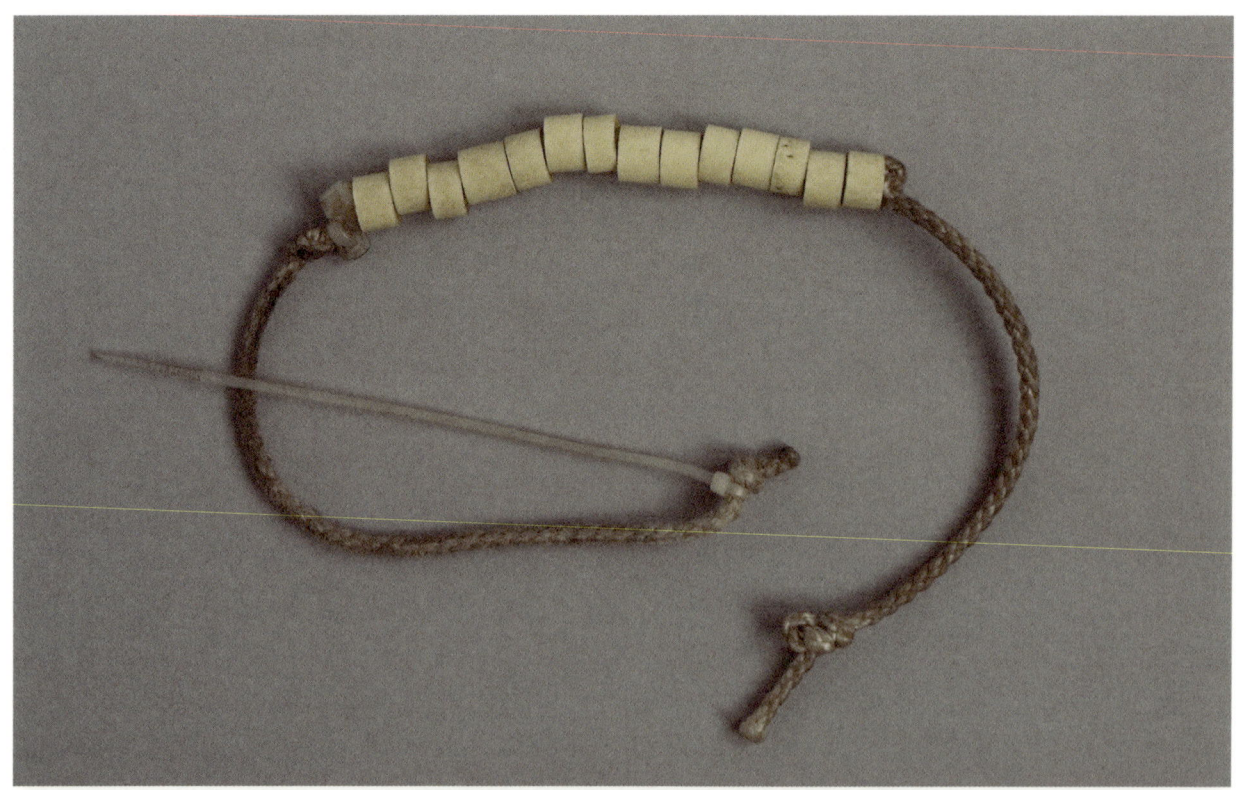
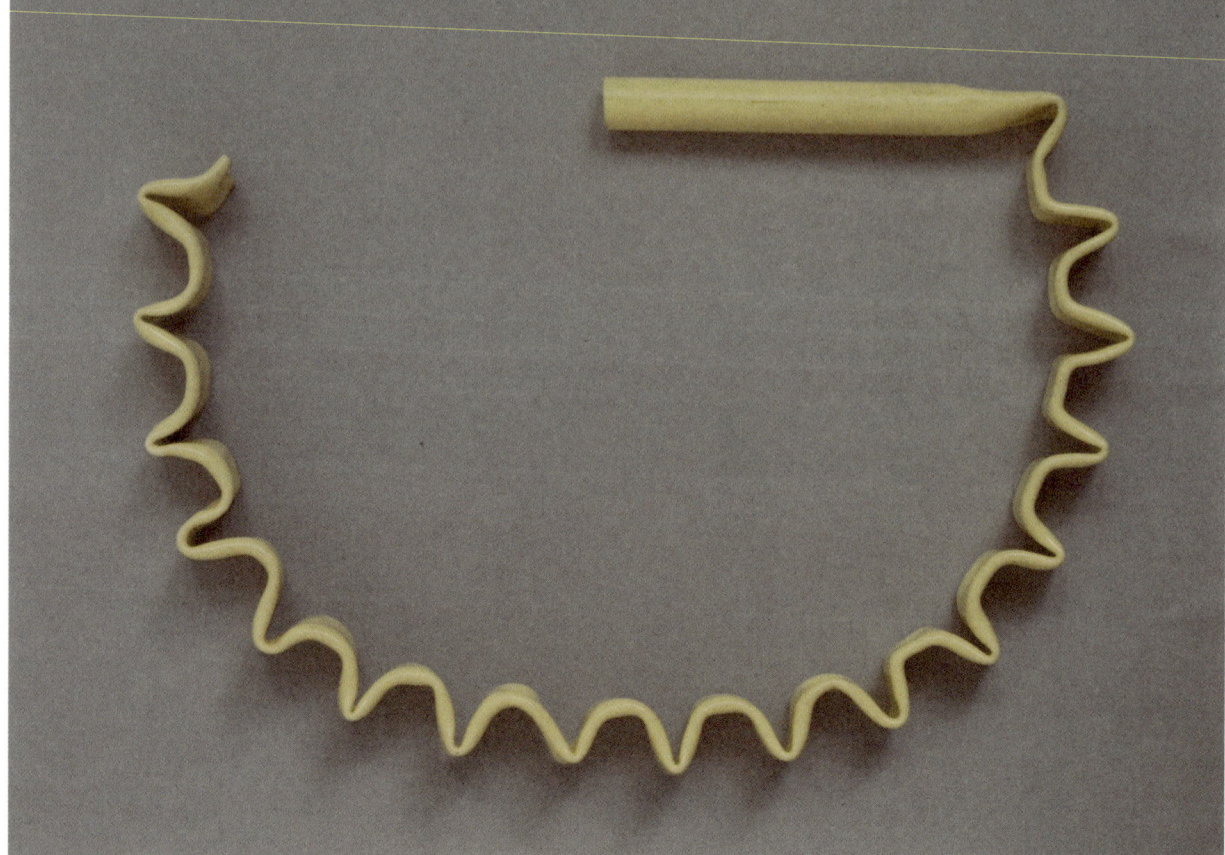

**Step-counter,
cog ribbon**

describes the other two sides of the triangle. During that time the animal is supported by the other legs which at this stage are on the ground.

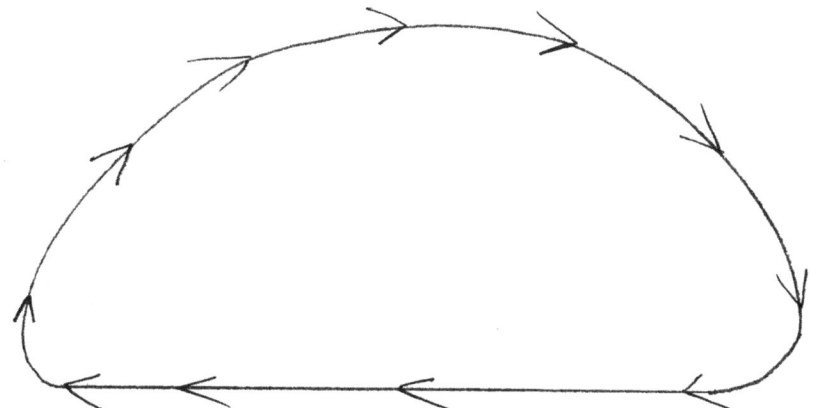

the ideal walking curve; flat base and rounded corners

The curve this produces is dependent on the ratio between the lengths of the 11 small rods. Another ratio gives an entirely different curve, a figure 8 for example. Of course, I had no idea beforehand which ratio between the lengths I needed for the ideal walking movement. Which is why I developed a computer model to find

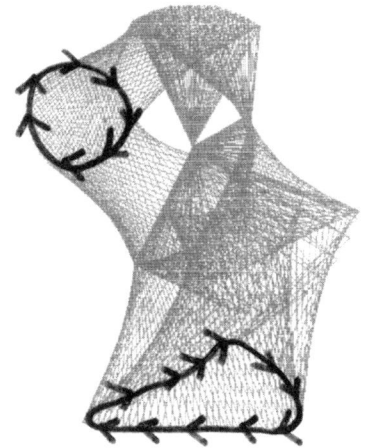

walking curve in thirty stages

this out for me.

But even for the computer the number of possible ratios between 11 rods was immense. Suppose every rod can have 10 different lengths, then there are 10,000,000,000,000 possible curves. If the computer were to go through all these possibilities systematically, it would be kept busy for 100,000 years. I didn't have this much time, which is why I opted for the evolutionary method.

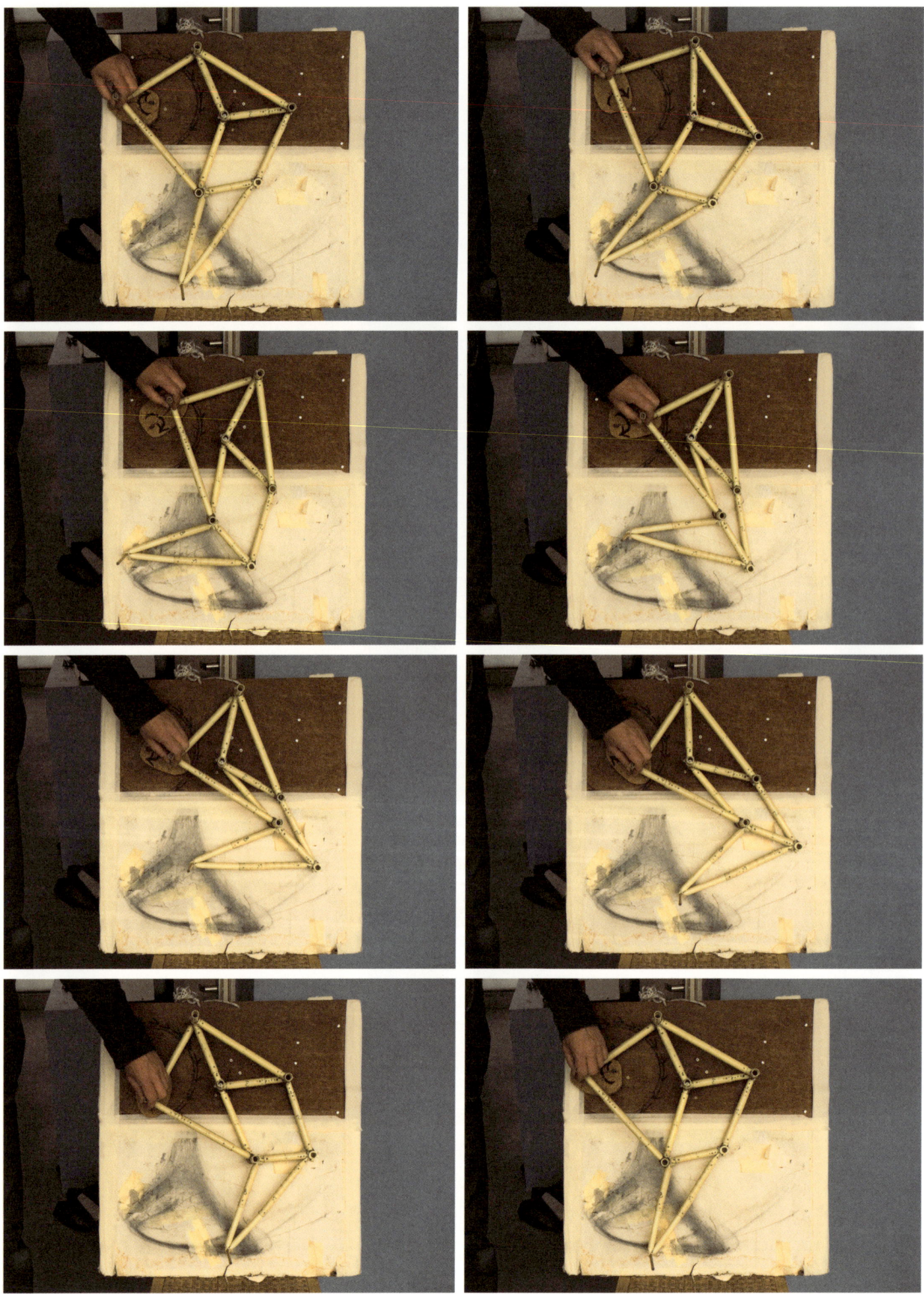

leg demonstration model

Twelve holy numbers
Fifteen hundred legs with rods of random length were generated in the computer. It then assessed which of these approached the ideal walking curve. Out of the 1500, the computer selected the best 100. These were awarded the privilege of reproduction. Their rods were copied and combined into 1500 new legs. These 1500 new legs exhibited similarities with their parent legs and once again were assessed on their resemblance to the ideal curve. This process went through many generations during which the computer was on for weeks, months even, day and night. It finally resulted in twelve numbers denoting the ideal lengths of the required rods. The ultimate outcome of all this was the leg of Animaris Currens Vulgaris. This was the first beach animal to walk. And yet now and then Vulgaris was dead set against the idea of walking. A new computer evolution produced the legs of the generations that followed.

These, then, considering the radius of the crankshaft of 15 (m=15), are the holy numbers: a = 38, b = 41.5, c = 39.3, d = 40.1, e = 55.8, f = 39.4, g = 36.7, h = 65.7, i = 49, j = 50, k = 61.9, l = 7.8.

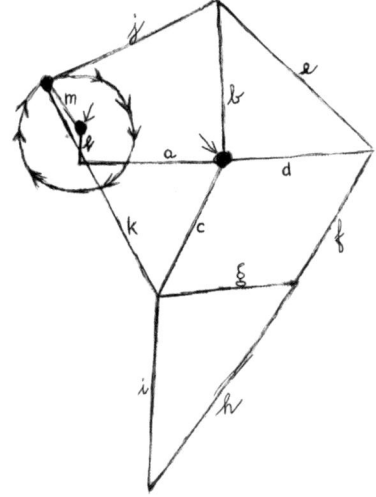

eleven holy numbers

It is thanks to these numbers that the animals walk the way they do. Their locomotion resembles that of a real animal. People often ask me how on earth I managed it. Needless to say, I didn't. The computer came up with the movement. I do have an idea why it is that the beach animal's gait resembles that of a real animal. Until now, I have said nothing about one of the criteria by which the legs were selected in the computer. Another assessment besides the shape of the walking curve was the amount of time the leg spent in the air. After all, a leg

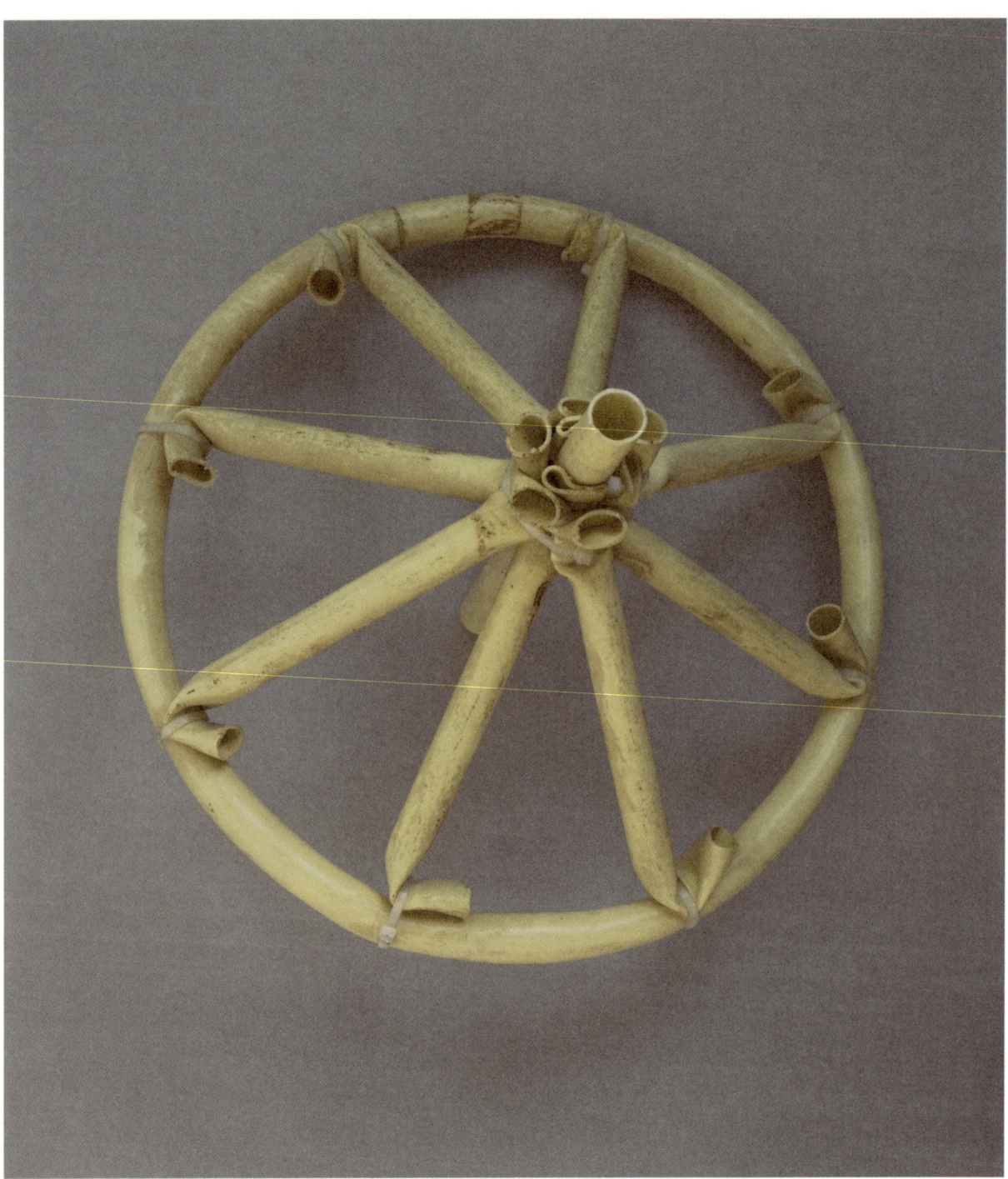

Animaris Properans, propeller front

hanging in mid-air serves no purpose. It doesn't contribute to supporting or balancing and is in fact dead weight. Legs that spend relatively little time in the air are more likely to pass the reproduction test. Over the generations, the movement became more and more characteristic of a leg. The acceleration the moment the foot leaves the ground is typical of leg movements, those of real animals too. Walt Disney had understood this completely when he animated Bambi. He exaggerated the principle of through-the-air-quickly, so that Bambi's motion became thoroughly organic. The fawn's endearing movements have a purely mechanical basis. Someone with a limp exaggerates the principle too. Their good leg shoots forward to regain the ground quickly, so as to burden the impaired leg at little as possible. We all limp slightly, if the truth be told.

Legs
Shifting objects across the earth's surface is no easy matter. The moon rushes through space, as do the satellites. Of their own accord, without an engine. Beyond the atmosphere it seems anything is possible. There is no friction, no air resistance. The bounds of the atmosphere are smooth but implacable. Within these bounds, it costs sweat and toil to move things. There is something unfair about this. But fairness apparently means nothing in nature's book. We, people and animals, must toil. Legs carry our bodies from one place to the next.
The mechanics of the leg are quite simple. Its duty is to carry the body while walking. Not that it does this all the time. There are moments when the leg is carried forward to make the next step. The leg is a tool you have to carry; you have to have it with you. Which is why it's attached to the body. The leg must be thick enough to carry the body's weight but thin enough to be carried itself. Each body needs legs of a thickness that represents it best. Elephants have thick legs and spiders thin ones. This optimum condition has a long and repetitive history. If the load is too heavy, the leg must be made thicker. But if the leg is made thicker, the load becomes heavier because of having to carry a thicker leg.
The leg produced in the Chorda is composed of triangles. The triangle is the basis of all truss constructions. Power pylons, the Eiffel Tower, there are triangles all over them. A square of rods can hinge at the corners; not so a triangle, a triangle is rigid. There is relatively much air and little material in the leg. It can support by virtue of the triangle, and is easy to lift and move forwards by virtue of the lightweight air. This is an ancient principle but it only dawned on me in the Chorda. In my self-chosen isolation I went without knowledge. I wanted to

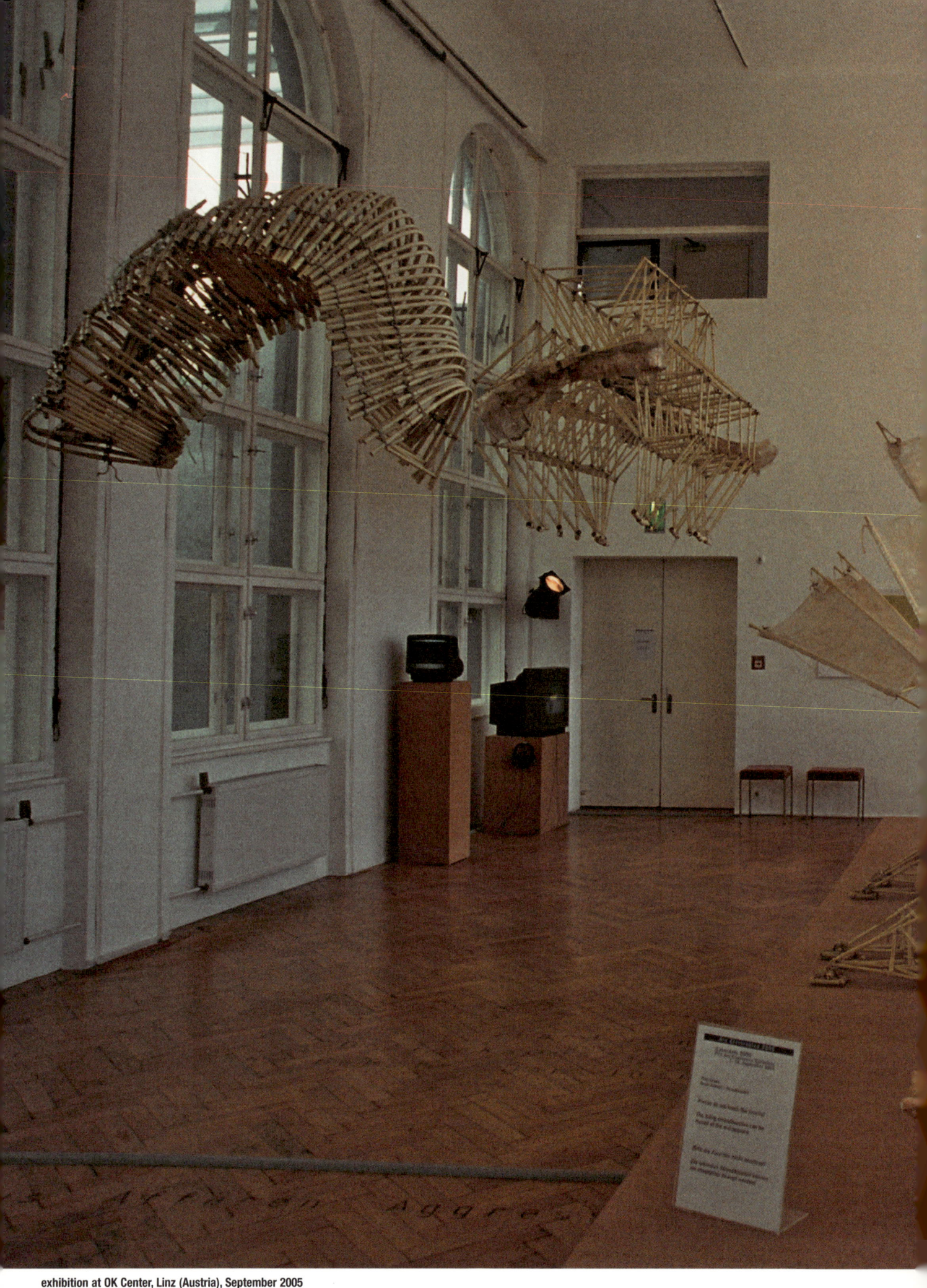

exhibition at OK Center, Linz (Austria), September 2005

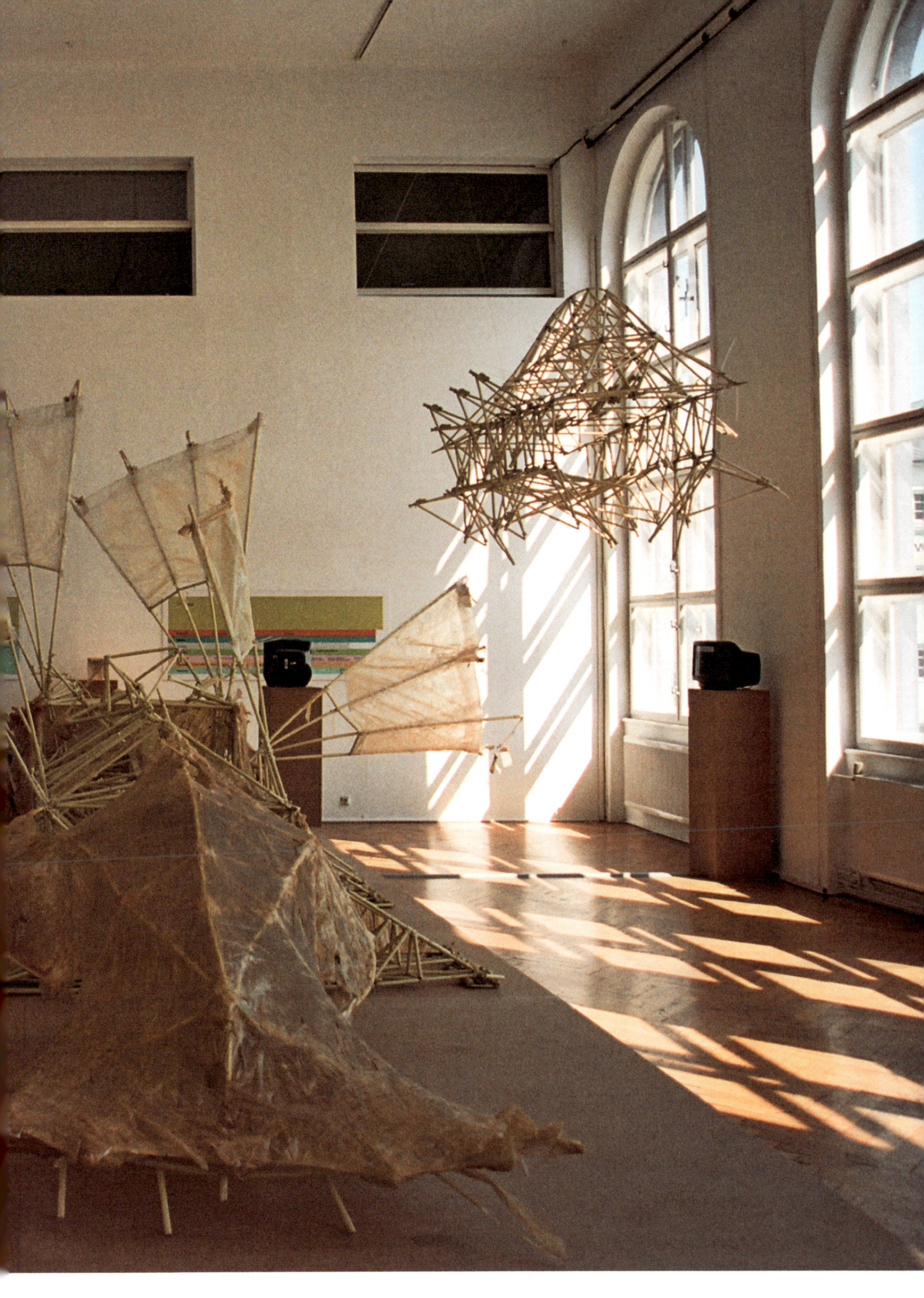

Machete

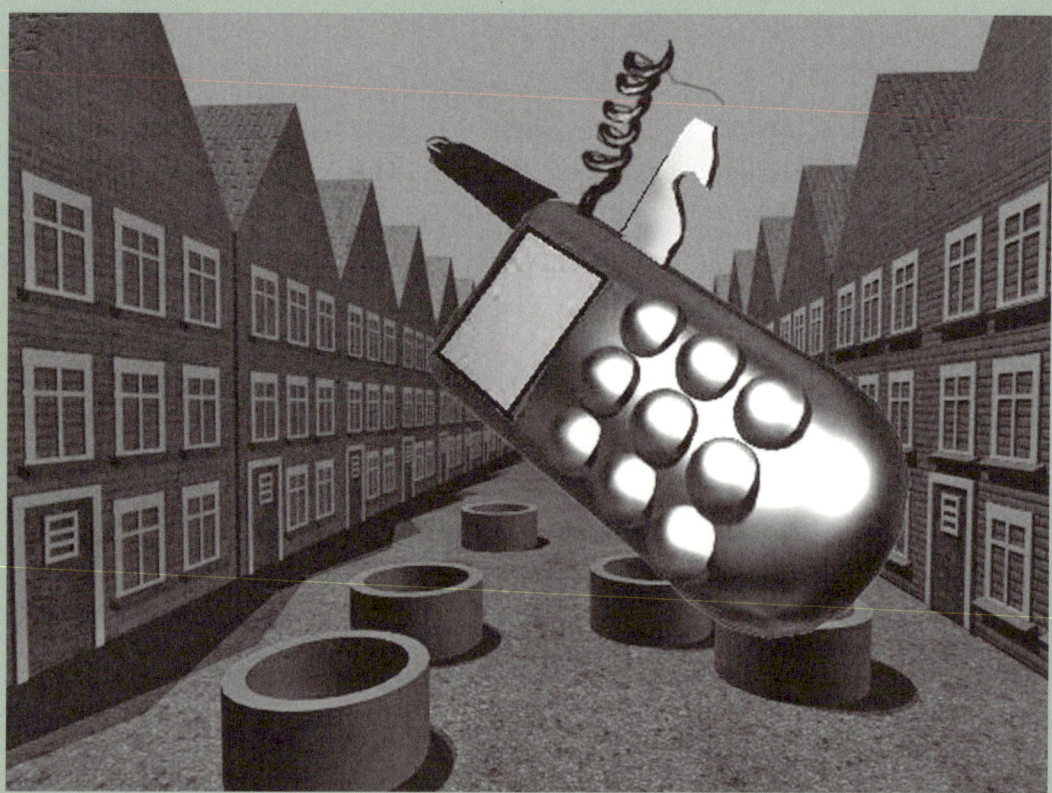

In the city where I live we have a legalized prostitution zone, a parking zone, an industrial zone and a begging zone. Begging is an innately human act. Not exclusively to alleviate need, but also to make clear that such needs exist. The begging zone takes up one street in the city centre. But the beggars there are not drug addicts or homeless but young students in nylon jackets. They want you to donate to Greenpeace, the Multiple Sclerosis Society, Amnesty International, Oxfam Netherlands. Every one a cause I support wholeheartedly. Still, I need to get through that street...

It's spineless of me, I know. But I do it all the same. I get out my mobile phone and cross the street at a brisk pace. All I have to do is hold my mobile against my ear to keep the nylon jackets at bay.

I've been doing this for a couple of months now and I see more and more fellow pedestrians rushing through the street, mobile at the ready. So I'm not the only one to use a phone as a machete to hack their way through the acquisitive undergrowth.

In the city where I live there's another street, where cars are allowed but only just. Nothing is spared in the campaign to make car drivers feel unwanted there. Speed bumps everywhere, chunky flower pots in the middle of the paving. They all say the same thing: get that car out of here.

And I agree with that wholeheartedly; the central-city area is for pedestrians. But there are times when I have go through that street by car. And then I do drive very slowly and very carefully. Still, the police stopped me on one occasion.

'Well, lads, what's the problem?' No, I'd never dare ask that way. I'm terrified of anything wearing a cap. 'Is anything the matter?' That was how I put it, trying to look as accommodating as I could.

'Yes, guv, your hand.'

'What's the matter with my hand?'

'There's a mobile phone in it, guv.'

Indeed, my girlfriend had just rung me.

That cost me twenty minutes and 180 euros. I could have subscribed to Oxfam Netherlands for eighteen months for that amount.

It was now clear that my mobile had failed me in its role of machete. Each job evidently requires its own tools.

I've pointed out the possibilities of the mobile phone as a multi-purpose tool on other occasions in this column. Let me just mention here the mobile with an aerial whose end is a crosshead so that you can use it as a screwdriver. A mobile can be as multifunctional as a Swiss pocket knife. I'm very pleased to see that the industry has realized this too. Now you have mobile phones that include a camera, a diary, a notepad, a tin opener, a bottle opener and, as we have seen, a street opener.

invent it all anew. And that takes time. That isolation was not a deliberate choice. I'm just not the type to visit the library before embarking on a project. This tendency to go it alone is more a question of indolence, or shyness.

The benefits of speed

Why should beach animals move about? Moving about is a property of animals. For example, it's a good thing sheep have legs, otherwise they'd be pathetic balls of wool lying around fields. The grass close by them would soon be eaten roots and all. The fact is, sheep simply have to move about. So it's a good thing they have legs.

It's different for a wolf. To a wolf, a sheep is something like a plant. A plant has to stand still, otherwise you can't eat it. Which is why a wolf has legs so fast that the sheep seems to be standing still. So wolves need not just legs but extremely fast-moving ones. A plant can't walk, let alone run. Plants have another tactic to move from one place to another. They use pollen which gets blown in the wind. Sometimes pips from berries hitch a ride in birds' stomachs to take root elsewhere when expelled as waste. A sheep can move faster than a plant and a wolf moves faster than a sheep. But they all move. What I'm saying is, that getting about is a characteristic of, and a necessity for, life. If I see someone jogging, I know that he or she is rerunning an old program. Running a marathon boils down to cultivated fugal behaviour. Skating and cycling too. Motoring events are called runs. We don't need to run, but we love doing it all the same. All hobbies can be traced back to old programs. Music is a mating call that got out of hand, dancing an over-the-top courtship ritual. Fishing and hunting have lost their economic benefits, although rerunning these programs makes a whole lot of men happy. Of course, one of humankind's special aptitudes is being able to think. You're happy if you think, especially if you think you're happy.

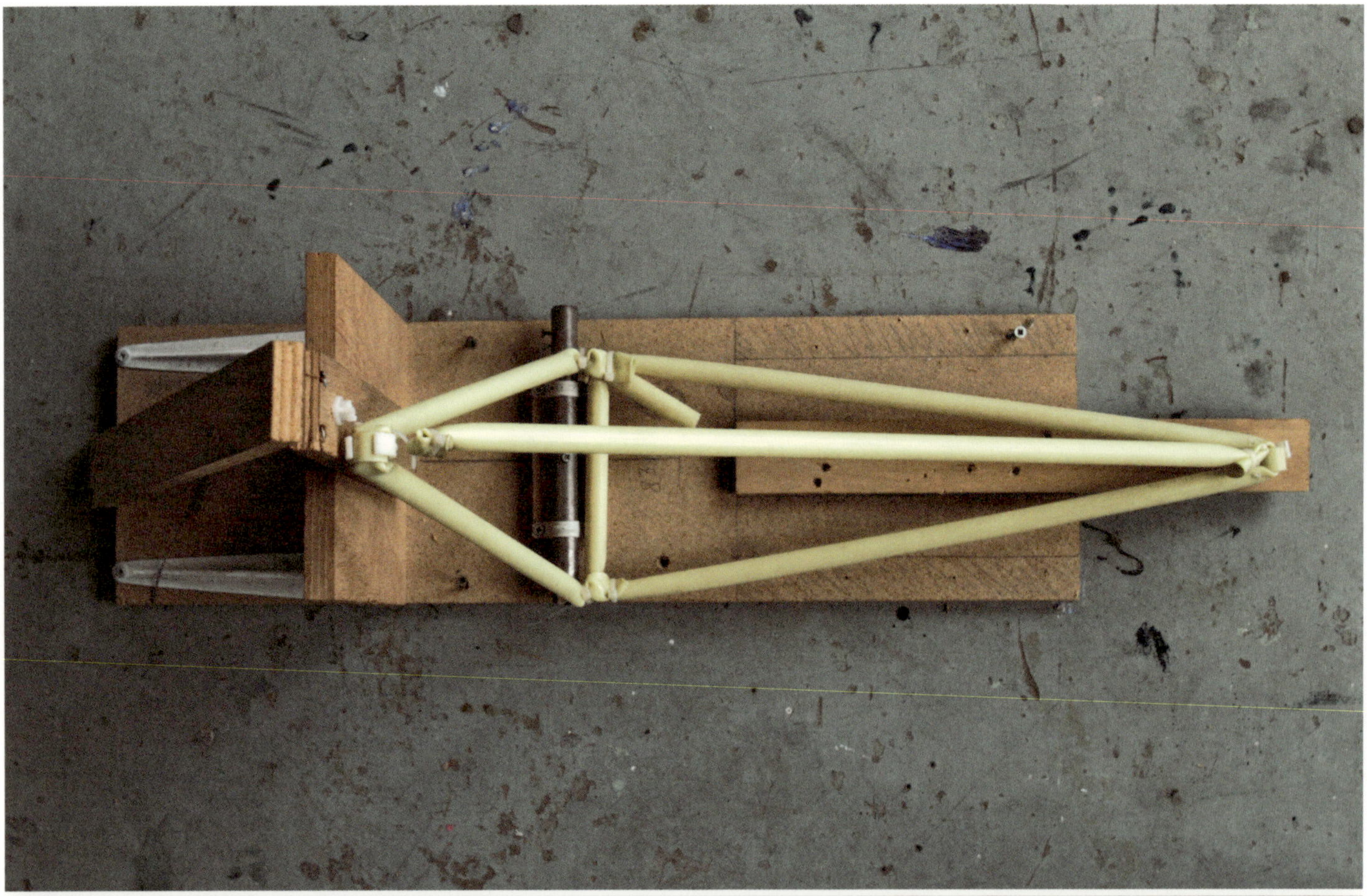

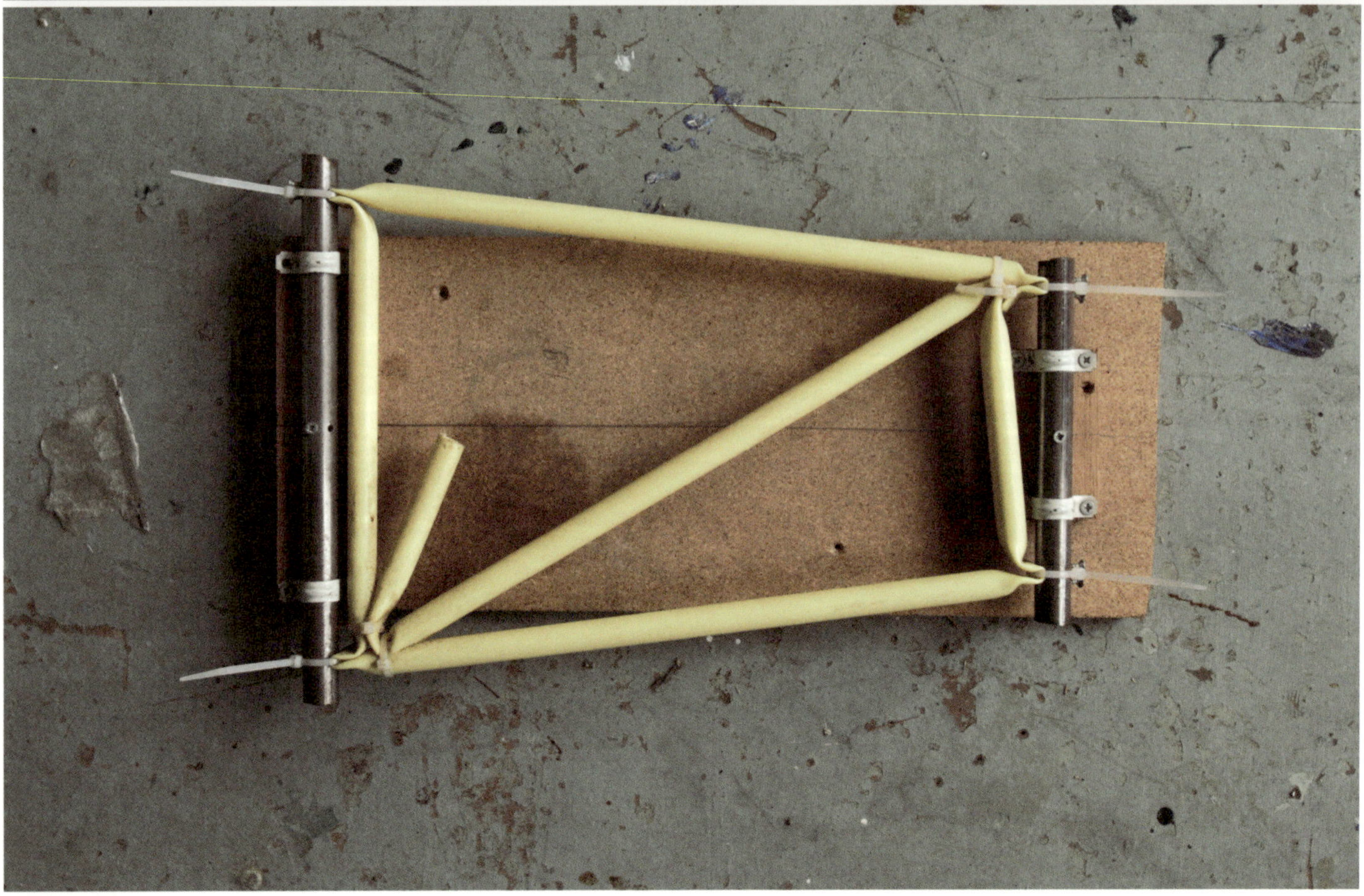

**lower leg mould,
upper leg mould**

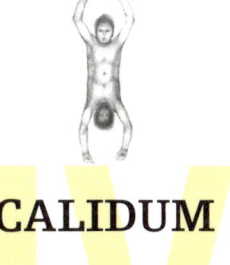

CALIDUM
the hot period (1993-1994)

Heat gun

The major find during the Calidum was the heat gun. In day-to-day life this is used to soften old layers of paint so that these can be scraped off. If you point this appliance at a plastic tube, this becomes as flaccid as a garden hose after a time. The tube can then be bent into shapes of all kinds and keep this shape on cooling. The heat gun proved an excellent tool for making the vertices of the assemblies tighter and tougher, giving a much more rigid skeleton. I discovered the heat gun only after three years. Obviously I knew of its existence but I didn't know you could soften plastic with it. People I know wonder how on earth it took me so long; they could have told me straight off. Oddball situations such as this occurred with greater frequency later on in beach animal history. You spend years frittering away time on something and then a possibility hits you that could have saved a lot of suffering and disappointment. If only I'd known. The heat gun proved equally good at making components identical. Just like biological life-forms, beach animals consist of numerous identical components. Each component had its own mould. This was made of wood with small metal tubes fixed in it with the same diameter as the plastic tubing.

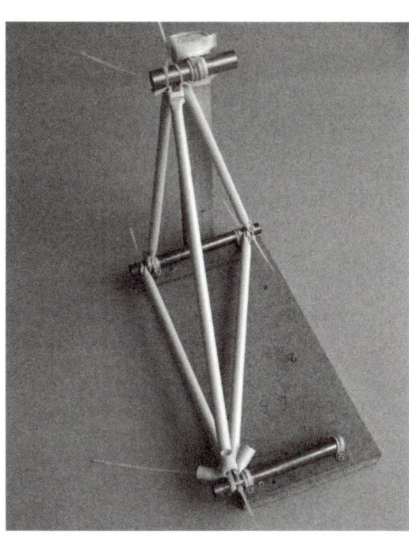

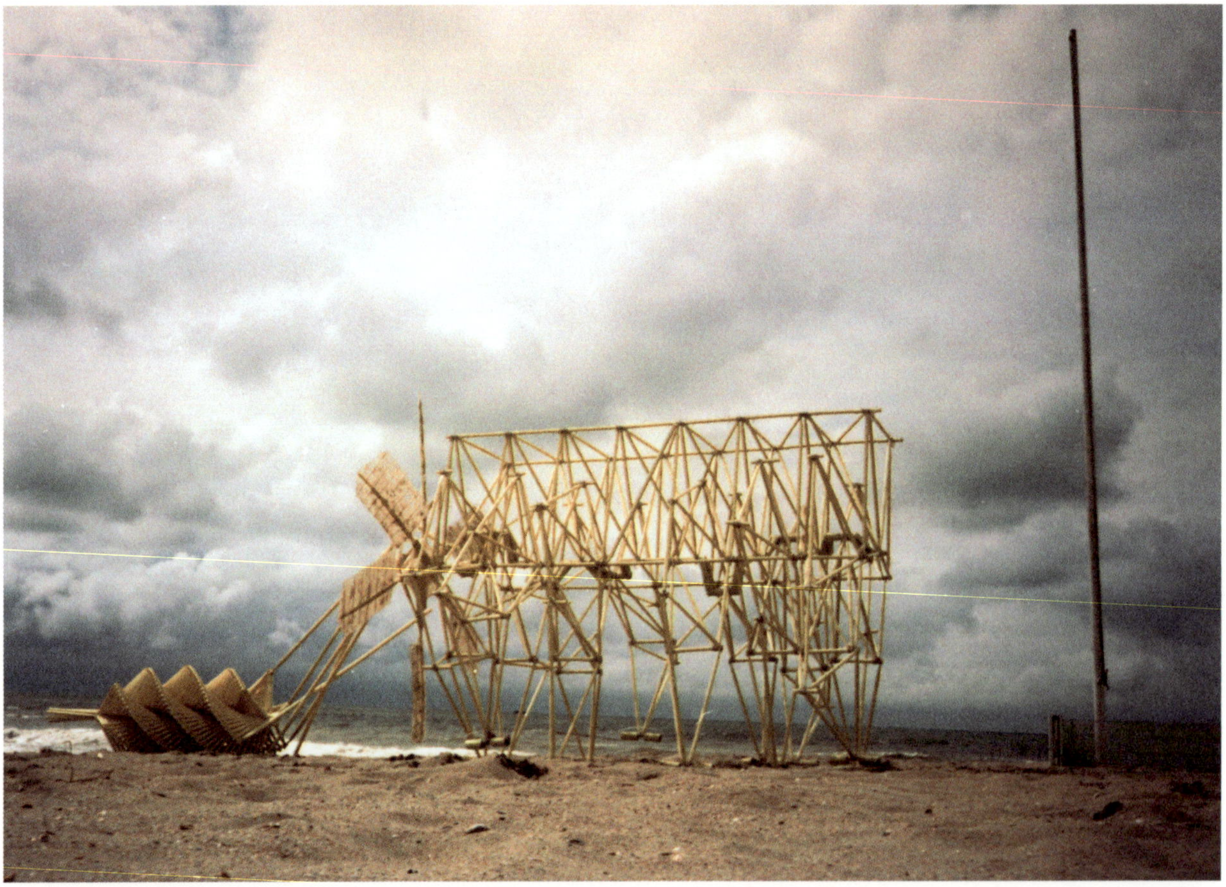

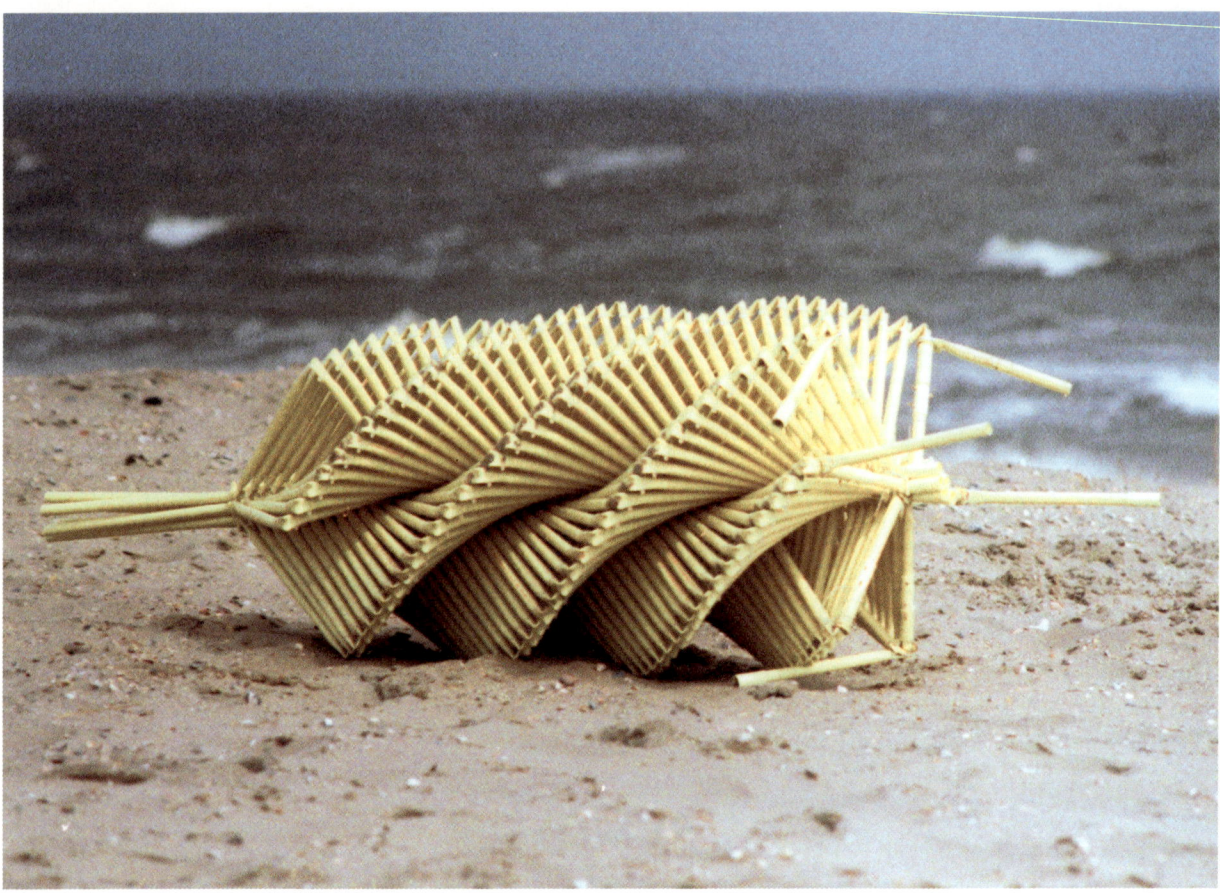

Animaris Ancora

The tetrahedron is one form that kept cropping up in the beach animals. It was assembled by heating tubes at their extremities and attaching them to a small piece of metal tubing with a cable tie. The cable ties were removed by cutting when the shape was complete. Later the tetrahedron was incorporated in the structure as a whole using new cable ties.

The tube-cutter was another late find. This is a special pair of scissors able to cut through a tube in a fraction of a second. You can cut a hundred tubes in two minutes. It only entered the picture at the end of the Chorda. The metal saw that had served me until then was primarily a source of irritation and minor wounds. The heat gun has a button with two positions, one for high-temperature use and the other for low. It transpired that the high-temperature position was suitable for stripping off old layers of paint and the other for softening up plastic tubes. I hadn't been aware of this difference at first. I had always used the gun on 'hot' as it did the job quickly. After a year, joints made this way were found to be brittle. Animaris Currens Ventosa was even no longer capable of standing up. It was suffering from bone porosity. At exhibitions, it was attached to the ceiling with thin cords to prevent it collapsing. Animaris Sabulosa is only able to lie there, spreadeagled, for the same reason. All of which is why the Calidum is also known as the hot period. The Tepideem by contrast is the less-hot period. The two names derive from the temperature of the heat gun at the time.

Animaris Currens Ventosa and Animaris Sabulosa were the two inhabitants of the Calidum. A. Currens Ventosa was the first beach animal to extensively incorporate the eleven holy numbers. Its legs could be made quickly using the moulds. Ventosa began with 56 legs. In the end only 48 remained. It proved easier to manage with less legs. Ventosa had two large wings that were actually too heavy for it. It could barely turn. One day I set off for the beach at Oostvoorne in a lorry driven by my brother Kees. This was to be a red-letter day for Animaris Currens Ventosa, the only day it has actually walked on the beach. It was also a crucial day for publicizing the beach animals. I had help from a number of students. Into the hands of one of them (Adriaan Kok) I thrust a camera. He took the photo that would make its way around the world. It was published in countless magazines and Animaris Currens Ventosa became the figurehead of the beach animals. There was almost no wind that day. The wings were unflapping and seemed destined to stay that way. There was a moment of panic as the tide suddenly began

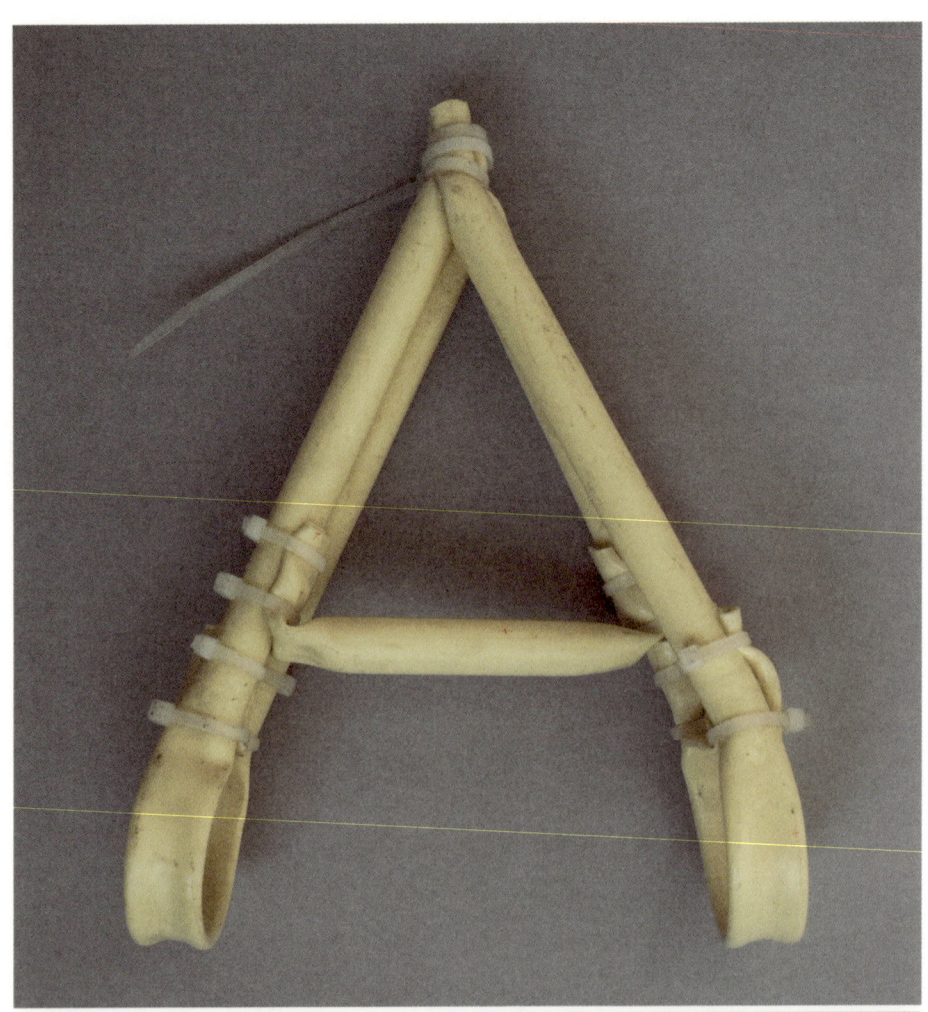
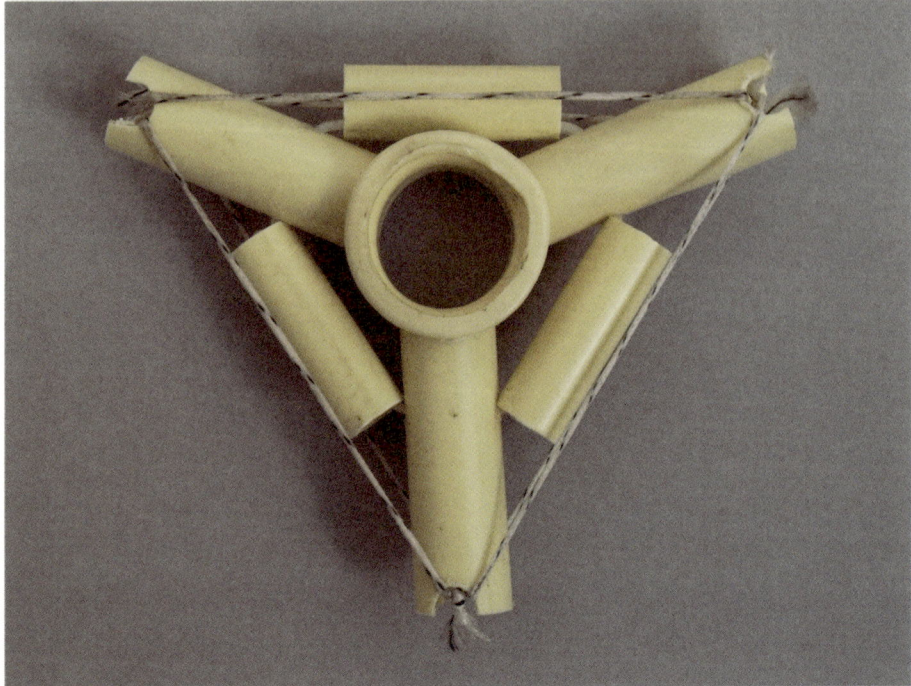

**pry-bone of Animaris Properans,
guide-piece (small)**

coming in, all the equipment had to be hurried off the beach and the animal itself saved from certain death by drowning. Then, during all this, a miracle happened. The wings began to move as if driven by a divine wind. They may not have flapped more than ten times altogether, but they had flapped. We had to fake the footage. I'm in it but invisible, tugging and shoving at the wing pins. And the animal had to be pulled by a thin cord for it to walk. Once out of the frame of the video camera, the beast collapsed under the sheer weight of its wings. Talk about timing. That image of the walking, flapping A. Currens Ventosa cleared the way for the other beach animals. That brief video got me a grant from the Netherlands Foundation for Visual Arts, Design and Architecture. A. Currens Ventosa is now hanging in the Haagse Hogeschool as an extinct species. It only managed one eventful day on the beach. Animaris Currens Ventosa means 'beach animal running on wind'. Animaris is a corruption of the word animal and *mare*, sea. So they are in fact sea animals. The names of the beach animals reflect the capacities they should have had and not their performance in practice. Animaris Currens Ventosa for one could not run on wind power, at least that's the general assumption. Actually we don't know because it has never been tried. The one day it was on the beach there was scarcely a breath of wind. Animaris Sabulosa did run on wind, sideways. It also gathered sand (hence the name Sabulosa) in minute quantities.

I had a container at my disposal on the quiet beach at Scheveningen, south of the harbour. The container belonged to Aphrodite, a seaside refreshment hall, on whose terrace I drank coffee between operations, gazing out at the horizon. I could work there in peace in the springtime. It was much busier there in summer. Then I was on holiday myself, returning only at the end of August. A sea container is not exactly a workplace, more somewhere to store the animals at night. This was essential at the time. They were so delicate in those days, I couldn't leave them unprotected. They would collapse in the wind. That happened enough anyway. The legs usually broke at the joints, just because of the wind. It was only in 2004 that the animals were strong enough to stand unattended in wind and weather. Even then, they were anchored with a tent peg. They could hardly be let loose yet. First they needed a little training. They had to seek out the dunes themselves at high water, anchor themselves to the beach if a storm threatened – and then pull themselves free afterwards and turn back into the wind. It could all be done, and in time it would be.

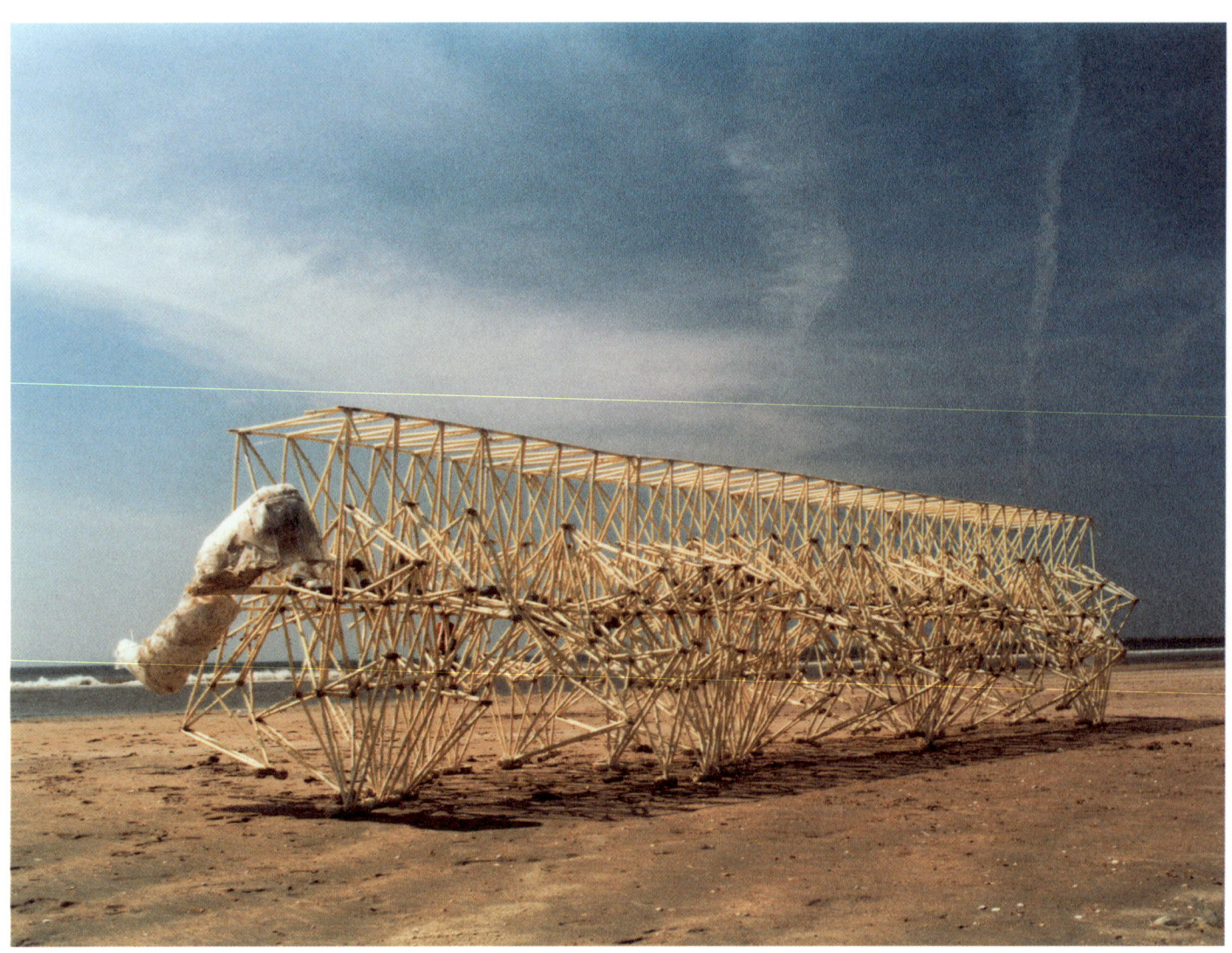
Animaris Geneticus

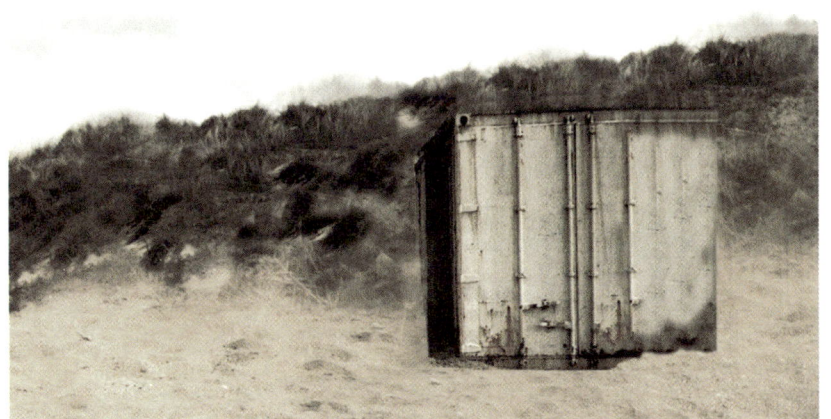

workplace on the quiet beach at Scheveningen

Every day I travelled the ten kilometres from Delft to the beach by bike. On the way, ideas would come to me. In the morning I'd cycle there bursting with optimism only to return in the evening in a state of depression. Asleep at night I was gripped by an inexplicable transformation – radiation from the cosmos or whatever, tossed on the waves of fantasy. At about six I would awake armed with new ideas.

Sand
Glutonia is a marmot-like creature the size of a badger. A native of Southeast Asia, it is remarkable in that it owes its camouflage to a resin-like fluid secreted by its skin. This causes leaves and twigs from the forest to cling to it. Done up in full camouflage kit, it rummages among the bushes. When danger threatens, its muscles tense so that it can't move. It stays there stock-still. It is only when danger has passed that its muscles relax. This was the scenario until recently. As you know, Southeast Asia is now undergoing deforestation on a massive scale, with just sand drifts in its wake. This means that for most animals their camouflage has lost its effect. Not so Glutonia. Instead of leaves and twigs, grains of sand now cling to its skin. A few examples of this transformation have been captured on film. I have seen this footage. Glutonia has become an entirely different animal, sand-coloured, something like a desert rat but much larger and less rapid. Patiently it goes in search of food. Brushwood, dead birds, anything will do. So you see, there are always a few survivors. And Glutonia lived happily ever after.

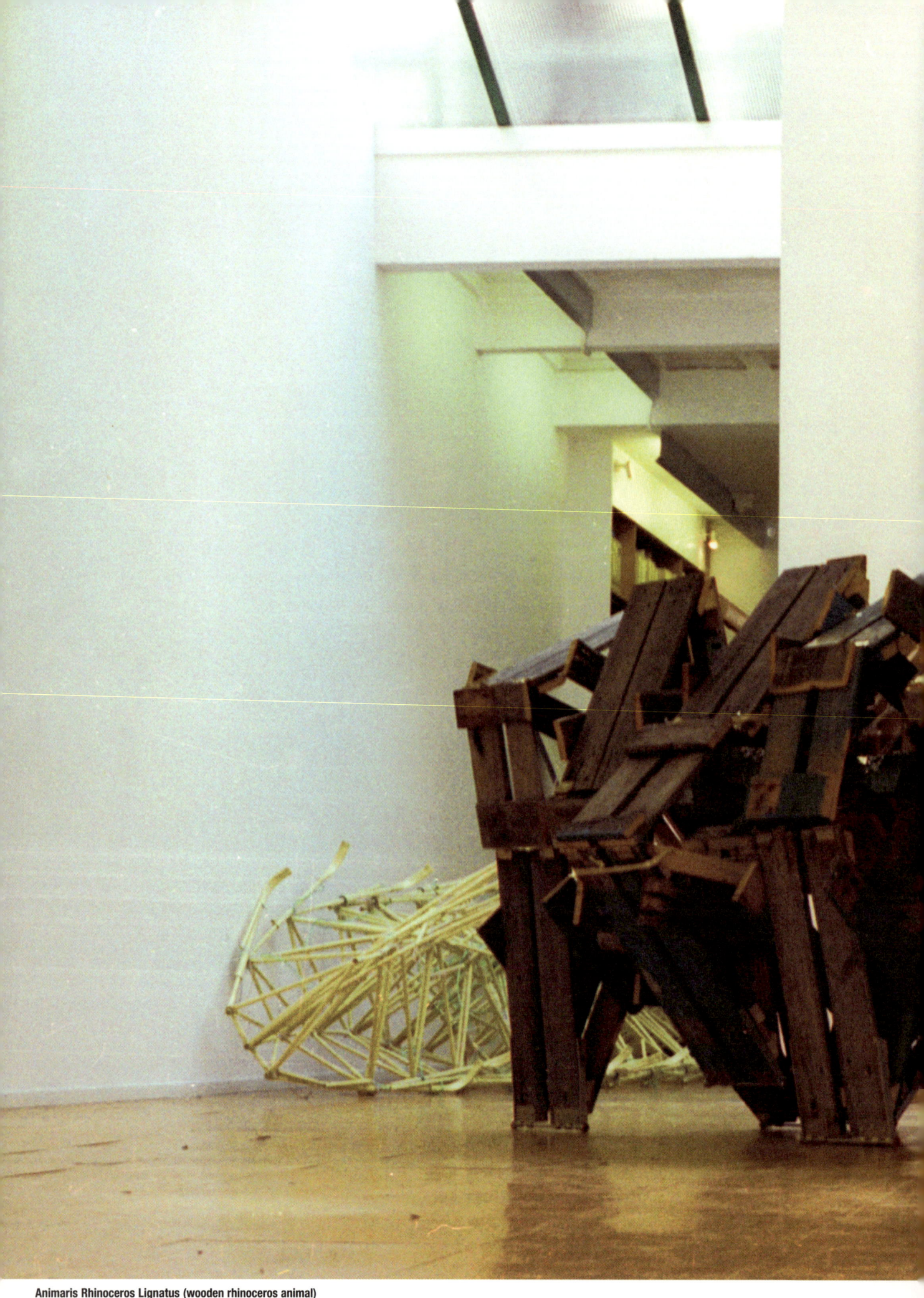

Animaris Rhinoceros Lignatus (wooden rhinoceros animal)

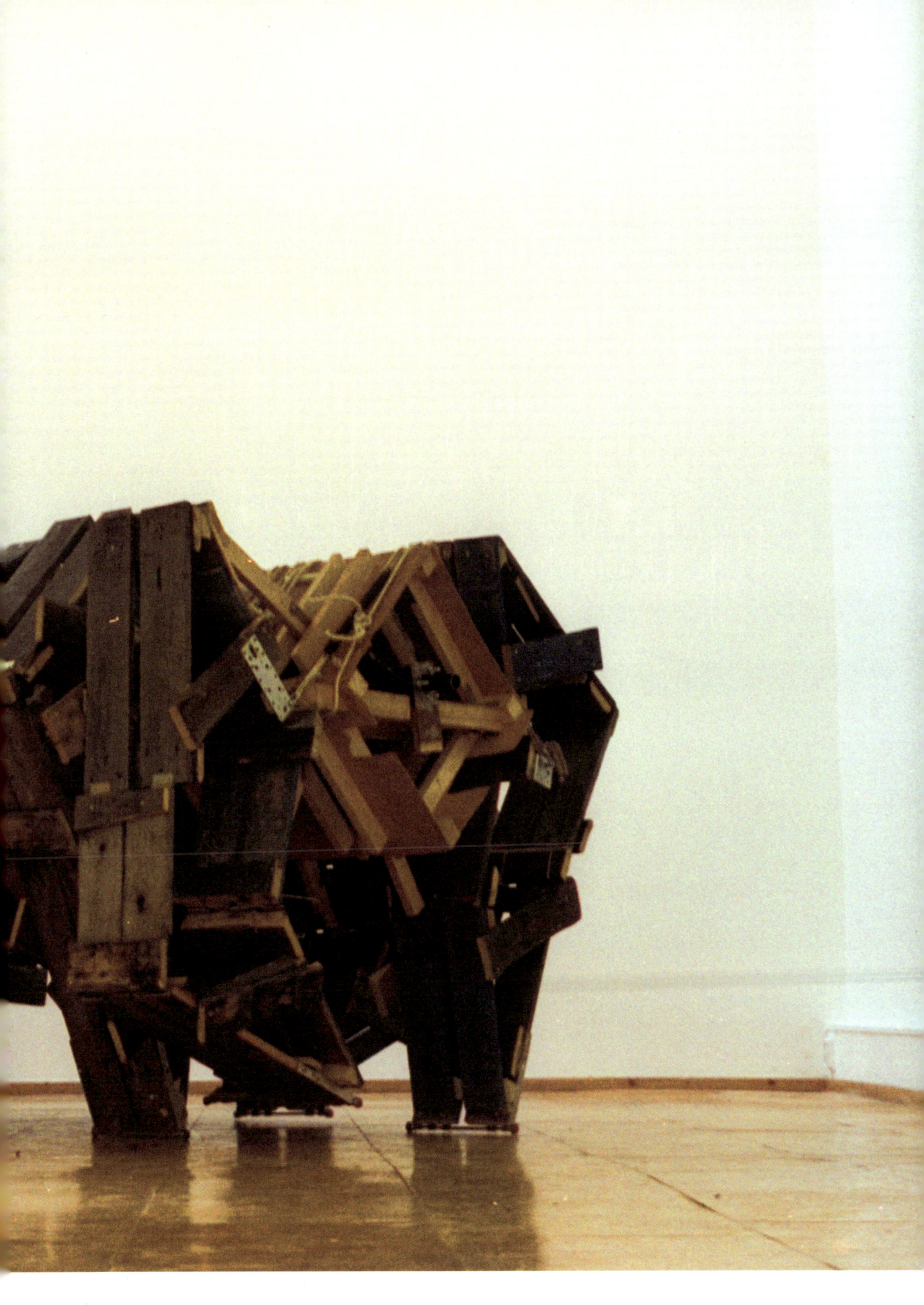

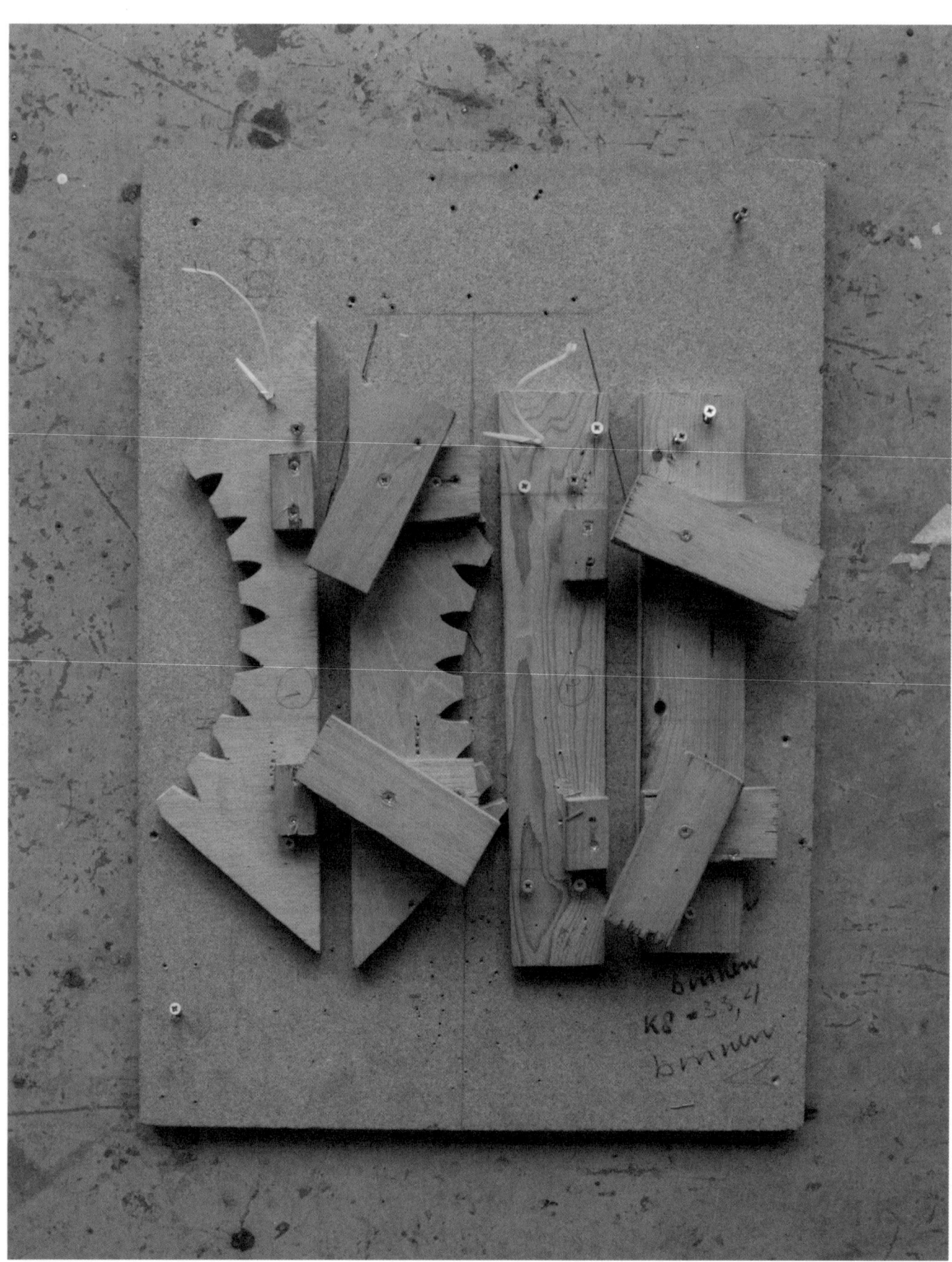

cog ribbon mould combined with crush mould

Dear reader, the above story is entirely made up. A product of my incorrigible imagination. But be honest, you believed it! In other words, the story could very easily have been true. There are some who believe that the current state of nature is merely the luck of the draw. Looked at that way, evolution could have taken an entirely different course. I don't know about that. There is so much variety in nature as it is, that I can hardly imagine another variety not overlapping with the nature we have now. What is certain is that nature has left a number of technical principles unexploited. The Creator left out wind as a source of energy. Let's say it didn't occur to Him. An animal species that requires little or no food because it gets its energy from wind is capable of lording it over nutrient-poor windy areas such as beaches. It doesn't need to compete with the rest of nature. Here resides a major advantage for the beach animals. Another is their unconventional means of reproduction, which we shall be looking at in the next chapter. I have tried to forget biological nature and genuinely create a new kind. Preferably using principles other than the existing ones such as food chains and procreation, in order to uncover the universal laws of nature. In this I have only been partially successful. Legs, for example, still prove to use the least amount of energy as a medium of travel. I have tried other means but haven't found anything as effective as legs. I had devised this animal, Emfime. Emfime uses its lungs to get about. It breathes in through its nose and out through its mouth. This mouth is huge and has its opening facing down like an air cushion on the ground. Each time it exhales, the mouth (and with it the entire body) is raised a few centimetres from the ground at which point the tongue sweeps back. This carries Emfime forwards like a hovercraft.

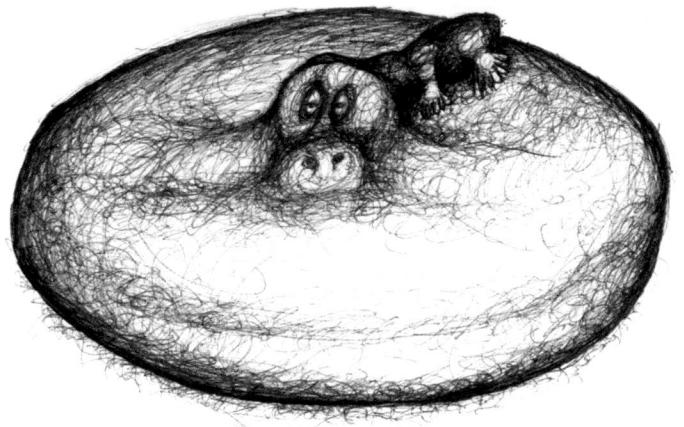

Emfime

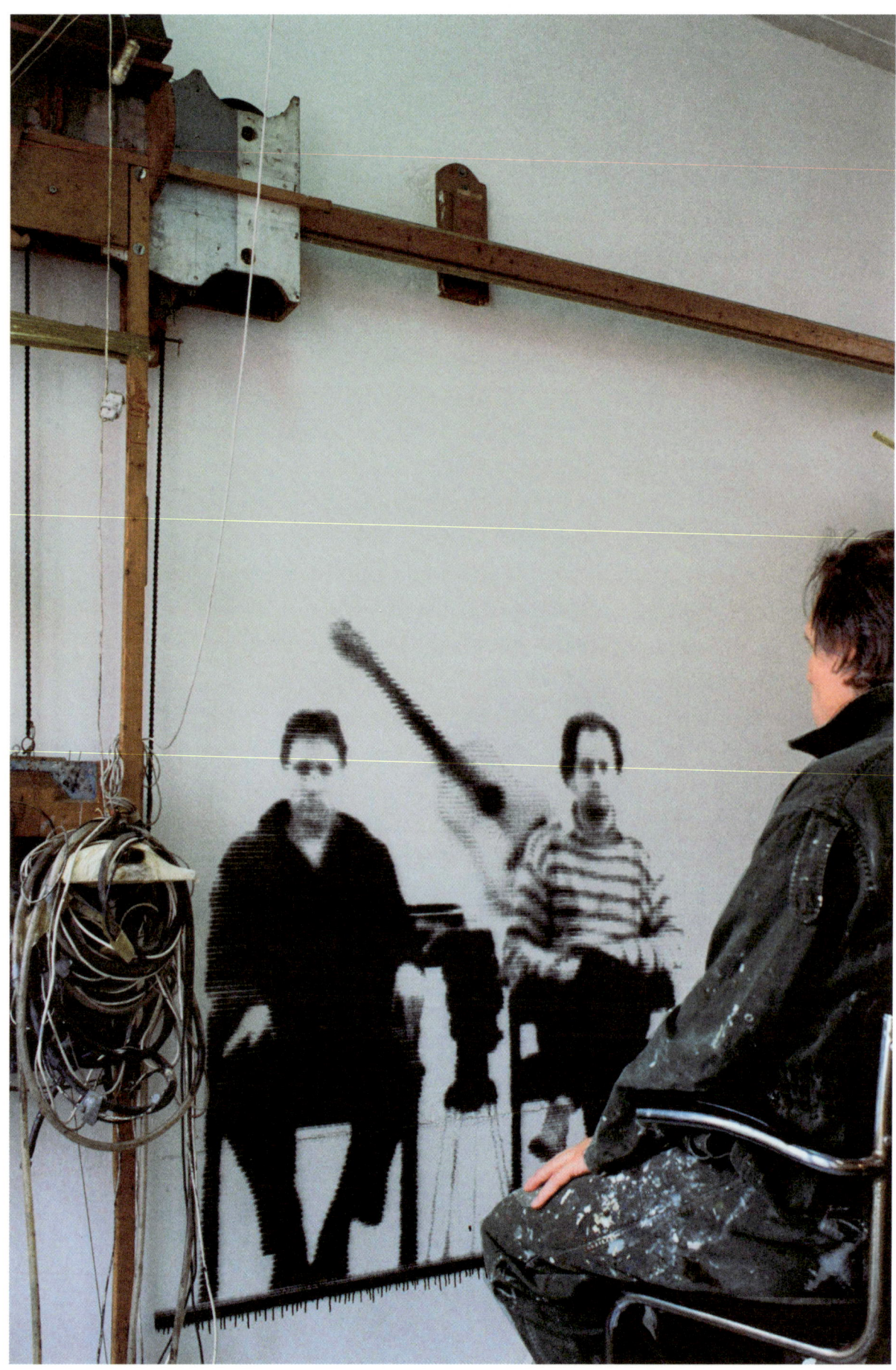

Painting-machine, 1984

The story of Glutonia suggested itself to me after an incident on the beach when making the spoiler for Animaris Sabulosa. First I made a frame of yellow tubing. Then I covered it with strips of adhesive tape. I had of course abandoned tape after Animaris Vulgaris with the advent of cable ties. Still, it came in very useful when making the spoiler. If you're willing to be uneconomical with it, you can make the most marvellous undulating forms with adhesive tape. The result on this occasion was most promising. But guess what? After the animal had stood there for a while, the wind blew sand from the beach against it and this stuck to the tape. This gave the beach animal a camouflage without this ever being my intention. I realized there must have been fortuitous occurrences of this kind earlier on, during biological evolution. Sweat, for example. Sweat is slightly sticky by nature. A mutation in which sweat is slightly more sticky than usual would seem quite feasible. All the same, it doesn't seem to have happened. At least, I've never heard of animals that secrete glue. Still, Animaris Sabulosa (sandy beach animal) owes it name to this chance occurrence. Sabulosa can also dig sand. There was a small spade on its tail that was moved by a nylon thread as it walked. This scooped small amounts of sand into the point of the animal's bent tail.

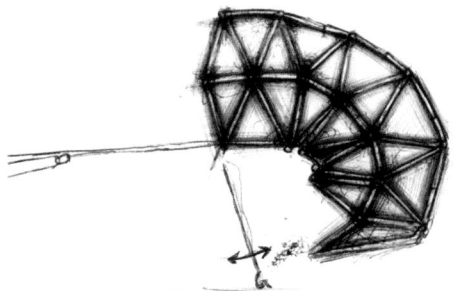

scooping sand

When the animal turned round, it stretched its tail and the sand fell out. The idea behind A. Sabulosa was that when oriented into the wind it would walk to and fro, crosswind, between the sea and the loose sand of the berm. Given the same wind direction, it would move along the same path and deposit the sand at the same places (the turnaround points). The sand dropped on the beach face would be washed away but that deposited in the berm would stay where it was. At least, that was the idea. It only worked now and then. I had to lend more than

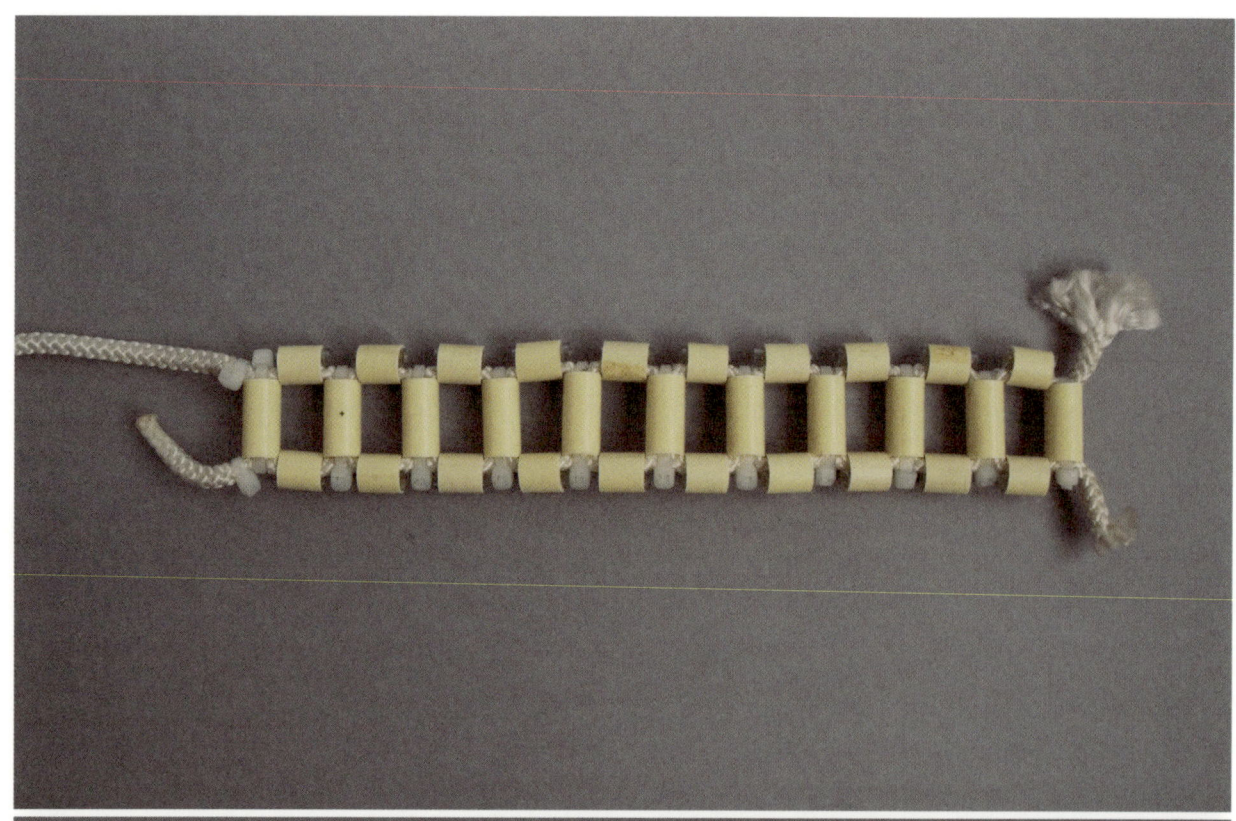
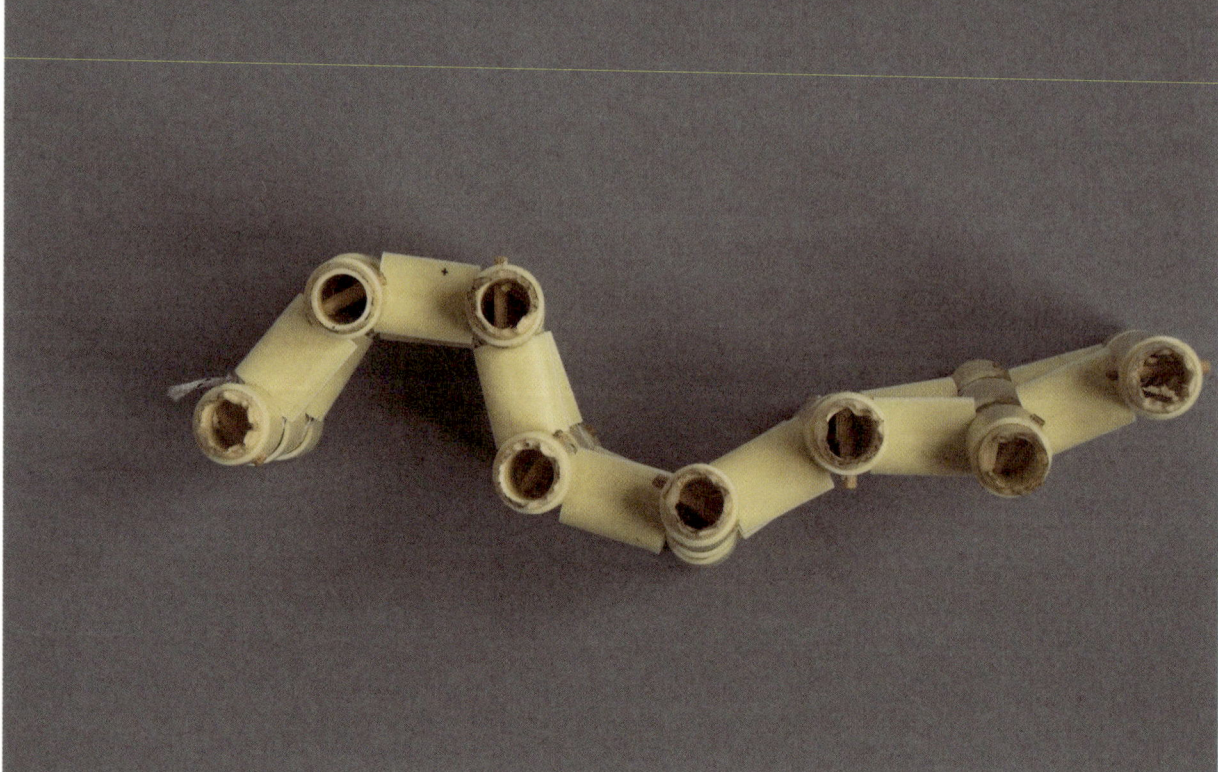

chains from Animaris Circodentis Primus
and Animaris Circodentis Secundus

just a hand. In the body of Sabulosa there was a gearbox that operated according to the principle of winding up and rolling off reels. Wind up a reel and its diameter increases due to the additional amount of cord on it. By the same token, the diameter of the reel from which the cord is unwound decreases. In other words, there is a change in the transmission ratio.

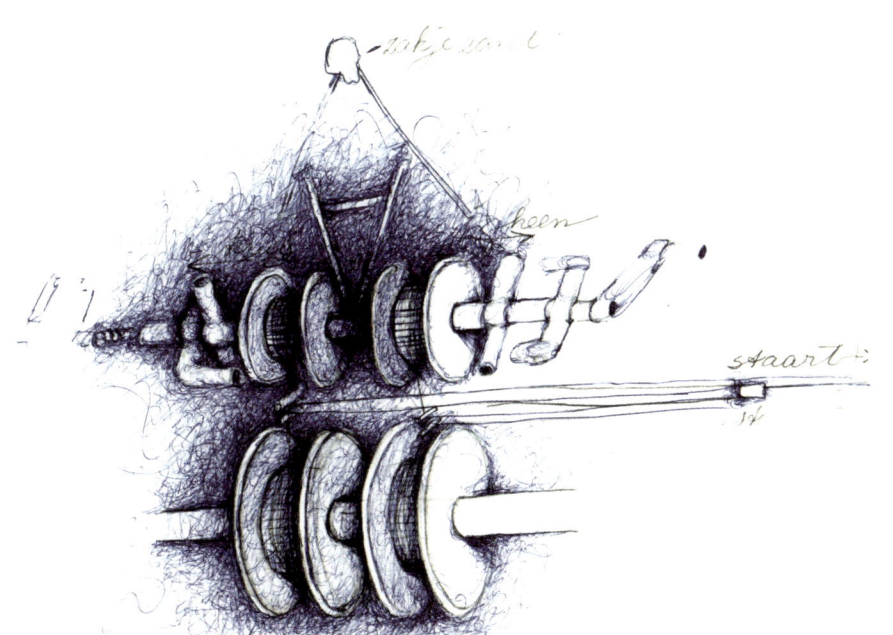

gearbox of Animaris Sabulosa

Sabulosa's gearbox worked like that of a car or a bike. Only it didn't change in stages; it didn't go from first to second, but shifted gradually during the walk with the changing diameters of the reels. The drive reel became steadily thicker and the roll-off reel steadily thinner. This meant that it travelled in a high gear – call it fourth – through the loose sand of the berm. The relationship between the reels changed when the animal reversed direction. The low gear speed of the new reel slowed up Sabulosa's pace although it did allow it to exert considerable power. Sabulosa was the first beach animal capable of walking at right angles to the wind. It also had primitive senses; two sensor wings at the front for determining whether it was in the berm or at the beach face. These wings flapped at just about the same frequency as the drive wings at the rear. Each pair of wings was lashed to its own crankshaft. The two shafts were aligned and tied together with a cord.

Floating

I hadn't been in a church in donkey's years. Nothing ever came of it. It took a funeral to get me back into God's house. Luckily the Catholic God is a merciful God.

First thing after that I went to Communion. As the host or consecrated bread mixed with saliva in my mouth, the memories came flooding back. Images from my childhood: white stockings, patent leather shoes, trousers by Terlenka, rosary in a leather pouch. These things had long remained caked on the margins of my memory. Besides offering a nostalgic backward look, they briefly placed my body and my life in the broad context of the universe.

Immediately on entering, I was struck by that distinctive smell of incense and candle wax. That aroma alone gave me a slight sense of floating. The Eucharist did the rest.

The Eucharist or Blessed Sacrament derives its powers to uplift from the blessings imparted to it during consecration. That done, this transcendent power is no longer influenced by the mass surrounding the host. Herein lies its effect of lifting up the believer.

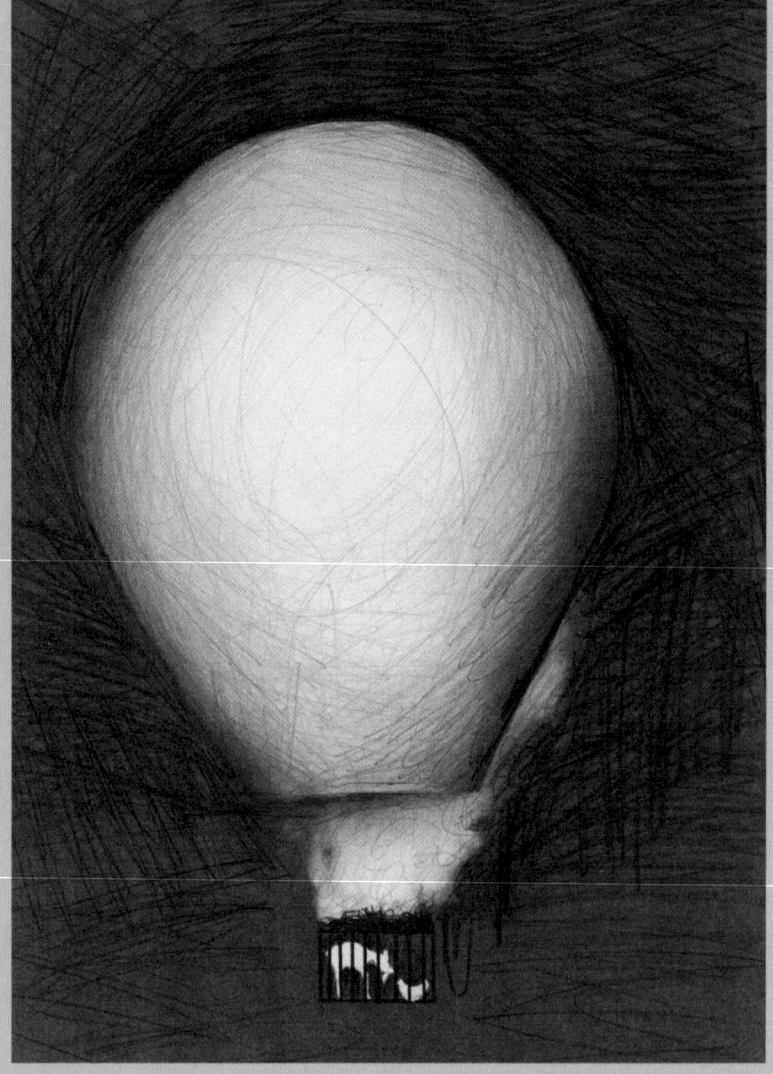

Having taken in and swallowed the host, you are then entirely free from the forces of gravity. Once in flight, you obtain another perspective on life, the becoming of the world, the heavens and all the other matters that are beyond our comprehension yet continue to fascinate us.

I would like to compare this rising up aided by the host with being enthralled by art or literature. You are seeing existence through the lens of imagination. Free of your ego. Free of the software of survival. But let us digress. 'Smoke rises.' Thus reasoned the Montgolfier brothers. 'So if you can capture the smoke in a balloon, it will rise.' And so it came to pass. In 1795 a balloon took to the air. A cage suspended from it contained a sheep, a duck and a rooster. Under the balloon itself was a fire of straw so as to produce as much smoke as possible.

We now know that the balloon rises not from the smoke but because of the heat from the fire. The Montgolfier balloon may have risen aided by erroneous reasoning, but rise it did.

The question is whether there is such a thing as correct reasoning. For in each correct argument resides one that is more refined. Not long after we discovered that it was the heat that made the balloon rise, it became clear that the upward force was caused by molecular movement. More molecules are colliding at the bottom of the balloon than at the top. This is what makes it rise.

There are bound to be dozens of better arguments residing in the one just named. We are unacquainted with them, which is why for the moment we allow the balloon to rise using the wrong argumentation. Phenomena like the rising hot-air balloon give us the impression that we can understand them. This is most accommodating of those phenomena but at the same time stupid of us to fall for it. Of course, we would be even more stupid if we didn't.

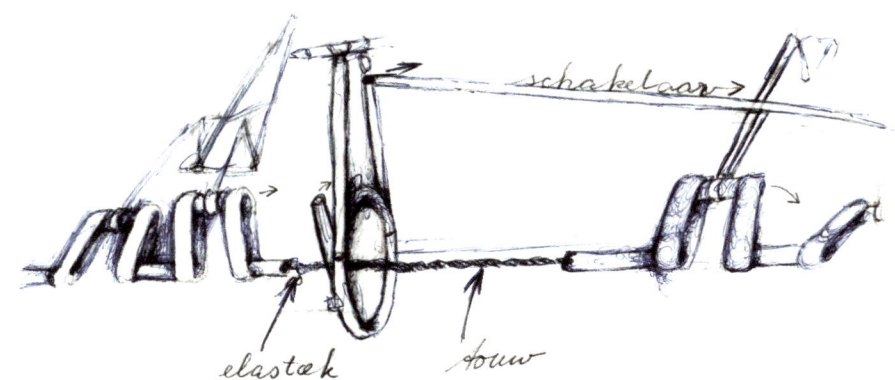

knotted cord is shorter

If the animal got bogged down in the breakers of the beach face or the loose sand of the berm, the drive wings stopped flapping since they were attached to the legs. The sensor wings, by contrast, flapped on regardless so that the cord between the axles became knotted and consequently got shorter. This activated a switch (a weight consisting of a bag of sand) causing the two reels to change places.

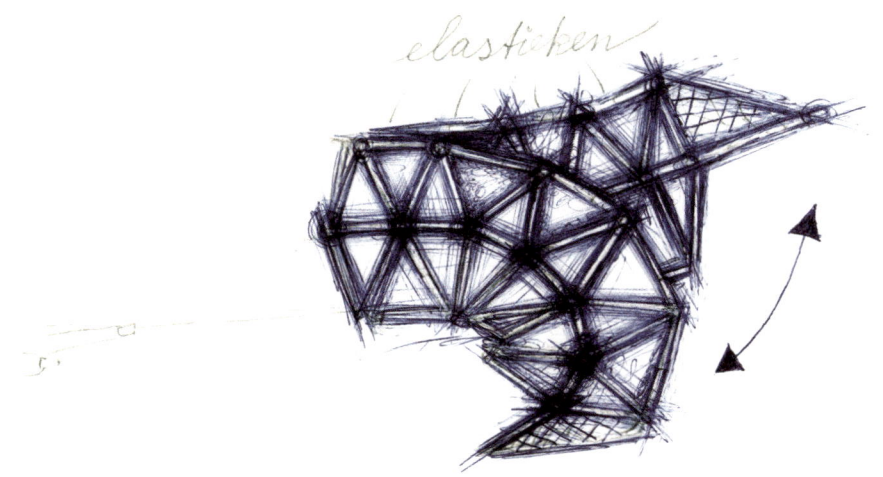

the tail curls

Sabulosa stretched its tail using a system of nylon strings. Its drive wings began to flap again. Then its tail curled and it walked in the opposite direction. In strong wind the drive wings had the tendency to flap too far. They then pulled on nylon strings in short tugs which drew in the plastic foil of the wings. Sabulosa reefed its wings, so to speak, to prevent them flapping until they broke.
Admittedly, the systems of reversing direction, reefing and digging described above worked in theory but exhibited shortcomings in practice. Later, in the

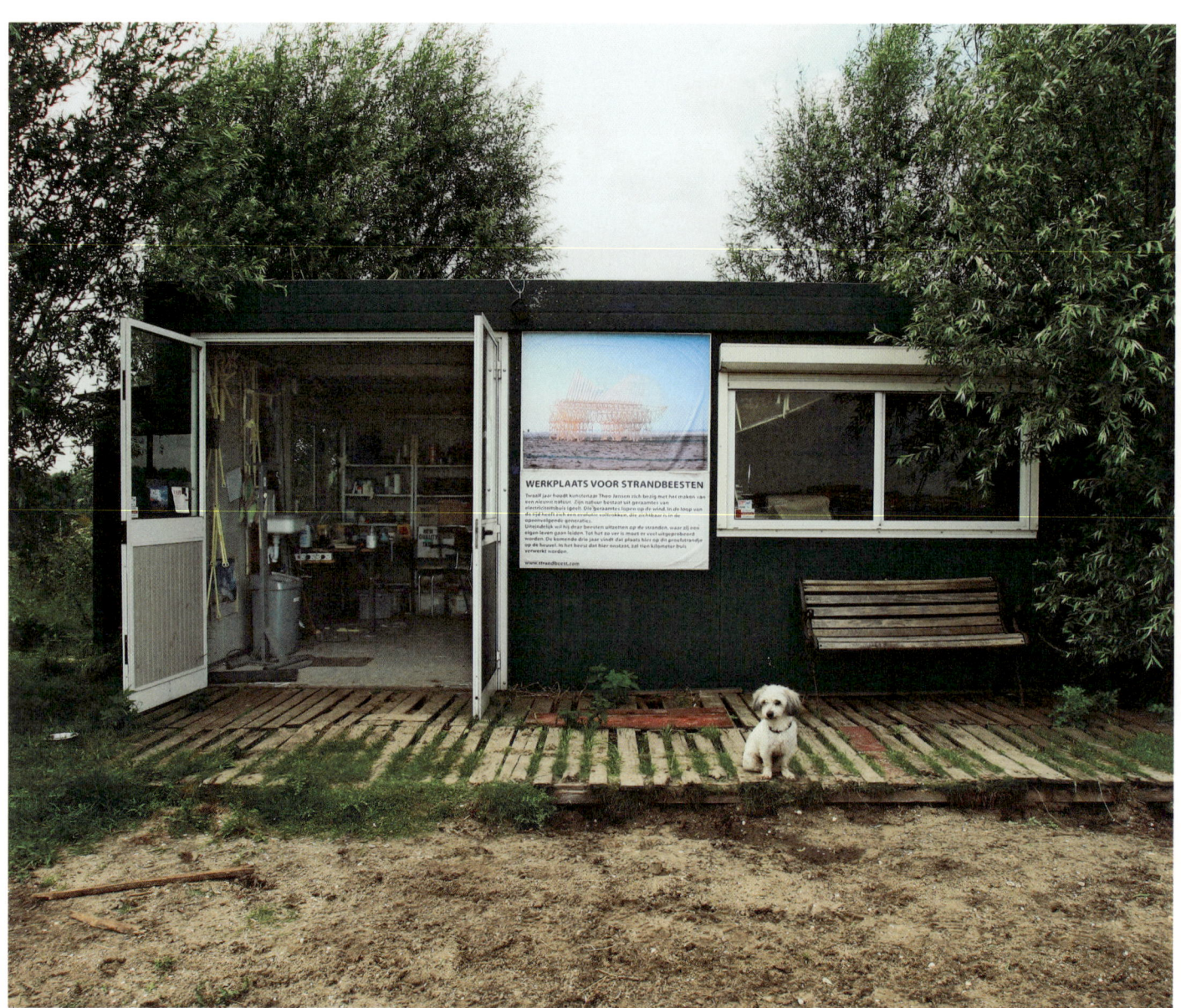

Mr. Murphy

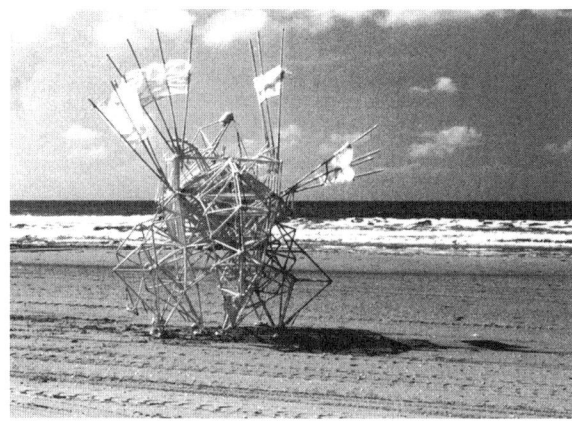
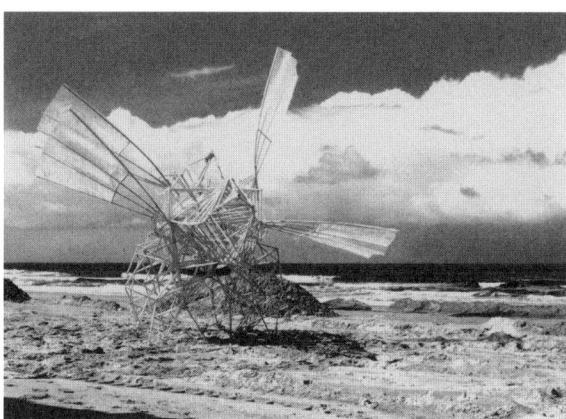

Sabulosa Adolescens reefed and unreefed

Cerebrum, the reefing system did become operational. Despite these repeated disappointments I worked with a boundless pleasure. I was disproportionately happy with the minor triumphs. Evidently I was less concerned with getting the things to work than with learning. My quest on the lonely beach was more a reconnaissance than anything else. With Animaris Sabulosa I was searching for ways for the animals to dig and reverse direction. I also addressed myself to the business of protection against storms. These were component issues, all of which needed tackling. It just required a little patience. Another thing: I've only been at it for seventeen years. Biological evolution took a great deal longer. Still, that must have had its own share of fumblings, particularly at the beginning. I sympathize with the Creator – assuming He exists.

Animaris Speculator
The outer edges of the beach are marked by sea and loose sand respectively. Animaris Speculator was to all intents and purposes a small, young beach animal that had remained attached to its mother by some kind of umbilical cord. This cord consisted of a tube containing a nylon thread which extended from a reel on the crankshaft of one animal to a reel on the crankshaft of the other. If one animal started walking, so did the the other, at about the same speed. If the little

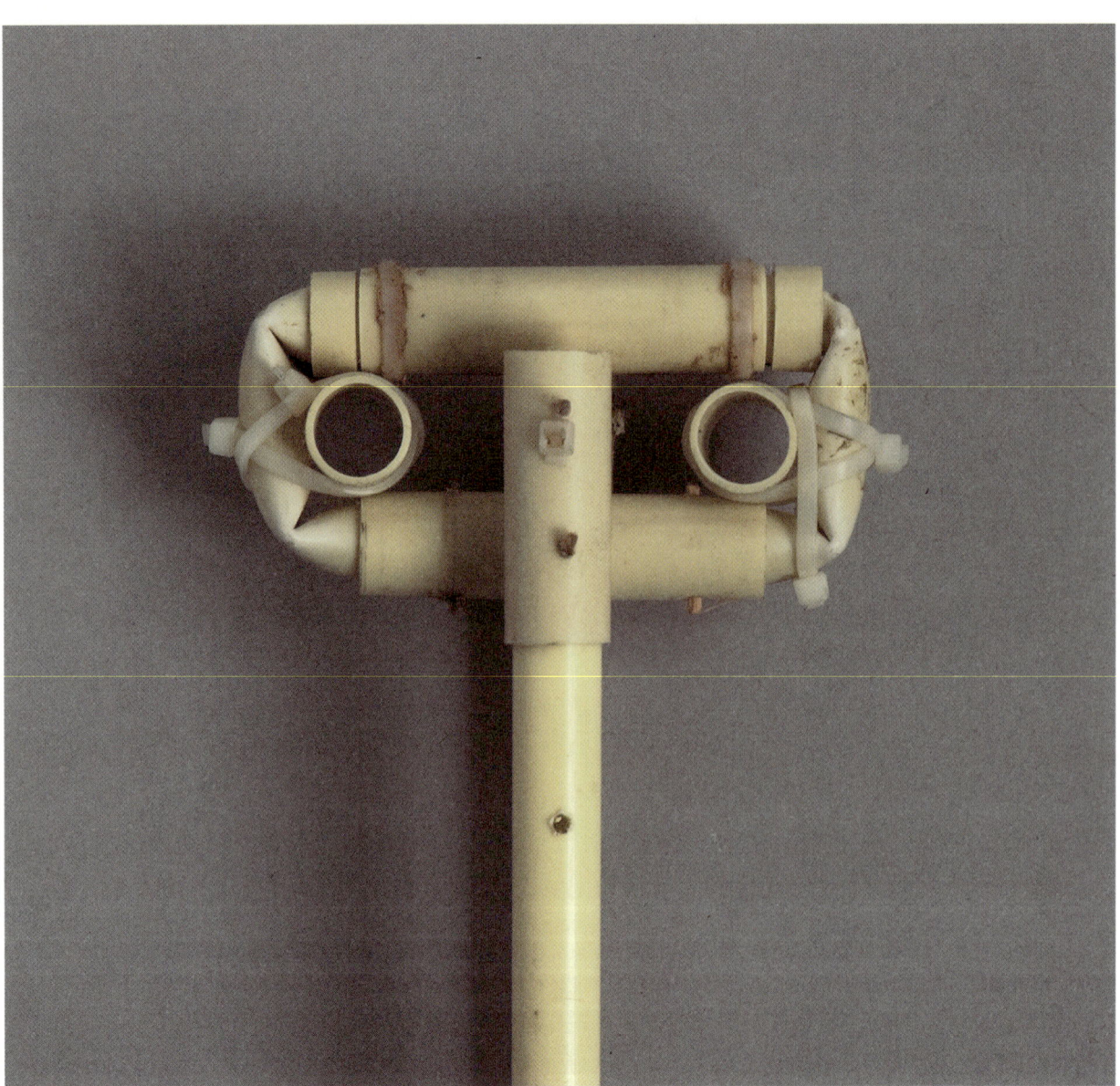
ankle of Animaris Sabulosa

one walked into the sea or into the loose sand, it had great trouble moving forward. The mother animal sensed this and because of her size was able to pull the little one back to safety.

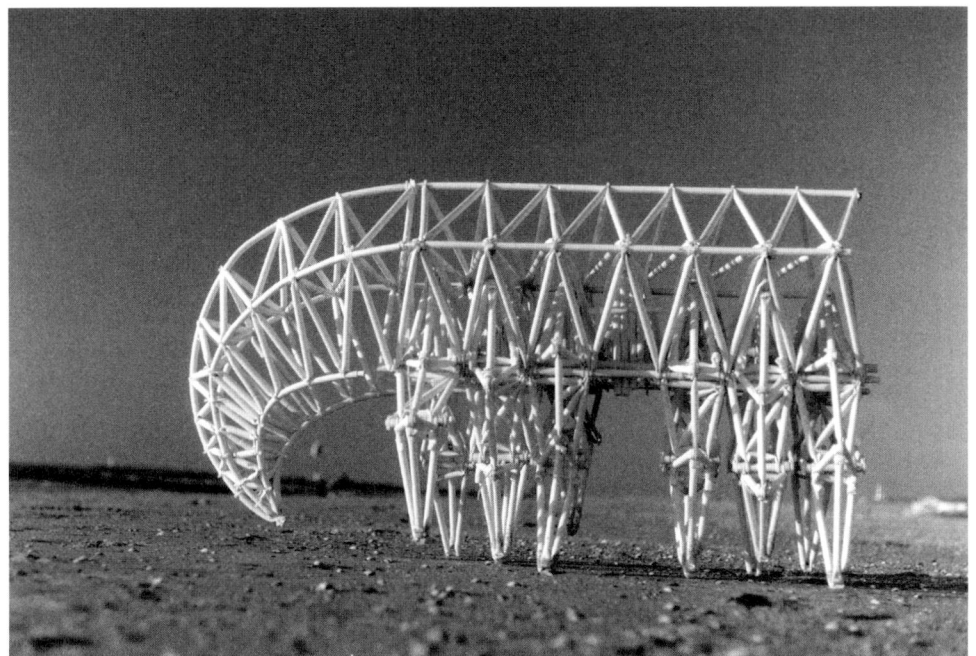

calf of Animaris Speculator

Animaris Speculator will go down in beach animal history as having introduced the concept of simulation. This was simulating reality in miniature, the way an architect first builds a scale model of his design. Speculator was sent as a scout with the following instructions: 'Give it a try; I can always pull you back.' A mutation in evolution is a kind of scout too. It's true that every mutation is occasioned by a mistake, an error, but you can also see it as a try-out. In nine cases out of ten, that try-out doesn't work. It means nine fatal casualties, which is a pity. In Speculator's case the try-out can be reversed; the mother animal can pull him out. The phenomenon of the moving amoeba, where the animal changes its shape within a generation, takes the form of 'trying' in Speculator's case. A single generation tries out different types of terrain. If something goes wrong, no matter. The same animal can try the same thing again somewhere else. In the case of a mutation, the animal dies. In the Cerebrum or brains period, such try-outs are simulated in the brain. No fatal casualties there. It is no longer a case of once bitten, twice shy, but just of thinking your way through it.

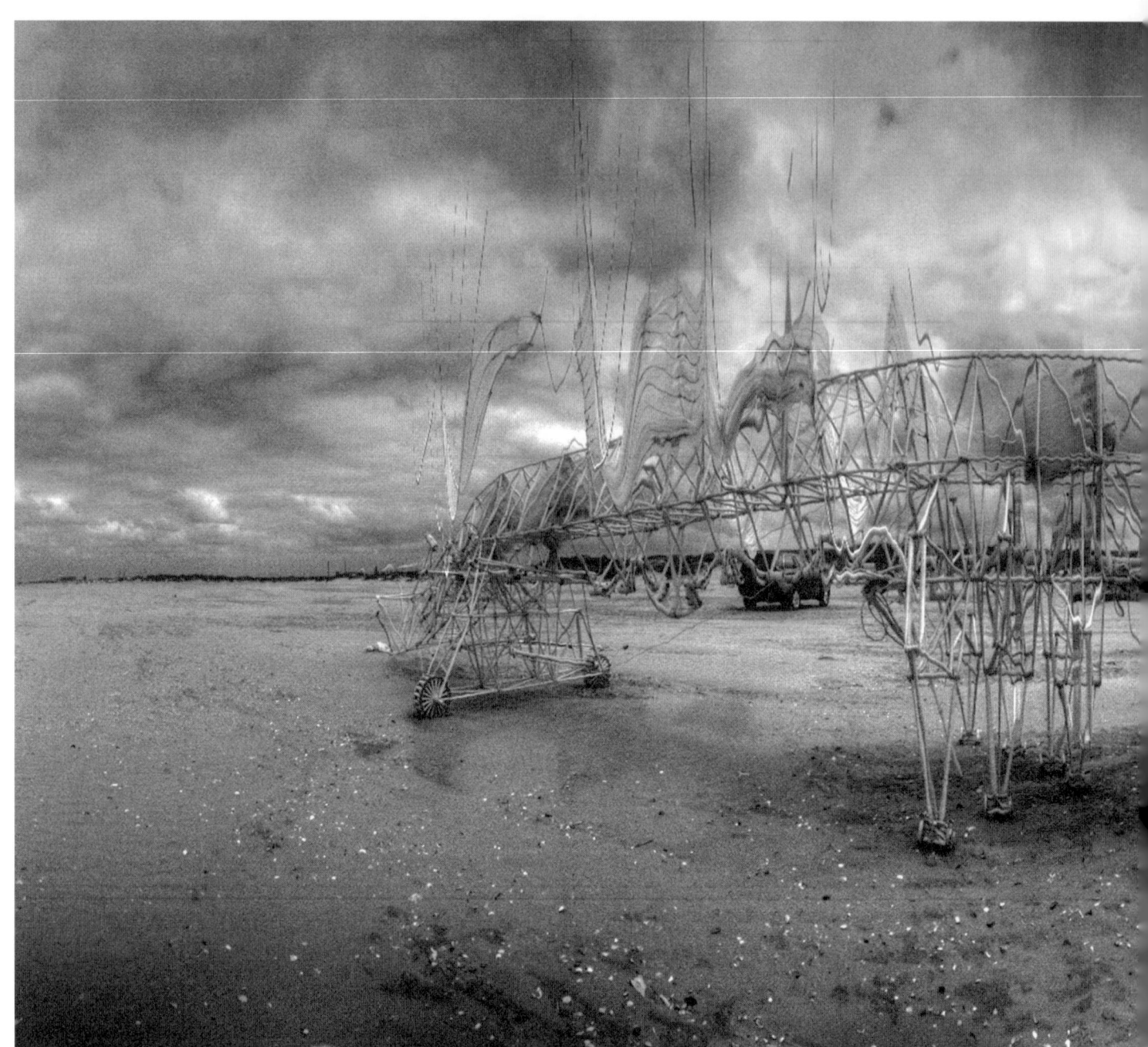

Animaris Rectus

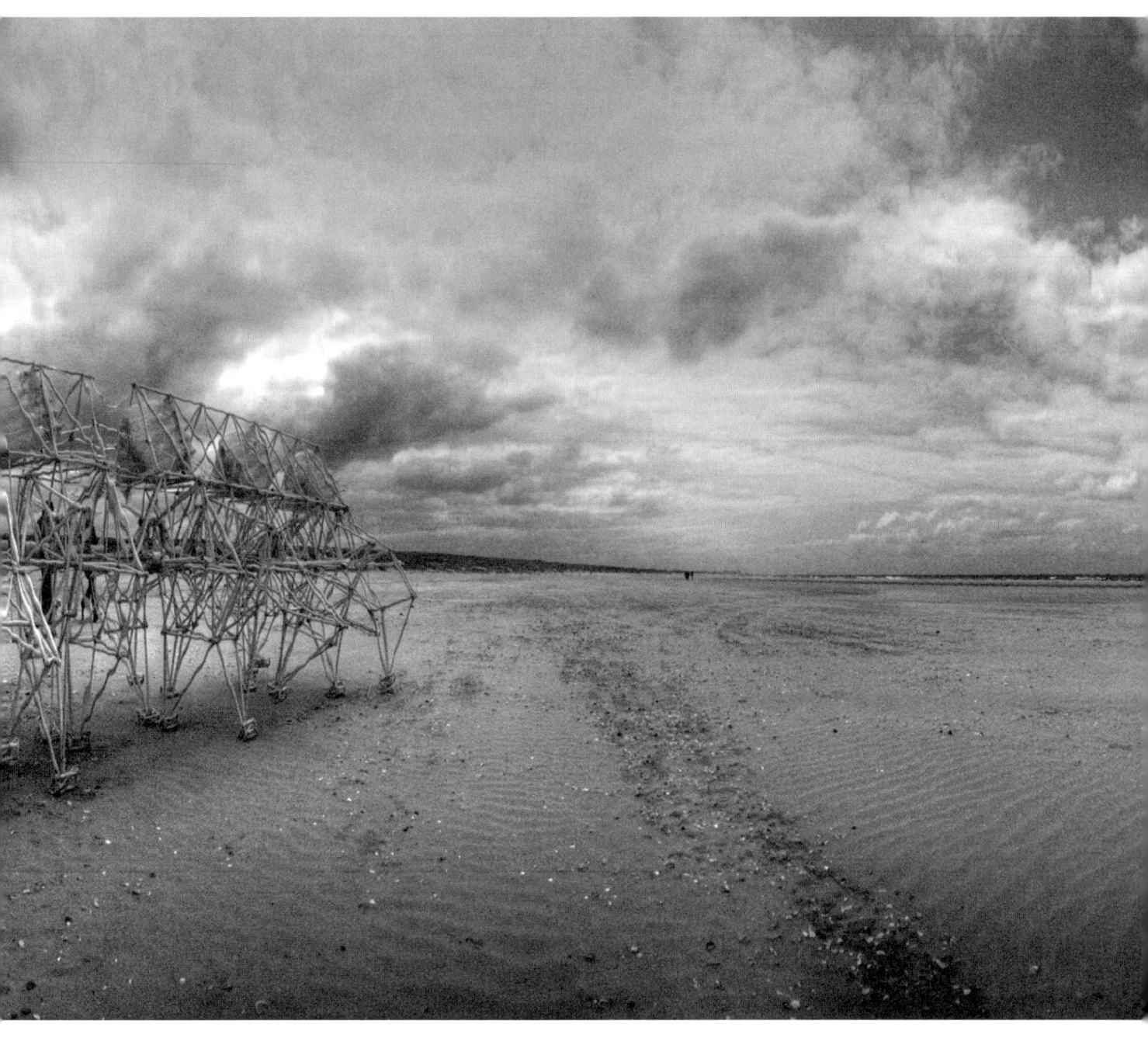

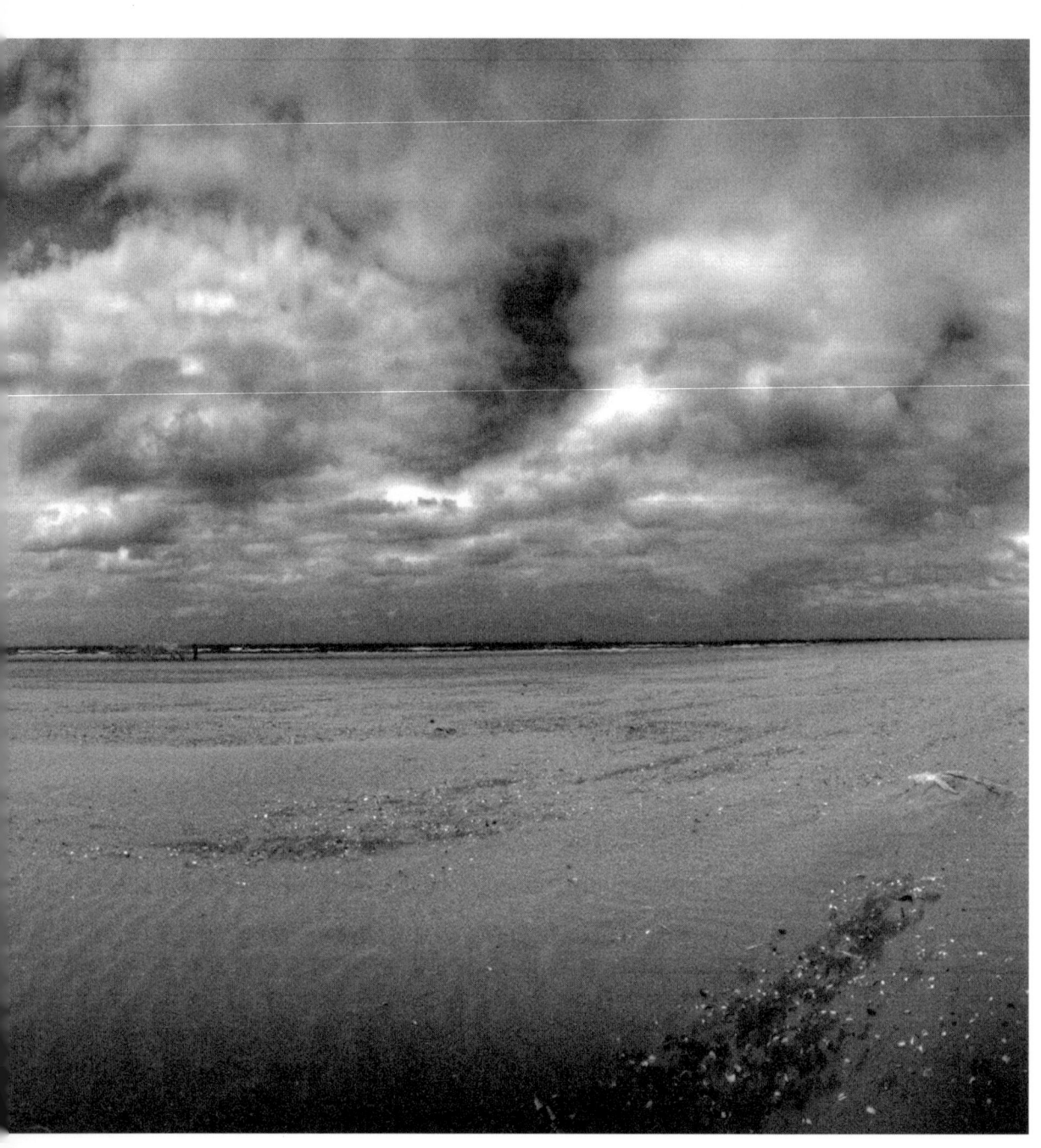

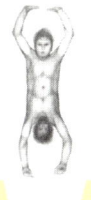

TEPIDEEM
the less-hot period (1994-1997)

Animaris Ancora

Anamaris Ancora had a heavy roller anchor attached to a propeller. This roller anchor was an anti-wind measure, so that the animal was less prone to blowing away. To help matters, Ancora automatically turned to face downwind, like a weathervane.

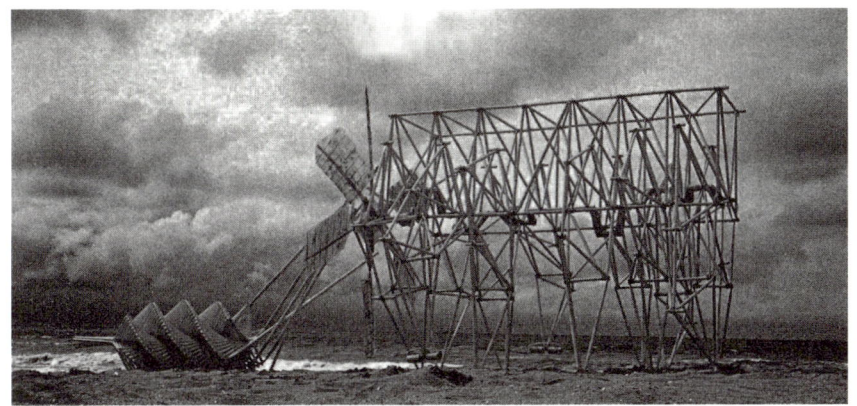

Animaris Ancora

The propeller animals were at their best travelling crosswind. They walked sideways. On reaching the sea or the loose sand, they had to reverse direction. Animaris Properans had a propeller whose blades could be reversed so that the animal then started walking the opposite way. But the propeller was heavy and didn't stay put for long. The crepe masking tape of the blades didn't last long either.

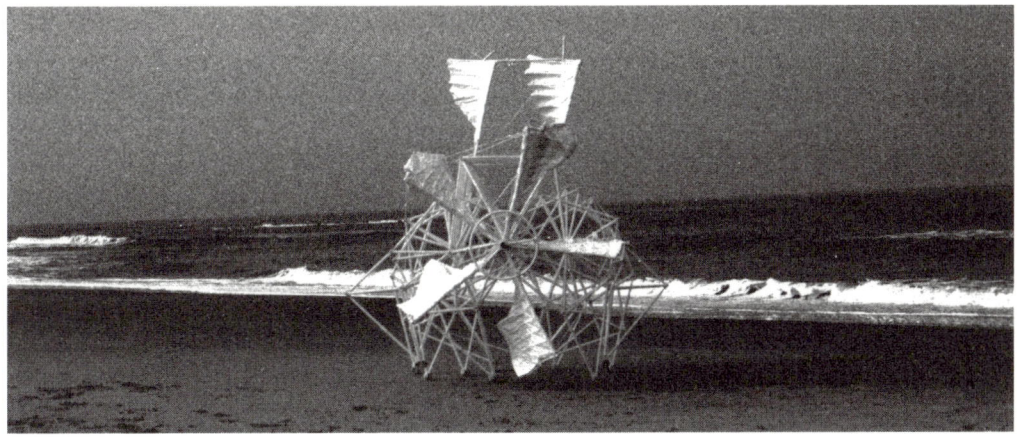

Animaris Properans

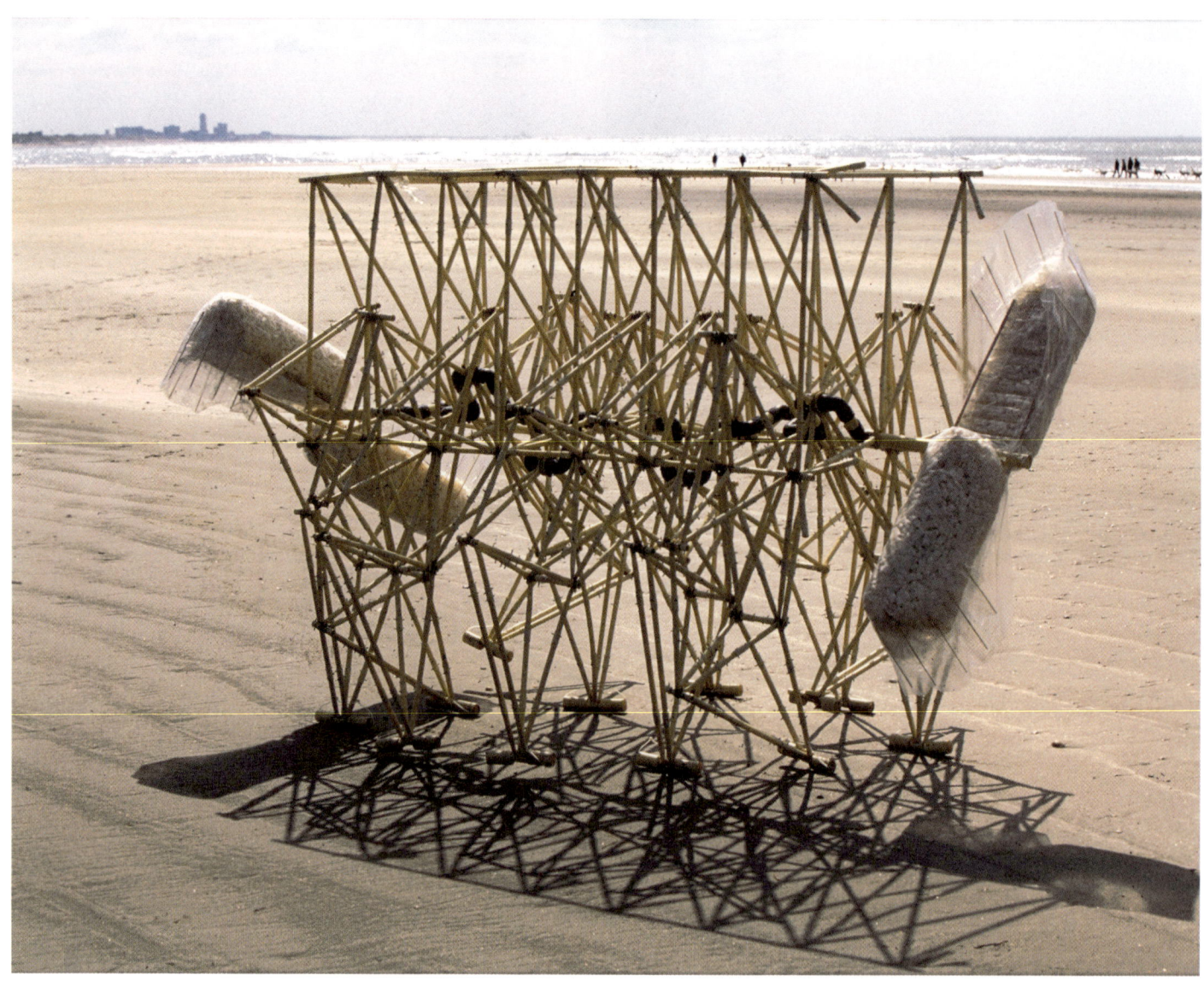

Animaris Geneticus

Animari Geneticae

It was in the Tepideem that the herd made its debut. Life in a herd has its advantages for beach animals, just as in the real animal world. Individuals can stand in the shelter of other individuals so that they don't blow away so easily. That advantage is admittedly at the cost of those catching the wind, but taken on average the chances of survival among beach animals living in a herd are slightly higher than those of solitary types. One disadvantage of herds of beach animals is that they can easily lock legs with one another and get entangled. Actually, something similar occurs in biological nature. There are times when tails of a nest of mice become entangled. Inextricably knotted together, they are fated to die.

Storms are the main adversaries of the beach animals. Many is the time their legs give way under them when the transverse forces brought on by the storm start playing up. The moment the wind slams into them, they first try to keep up with the higher speed, then they trip and begin to roll. There was one occasion on the beach at IJmuiden when a complete herd began rolling across the expanse like so many tumbleweeds. Now and again they would bounce a metre or two into the air. Then the wind caught them and slammed them down mercilessly several metres further along. It was a wonderful sight. After a kilometre or so they came to a standstill against the barbed wire of the fence protecting the dunes. They had to be dragged away one by one, as many of their legs had been rendered useless.

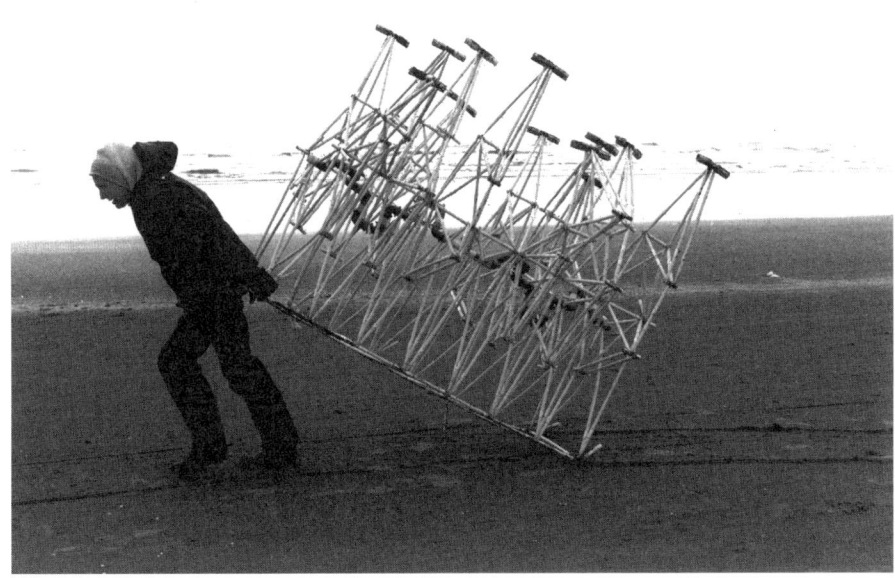

dragged away after the storm

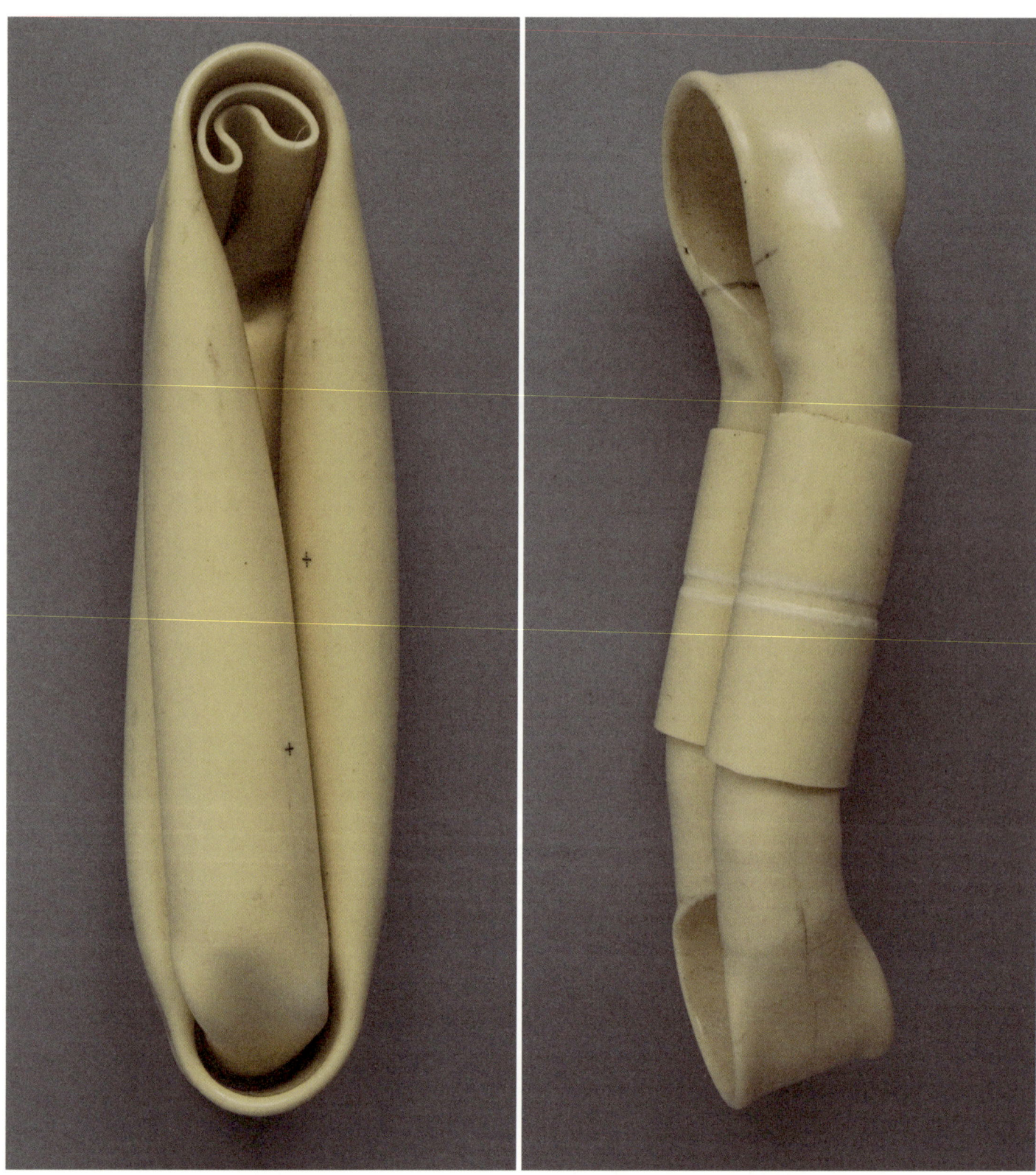
vertebrae of Animaris Rigide and Animaris Currens Vaporis

The Tepideem was all about reproduction. This was why Animaris Geneticus lived in a herd. It takes two to multiply. It was the first beach animal that could do that. Not that it amounted to much. Reproduction makes one immediately think of sex, of two beach animals doing it. A quite understandable conclusion actually. And they are indeed able to mate. I don't mean literally, they do something that encapsulates the *essence* of the act of mating. During insemination, genetic data gets passed on. A sperm cell contains DNA strands which carry the genetic code. The DNA strands are bearers of information, much like a CD-ROM or a DVD. The surface of a CD contains millions of tiny pits. Let's call this pit a one. The distance between ones is not uniform; the pits present an irregular pattern. The surface between ones is smooth and we'll call that a zero. These zeros and ones can be translated into music, text or film, whatever. There are no ones and zeros on a DNA strand but certainly something comparable. It contains a pattern of four types of molecules. In theory, you could record music on a DNA strand. The only problem would be to translate the pattern of molecules into sound. No audio equipment exists for playing back DNA. Or maybe it does. How about women? Women don't make music, they make babies. The man is considerate enough to transport the sperm cell to the woman and the sperm cell is prepared to swim with the DNA to the egg. This transportation business may seem to be no more than pleasantly passing the time. In truth, we are constantly engaged in transferring information. In our age, this information transfer is not exclusively by way of the genitals. We have computers, we have the internet and so an unbelievable quantity of data flows across the planet. From shallow streams to deep rivers, all of it data. Most data transmission via the computer is done using plugs. It's the simplest way. At the back of the computer you have a whole row of inputs and outputs through which the information flows. You can transfer information using light pulses, radio waves or sound; yet the simplest means is still the plug. We distinguish between male and female plugs. The male plug has a bit sticking out and the female plug or socket has a hole. This combination of protuberance and hole is a logical choice if two things are to be attached. Plugs are divided into males and females in every language. It has persisted through its effectiveness as a memory aid. Yet it is more than just a mnemonic. The most fundamental data transfer in the animal kingdom is the transfer of genetic information. This could have been done in so many ways but the plug is still the simplest way. We humans think we invented the plug but in

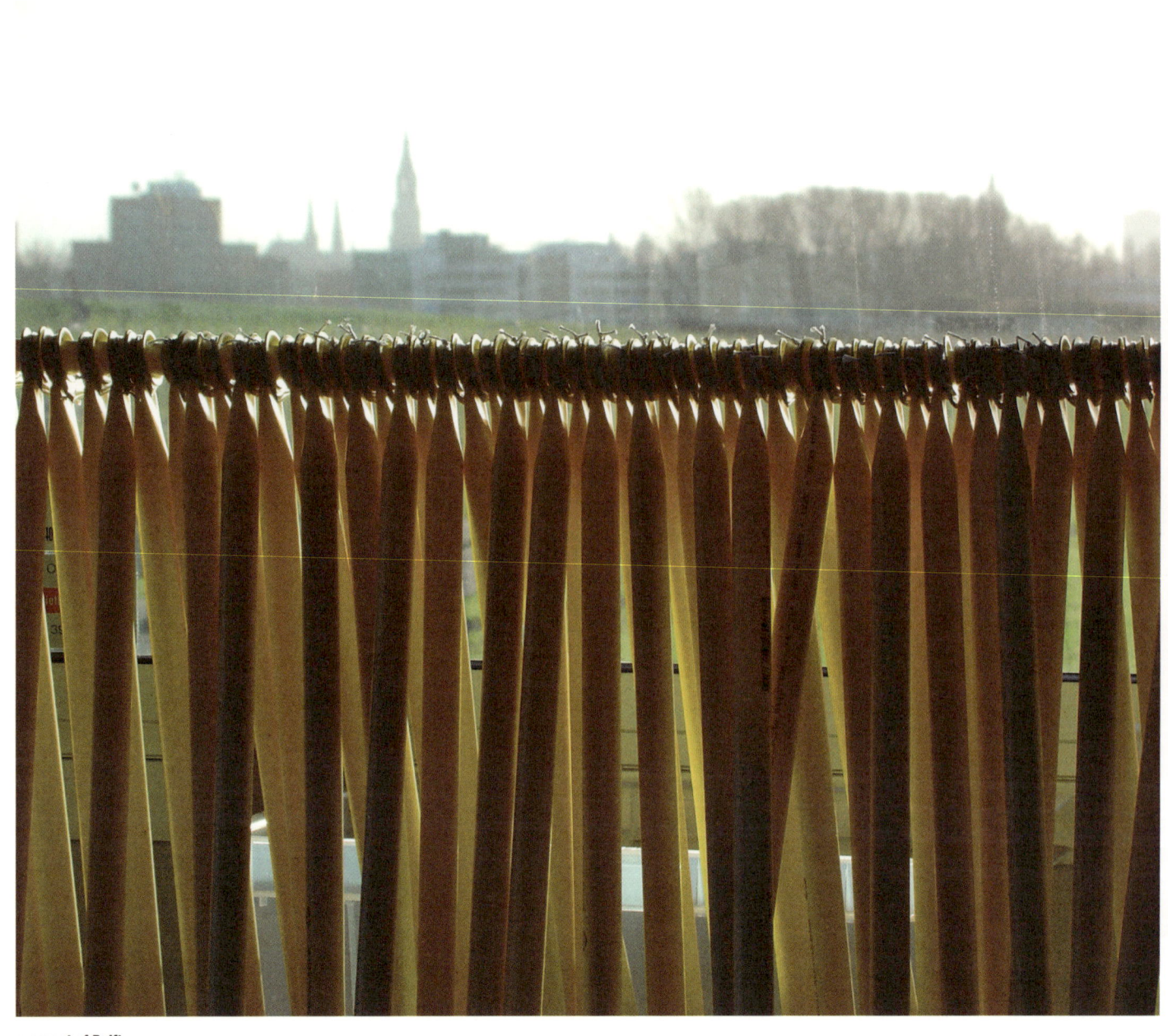

prospect of Delft

fact the principle came from nature. Just as a plane is an imitation bird and the camera an imitation eye, so the male plug is an imitation male and the female plug an imitation female. The plug of Animaris Geneticus is of course made from plastic tubing. It consists of tubes of differing length, short tubes and long tubes set in a row. At the end of the plug is a pattern of indentations or pits. When Geneticus plugs into another Geneticus, the genetic information can be read by that second animal. We shall return to this plug at the end of the Vaporum.

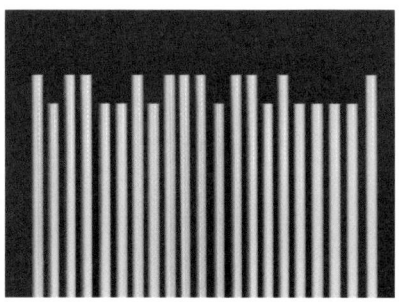

plug

In the Tepideem data transfer was still done by hand rather than with plugs. I still had to help with the reproduction process. And that's the difference between a real God and an imitation one like me: I had to help. If there is a real God, He starts things up that carry on of their own accord. Gulls and pigeons continue to fly about, rats still go about their business in sewers; they don't need our help. Not so Animaris Geneticus, certainly when it came to reproduction. And yet it really was able to reproduce. These were real genes that were transmitted, not imitation genetic data. Granted, they were not made from amino acids like ours, but from another material – you guessed it – plastic tubing. Indeed, the anatomy of Animaris Geneticus was exclusively geared to reproducing. It could walk but only in the direction the wind was blowing in and not at right angles to it like the other beach animals. Its genes hogged all the attention.

Genes of electrical tubing
The genetic information of Animaris Geneticus was defined by the rods comprising its skeleton. It takes the eleven holy numbers from the previous chapter a

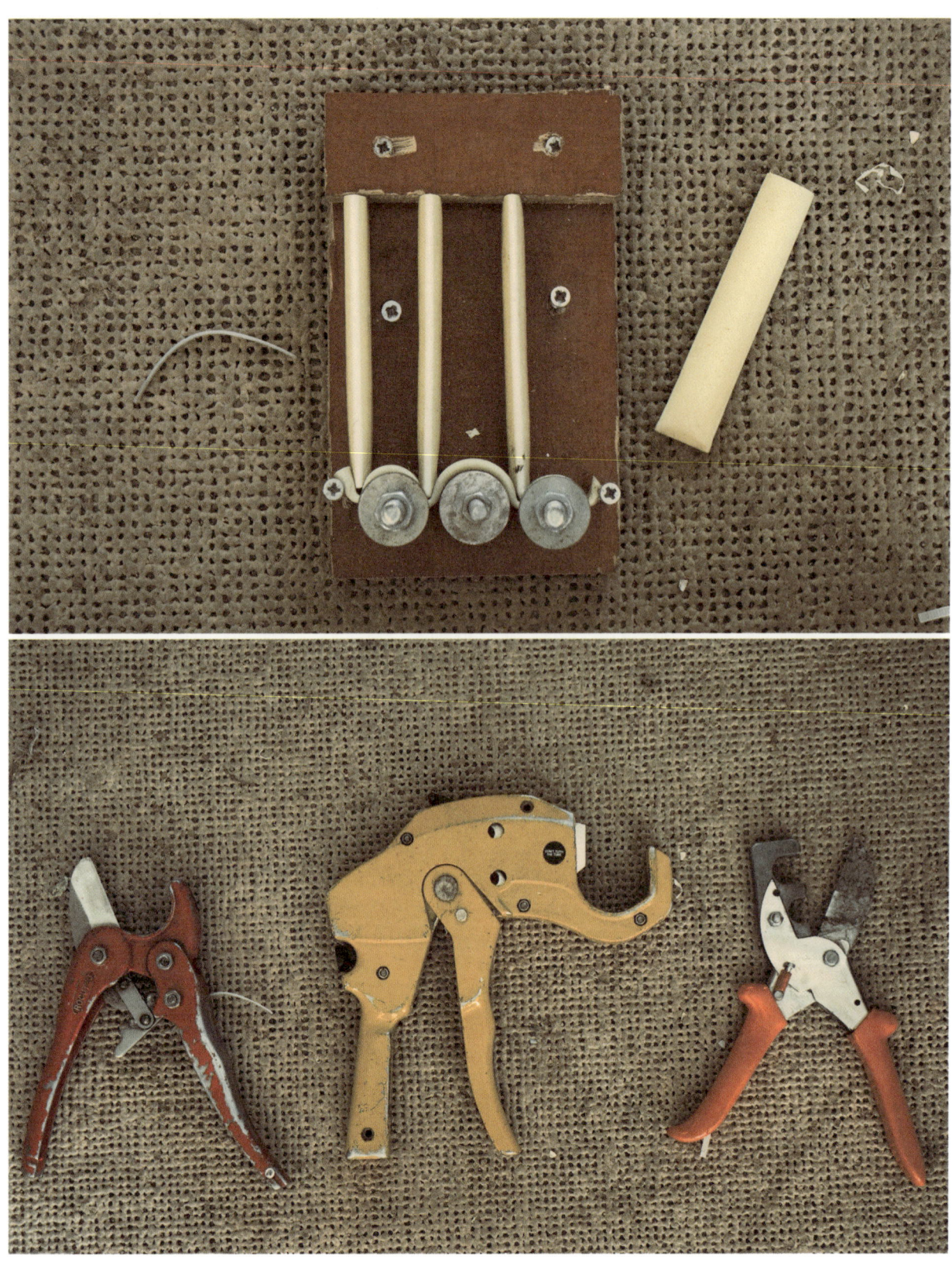

mould for muscle bundlers (Vaporum),
cutters

step further. These numbers represent the lengths of the rods in one leg. The ratio between those rods determined the quality of the leg. The ideal ratio was found through an evolution acted out in the computer. The eleven rods can be regarded as the leg's genes. Geneticus's skeleton was not just one leg, however, but a body and twelve legs. Altogether the animal had 357 rods – in other words, 357 genes. The lengths (the numbers) determined the animal's shape and size. If you were to double the length of all the rods in Geneticus, you'd get a version that was twice as big.

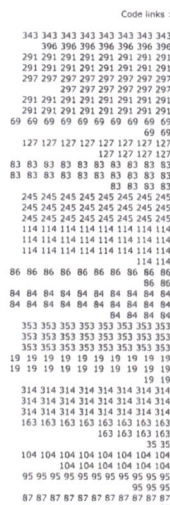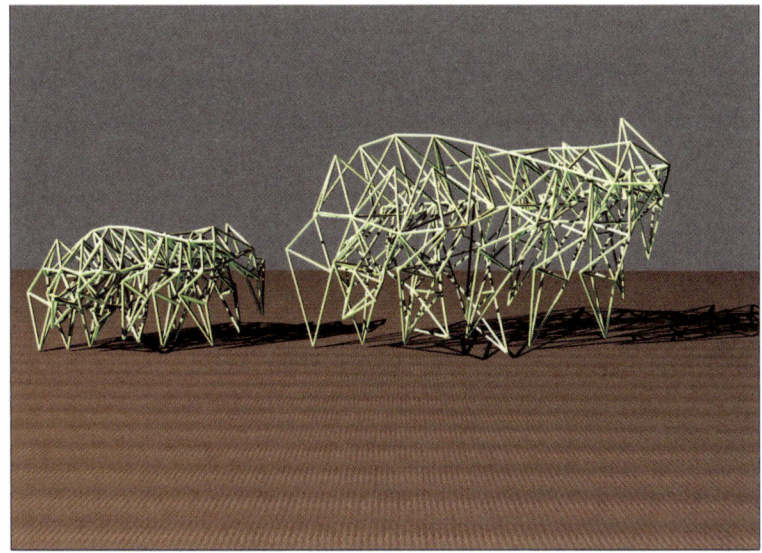

the rods of the animal at right are twice as long as those of the animal at left

This double-sized Geneticus would be pretty flaccid, which goes to show that not just the shape but also the walking properties are determined by the lengths of the rods, as were the walking properties of a single leg in the previous chapter. So the rods are the animal's genes. Its shape and walking characteristics can be rendered as numbers that in turn can be stored on a CD-ROM.

The handy thing about Geneticus's genes was that they were replaceable. Geneticus was compounded from connecting pieces or nodes into which the rods fitted exactly. A rod could be removed and replaced by one that was either longer or shorter.

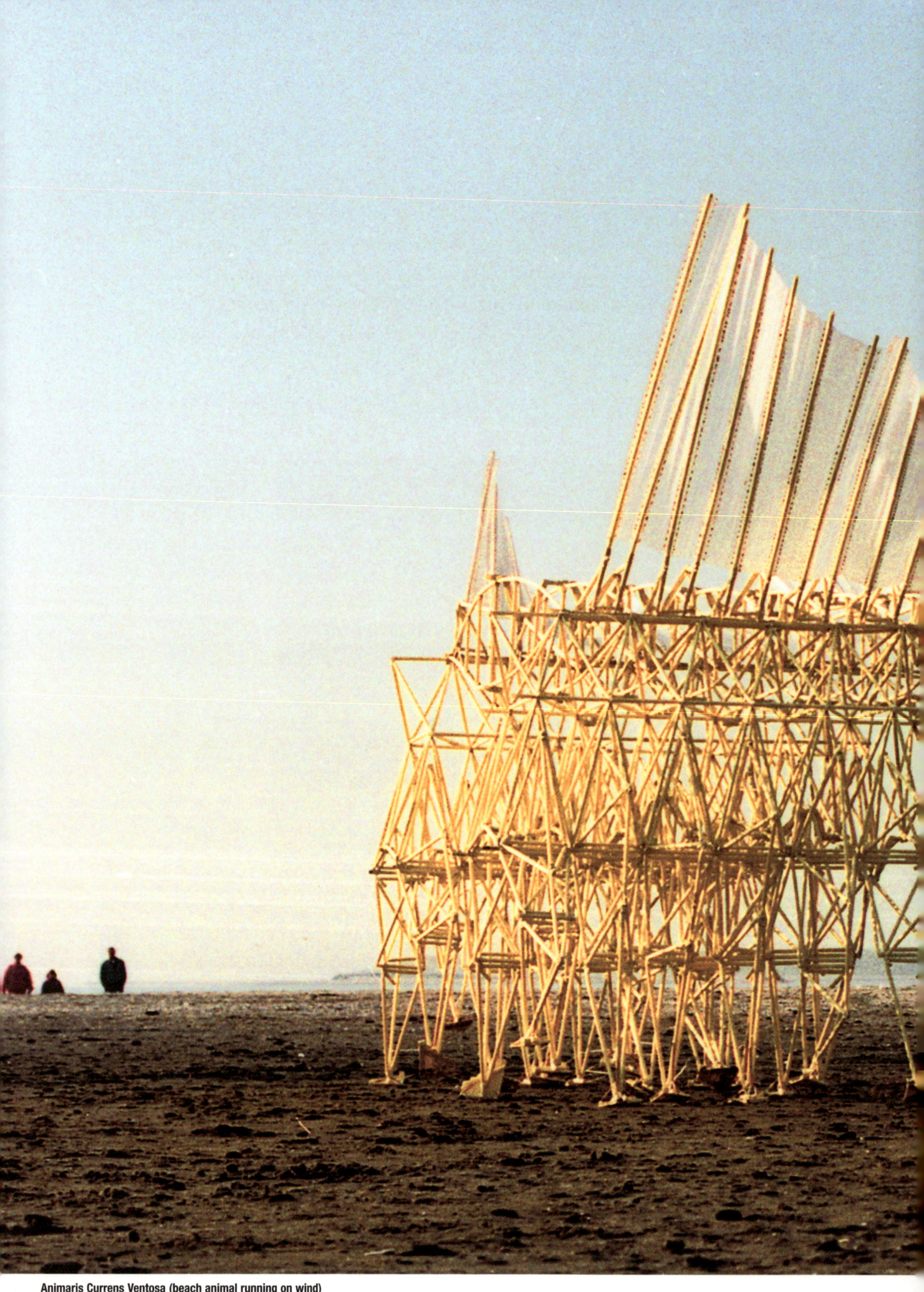

Animaris Currens Ventosa (beach animal running on wind)

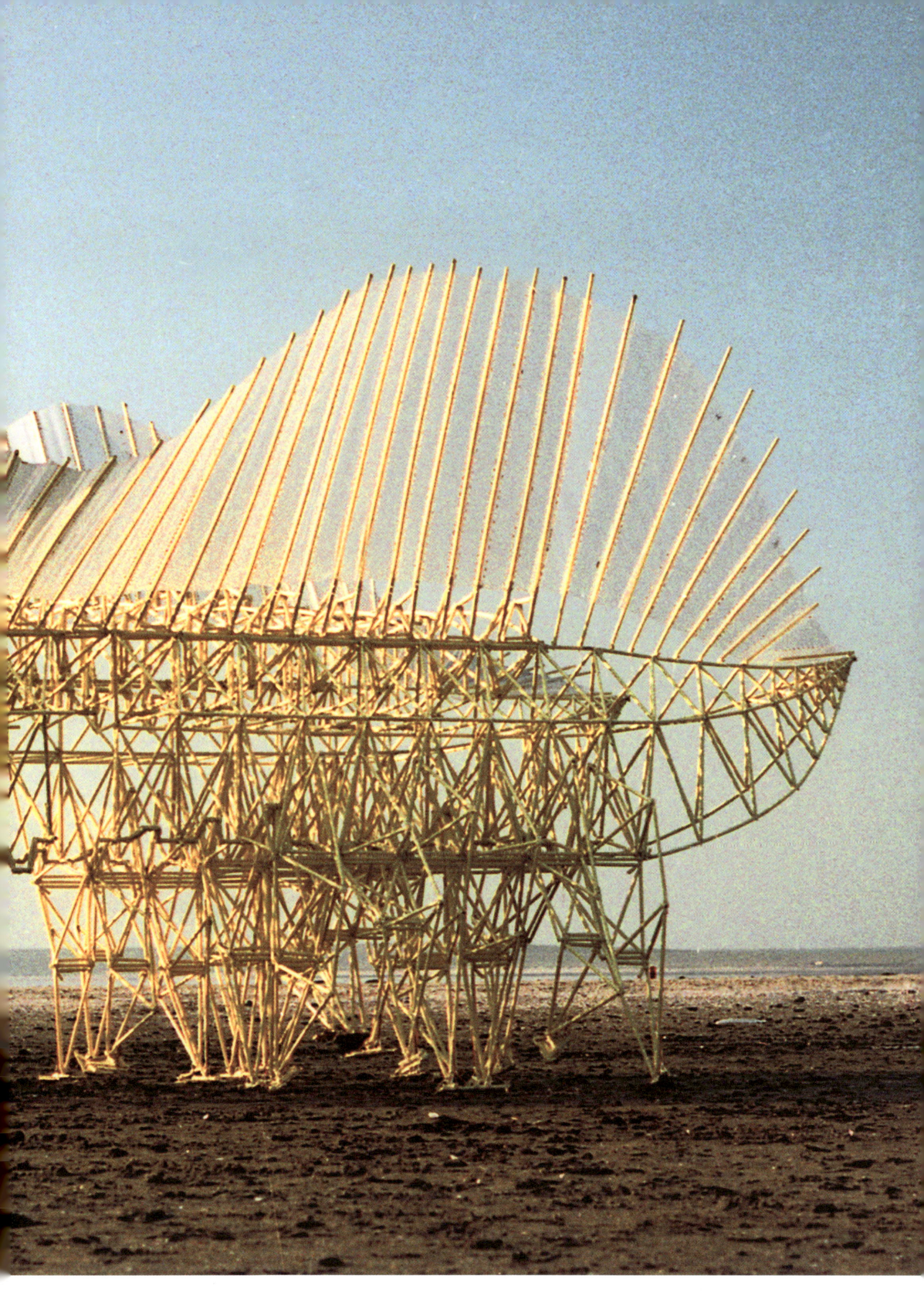

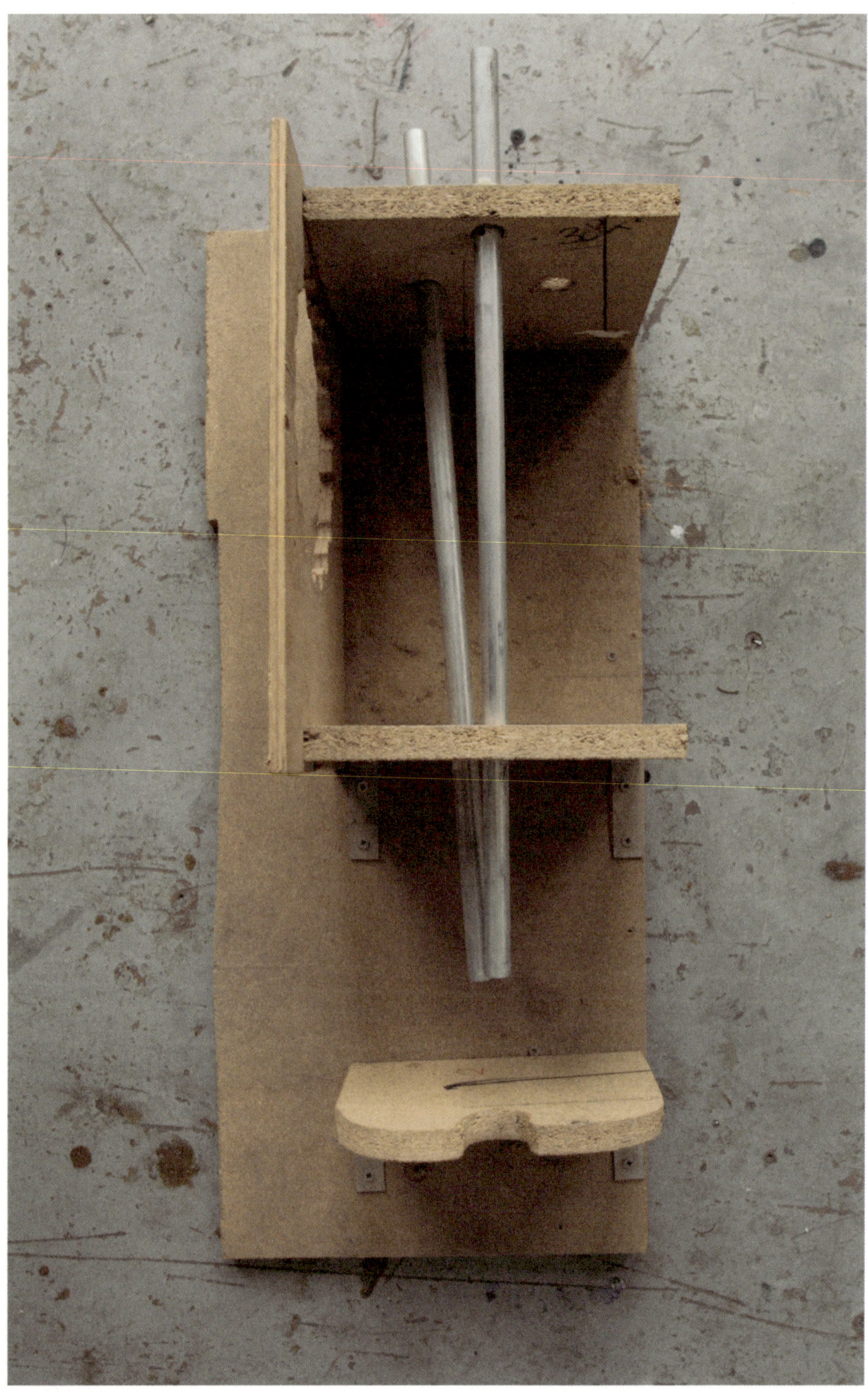

tri-split mould (minus one tube)

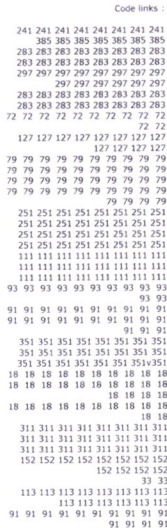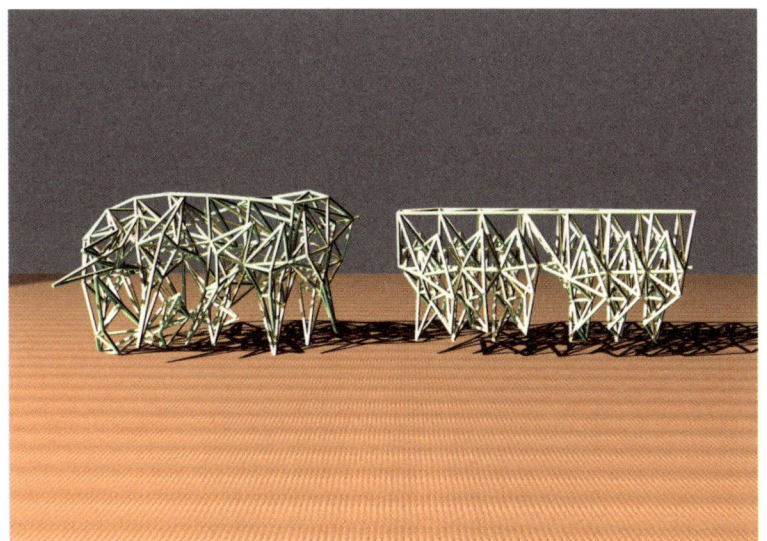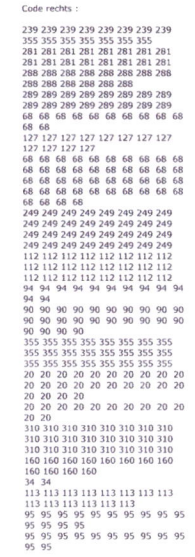

two animals with utterly different codes

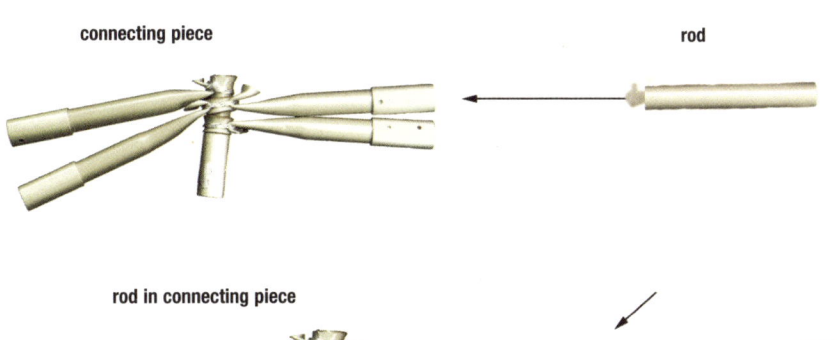

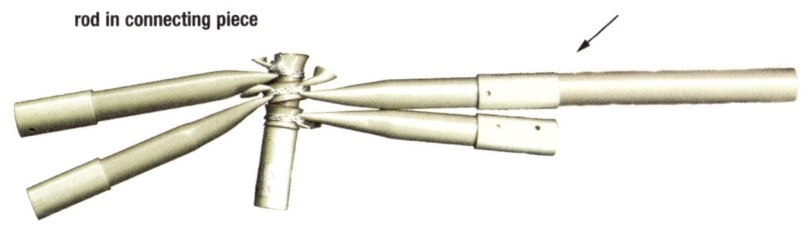

node to accept rods

So Geneticus's gene pattern could be changed at any time. This is where it differs radically with the selection of genes described in the previous chapter. There the evolution was acted out in the computer. With Animaris Geneticus, that evolution took place on the beach itself. A herd of seven Animari Genetici was let loose on the shore at IJmuiden. Each member of the herd differed slightly from the others in its gene pattern and therefore in its walking properties. They engaged in crosswind walking races. As they walked, they sorted themselves into fast and slow animals. Evidently one genetic code is better for walking than another. Then I entered the arena in my role as God. I awarded the fast-moving animals the privilege of staying alive; the slow ones were relegated to the condition of

Coupe de grace

I have written about the poet P.C. Welbeest in this column before. Then it was occasioned by his poems and his performances in trams and in crowds of shoppers. Whenever I wrote about him, I was invariably approached by people from the radio or the newspapers who wanted to interview Welbeest. He always refused point-blank, just as he refuses to go to a publisher to get his poems published. He publishes them himself. He sells his volumes of poetry privately. You needn't bother to ring me, since he forbids me to pass on anything about him. Not his telephone number, not his address, not even his real name.

Recently he has been working with videotape recordings, made using second-hand video cameras. These he buys for next to nothing and does things with them they usually don't survive. The videotape often does.

For instance, he is standing at a station. He holds the camera in front of him and films himself. You can hear the train arriving. He videos the train. And then casually tosses the camera in its path.

I know that Welbeest isn't planning to top himself. So he doesn't see it as a premonition, or an exercise. It's an excursion, it's just a camera after all. He threw a camera in the air and caught it. Juggled with two cameras. He didn't always succeed in catching them. But the film was always a success: it stayed in one piece.

Of course this can only be done with video cameras since when real film cameras break, the photographic material is immediately exposed to the light. At the moment of death, the foregoing images are erased as if the camera wants to prevent us from viewing those last seconds. Evidently, the film camera sees it as an intrusion into its private life. After all, you don't film a person when they're dying. Since the emergence of video nothing is sacred any more, not even the private life of a camera.

The piece with the baseball bat tugs at the heartstrings. In the film we see Welbeest waving the bat about. The first knocks the camera gets are slight. Then comes the *coupe de grâce*, followed by static.

I've seen every one of those images during editing. The last two pictures are interesting. These occur after the fatal blow and together last forty milliseconds. In other words, the camera wasn't broken immediately after the blow. It lived on briefly. Those must have been forty intense milliseconds, maybe intense enough to feel like an eternity.

As a rule, breakage or mechanical deformation means the end of the line. This holds as much for cameras as for the workings of the human body. A special property of video cameras is that you can examine their brains after they are dead. Maybe we'll be able to do that with people in the future. Nothing is sacred for long.

corpse. The code of the winners was copied, in other words there were new tubes cut to match the length of those in the winning animals. Next, the tubes in the corpses were replaced by the copies of the winning tubes. Now not all tubes needed replacing; a great many of them already matched. Only those tubes that differed from those of the winners were replaced. This saved me a lot of work. The corpses were reanimated, so to speak, with new genes. You can regard these new animals as the offspring of the winners. The winners were engaging in reproduction. And what is reproduction other than copying a genetic code? It seems like a new departure, reanimating corpses with new genes. But this happens in biological nature too. It happens during a woman's pregnancy.

Pregnancy and death
If you're healthy you don't feel bad.
If you're poorly you don't feel good.
If you're dead you don't feel, period.
In the Netherlands, pregnancy is sometimes referred to as 'the healthy disease'. And rightly so of course – new life is being created and what could be healthier than new life. Logically, one could then describe death as an unhealthy disease. A dead body is no longer capable of reproducing. So death is, above all else, a reproductive disorder. An incurable one too. And yet dead bodies do have their uses. I'll explain. The living body is a transformer; we take in food and fluid through our mouths. After transformation this matter leaves our body as urine, faeces, wax, nails, tears, spit, hair, snot, sweat and toe-jam.

When tabulating forms of voidance one very important one is generally overlooked: the voiding of offspring. We overlook the fact that our offspring are made of food. During pregnancy, the food the expectant mother eats is turned into molecules which are cleverly reassembled in her stomach in the form of a baby. Babies are made of whatever the woman put into her mouth. That stuff, that food is in turn made of other forms of life, or rather other forms of death. Food is not made up of clay or stones, it consists of corpses from the plant and animal kingdoms. Dead bodies are used for new life. Reproduction boils down to reanimating corpses. From those corpses comes new life. The female body is a remarkable thing. It demolishes corpses and makes new life out of them. A remarkable transformer then. And a very healthy one. In the case of the beach animals, reproduction is much more unusual but more efficient. The corpses left by the losers of

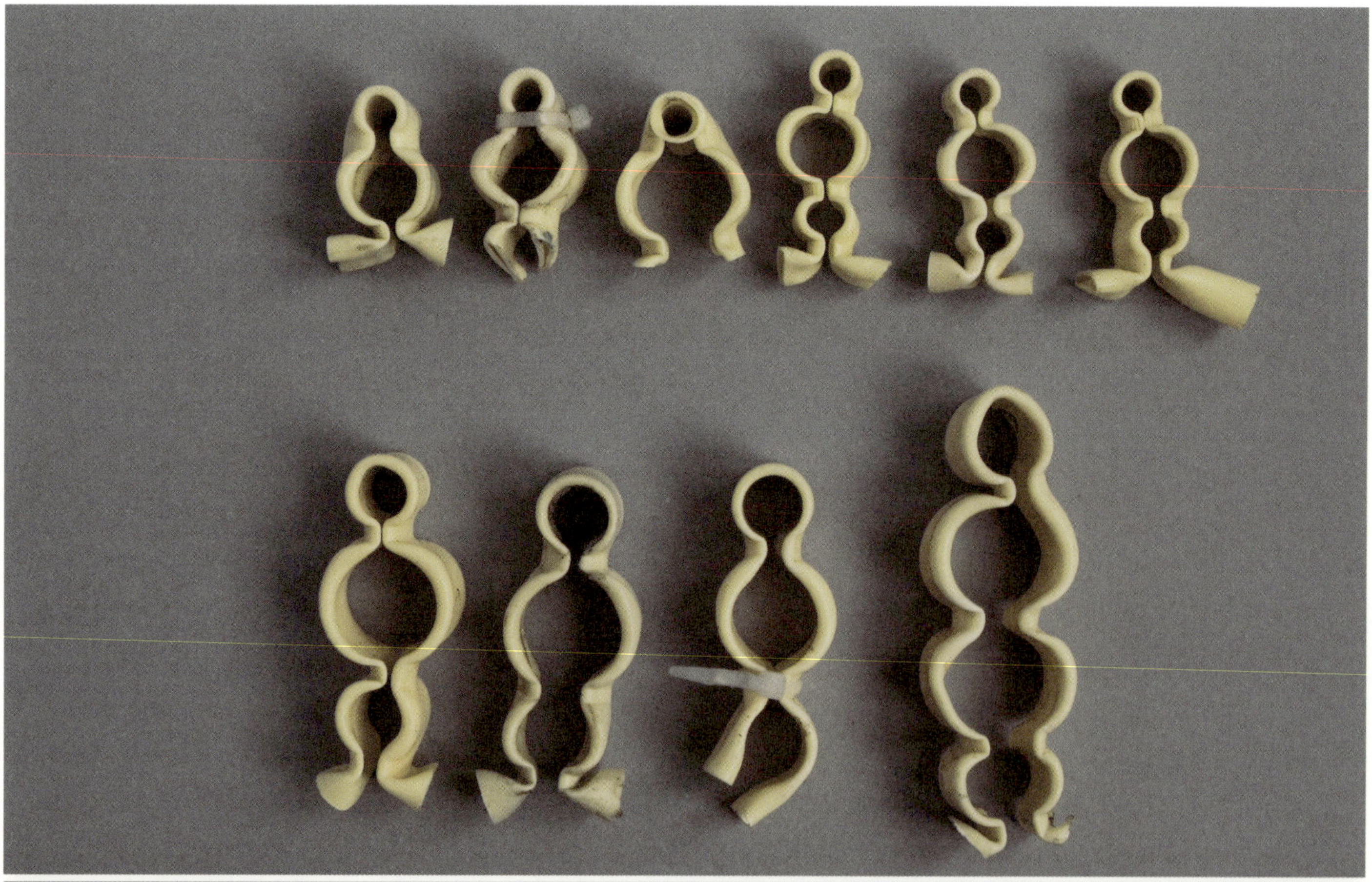

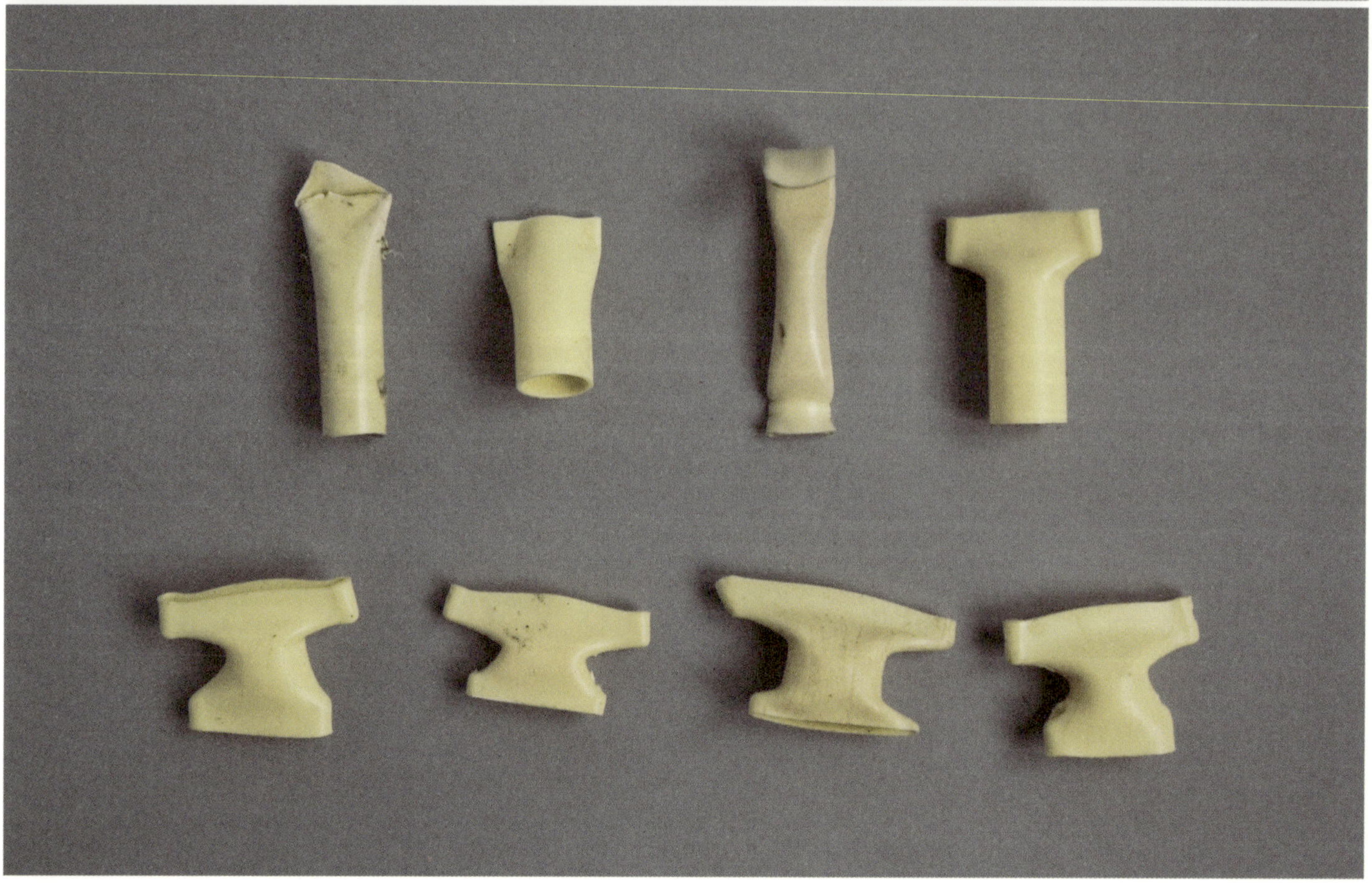

muscle bundlers,
teeth for sealing off hoses

the walking race on the beach weren't chewed fine. They remained intact for the most part. Beach animals have a much more economical approach to matter than we have. When all's said and done, they are cannibals, they exploit the corpses of their own species. Beach animals have a great respect for other life-forms. They only reanimate the corpses of beach animals.

The form of things
Elephants are objects. They obey the laws of mechanics. Dogs as well. The first astronaut was a dog. She was called Laika and was fired into space on November 3rd, 1957. She did it – that is, she stayed in one piece, she survived the journey. In those days I was catching flies, putting them in a shoebox with cellophane on the front and looking after them pretty much like pets. I kept them alive with chocolate sprinkles and jam. It was fun, though it did get a bit boring after a time so I thought up tests to do on them. Tests on animals are a bit perverse but a bit exciting as well. I sent them high in a matchbox attached to a kite – air travellers rather than space travellers – to see if they'd survive the more rarefied regions of the atmosphere. And they did. This living matter stayed the course. It's odd to regard an animal as an object, a piece of machinery – it works or it doesn't work. Dogs, elephants and flies are complex objects. Elephants have thick legs to prevent them sinking into the mud. Perhaps at one time there were elephants with legs as spindly as a sheep's but these weren't destined to live long. Their legs either gave way beneath them or sank into the ground. That's why elephants' legs are as thick as they are and why spiders' legs are as thin as they are. Living objects are different from dead objects. One difference is that dead objects get rounder and chubbier as time goes by whereas the forms of living objects just keep getting spindlier, with more and more bits sticking out. Throw a stone in the river and it will become smooth in time. All the sticking-out bits will disappear. Smash a glass bottle and the pieces will be smoothed out into globules in the fullness of time. Dead nature pulls out all the stops to create spherical shapes. Not just from bits of stone or shards of glass but from our planet as a whole. Mountains get worn down and the matter lost in the process fills up the valleys, the result being that the contours of the earth's surface are steadily disappearing. The very opposite happens in the case of living objects. An inverted erosion takes place.
Hundreds of millions of years ago, all living creatures had the same shape, that

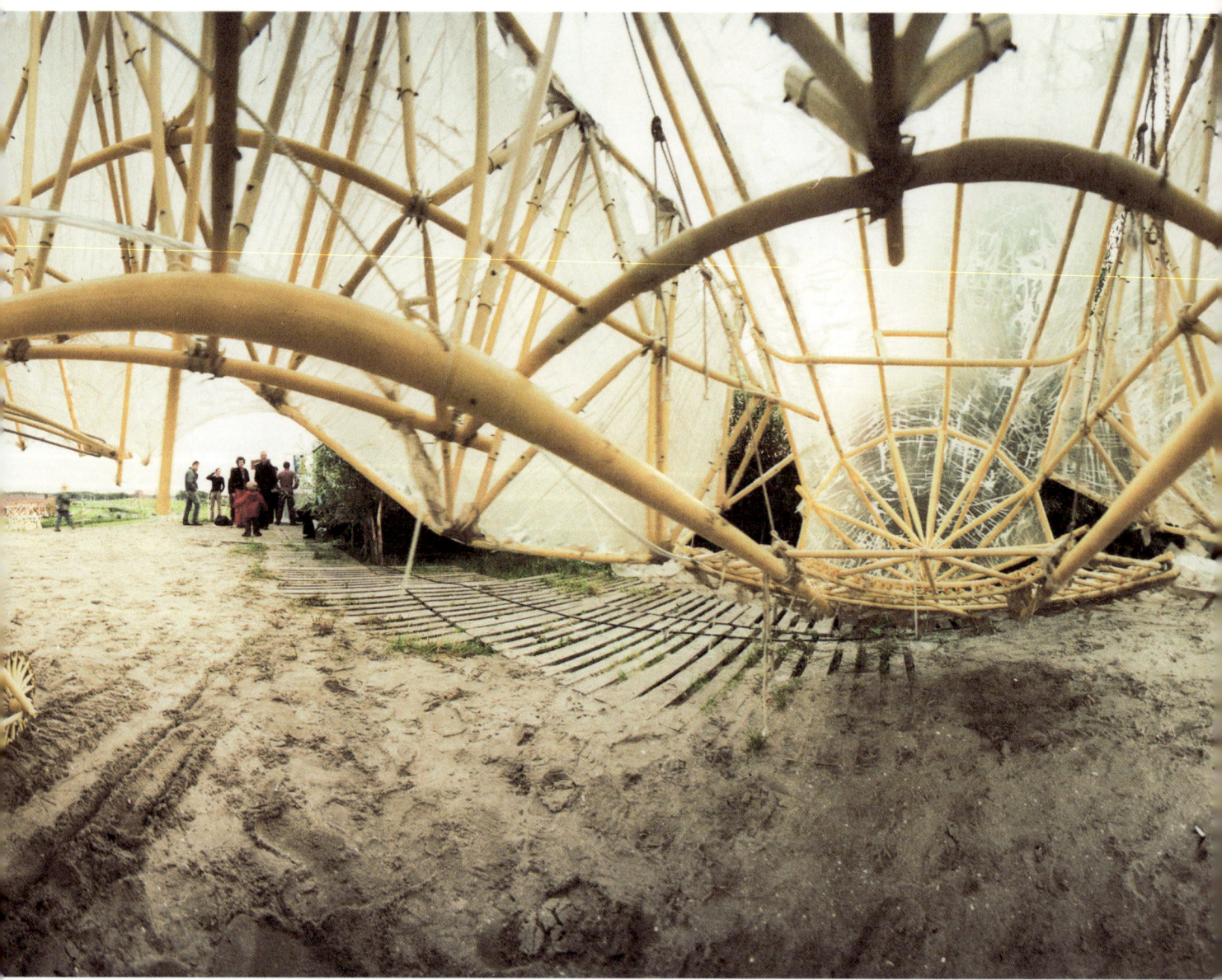
Animaris Percipiere, inside of nose

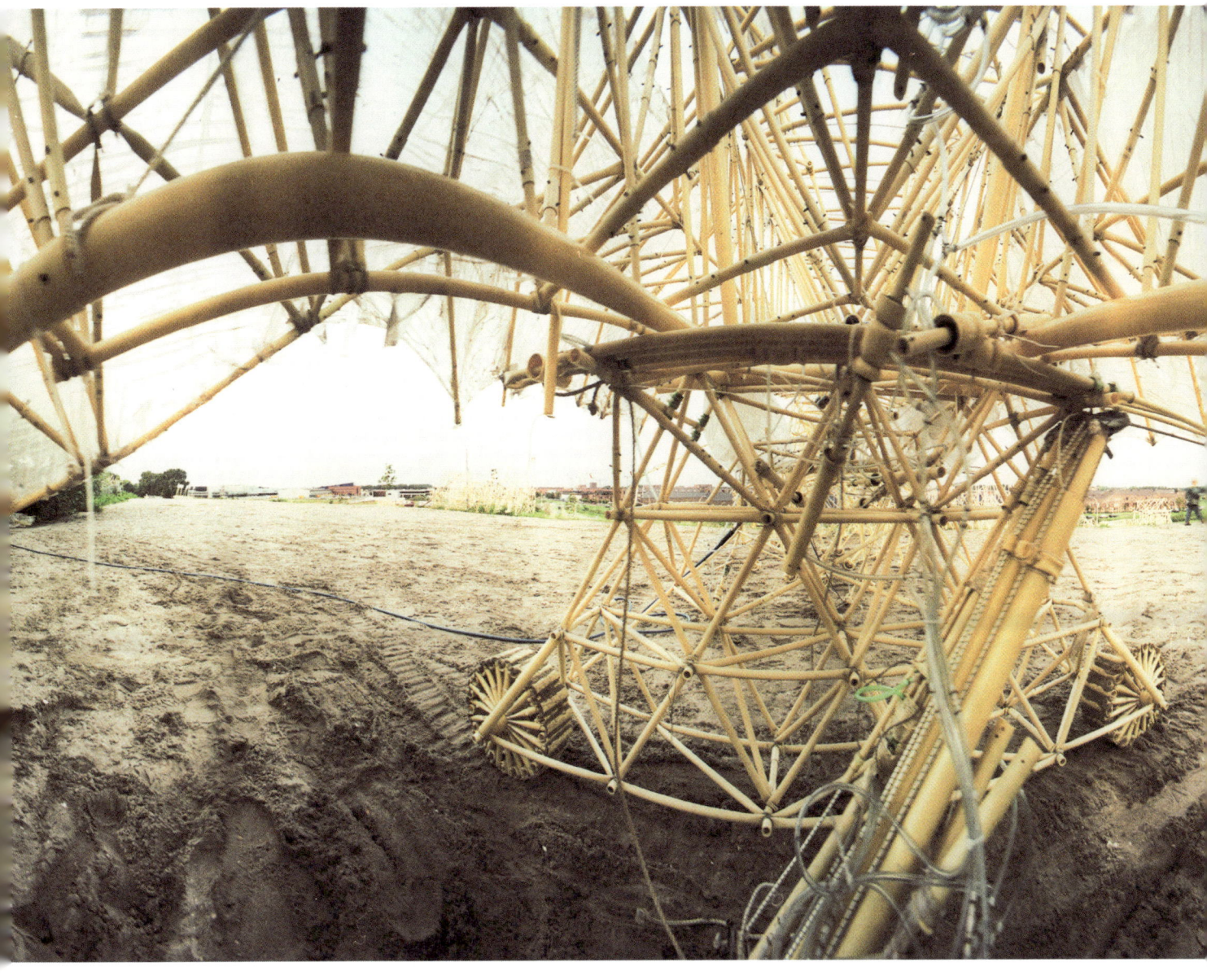

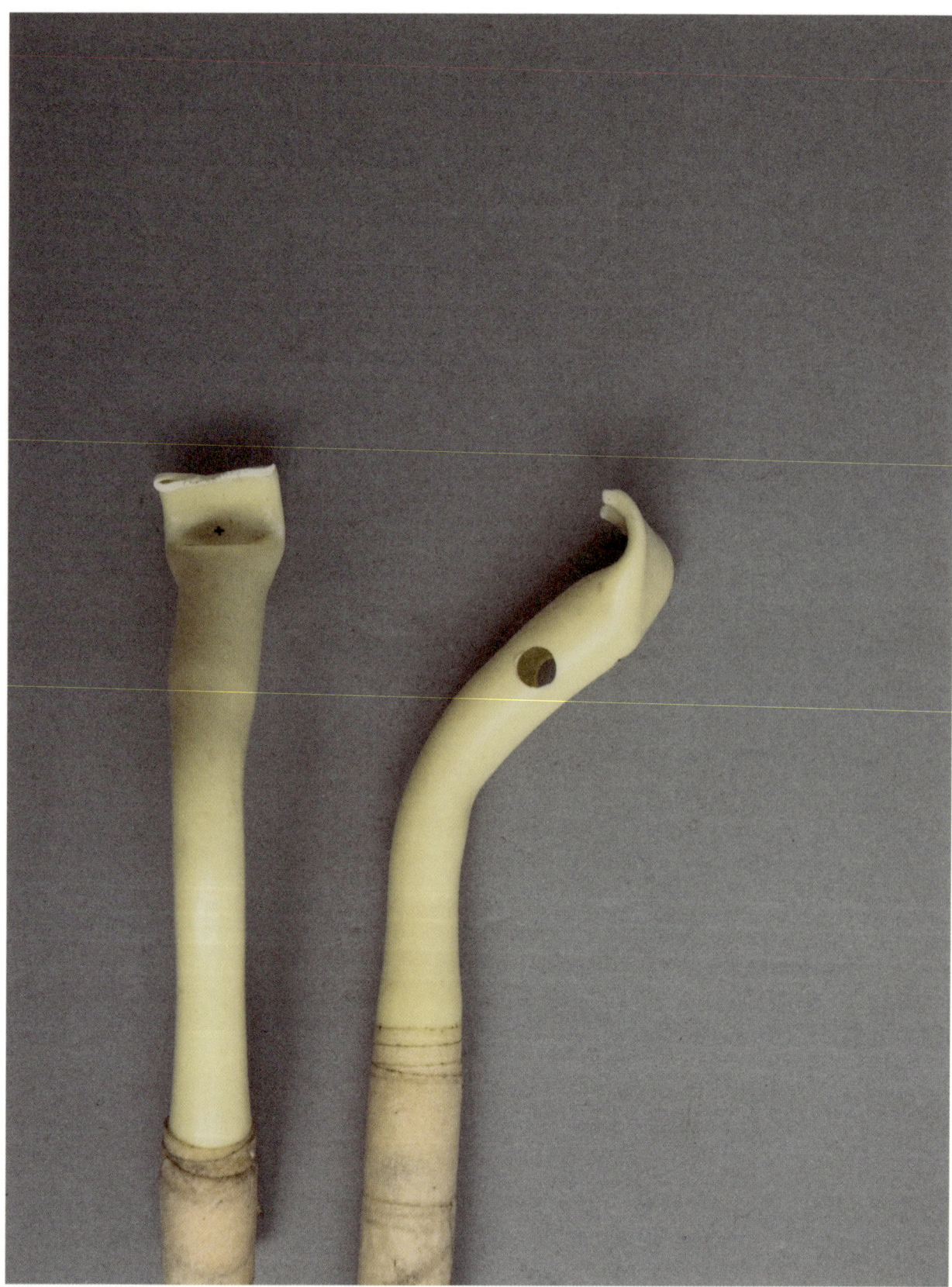

Animaris Percipiere Primus, teeth used in bending tubes (broken)

of the sphere. This stretched in the course of time into a shell-form, with an entrance at one end and an exit at the other. During the subsequent evolution, appurtenances – arms, fingers, legs, head, horns, spines, teeth, hairs, ears – made their appearance. Unlike the stone in the river, the surface of this shell actually became more rugged. Not that this erosion in reverse was fortuitous. The appurtenances sprouting from the original sphere-shape did so symmetrically. It seems that if an appurtenance sprouted from one side of the sphere, another would appear on the opposite side. Appurtenances are never alone, often being in pairs or four together. That to my mind is a result of growth through cell division. Cells divide symmetrically. My thoughts often wander to the time when man was just a shell and not much else. No limbs, no head, nothing. A small, weak-willed, listless shell.

So there was an entrance at one end and an exit at the other. The entrance was where the food went in and the exit was where the waste matter came out. That's how the body used to work and still does. All trimmings and trappings which

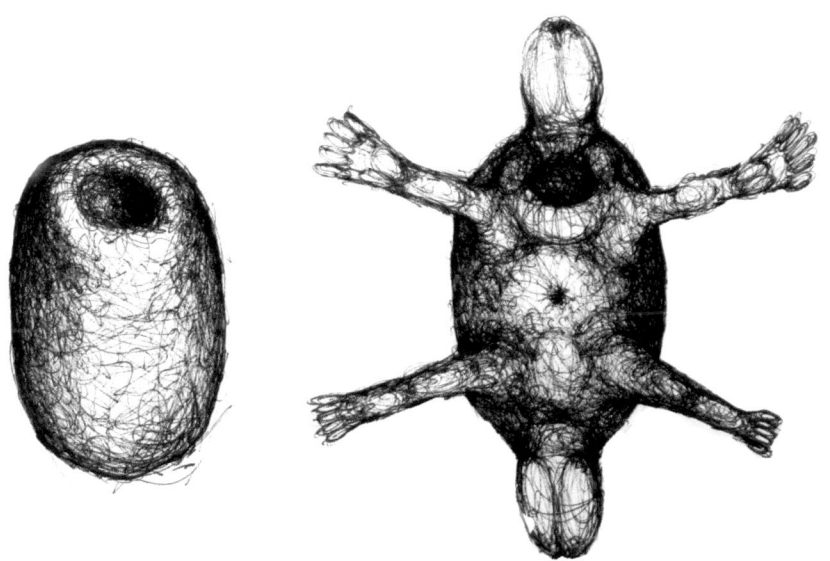

shell and shell with appurtenances

came later in evolution – eyes, ears, arms, legs – have changed nothing of the body's basic build. Even today's economics originate with that shell. Indeed, according to the principles of economics, you and I are driven by scarcity. This

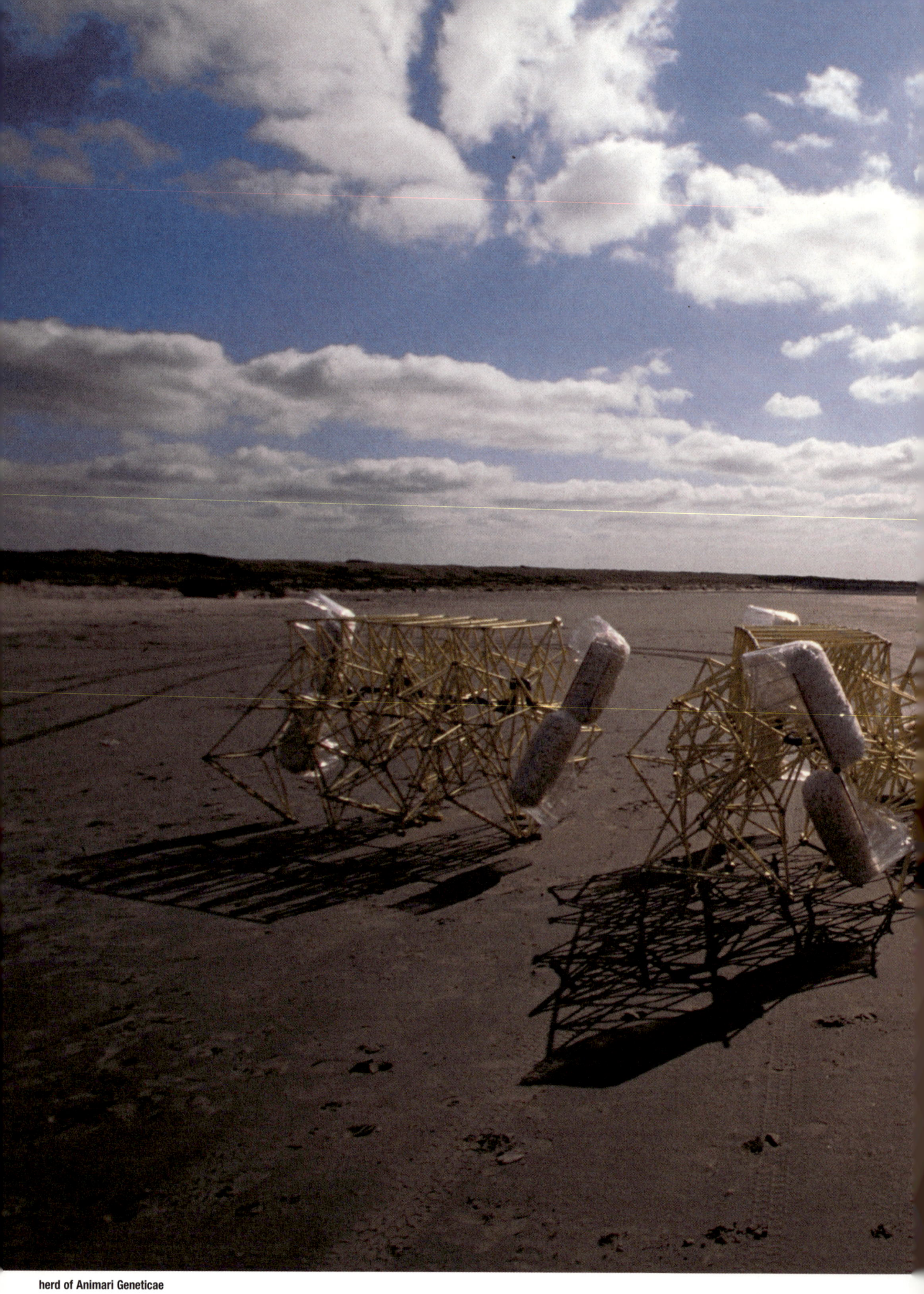
herd of Animari Geneticae

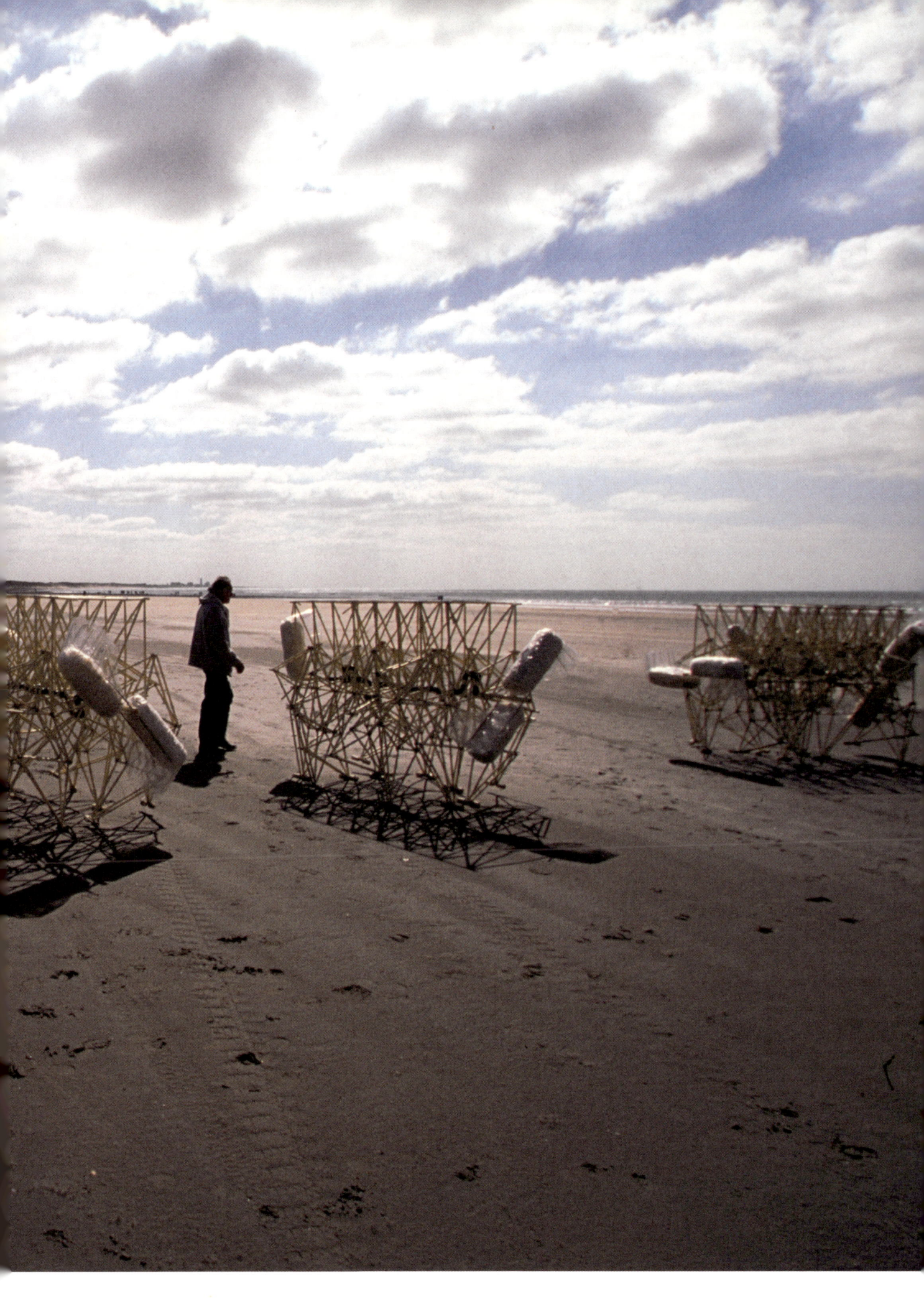

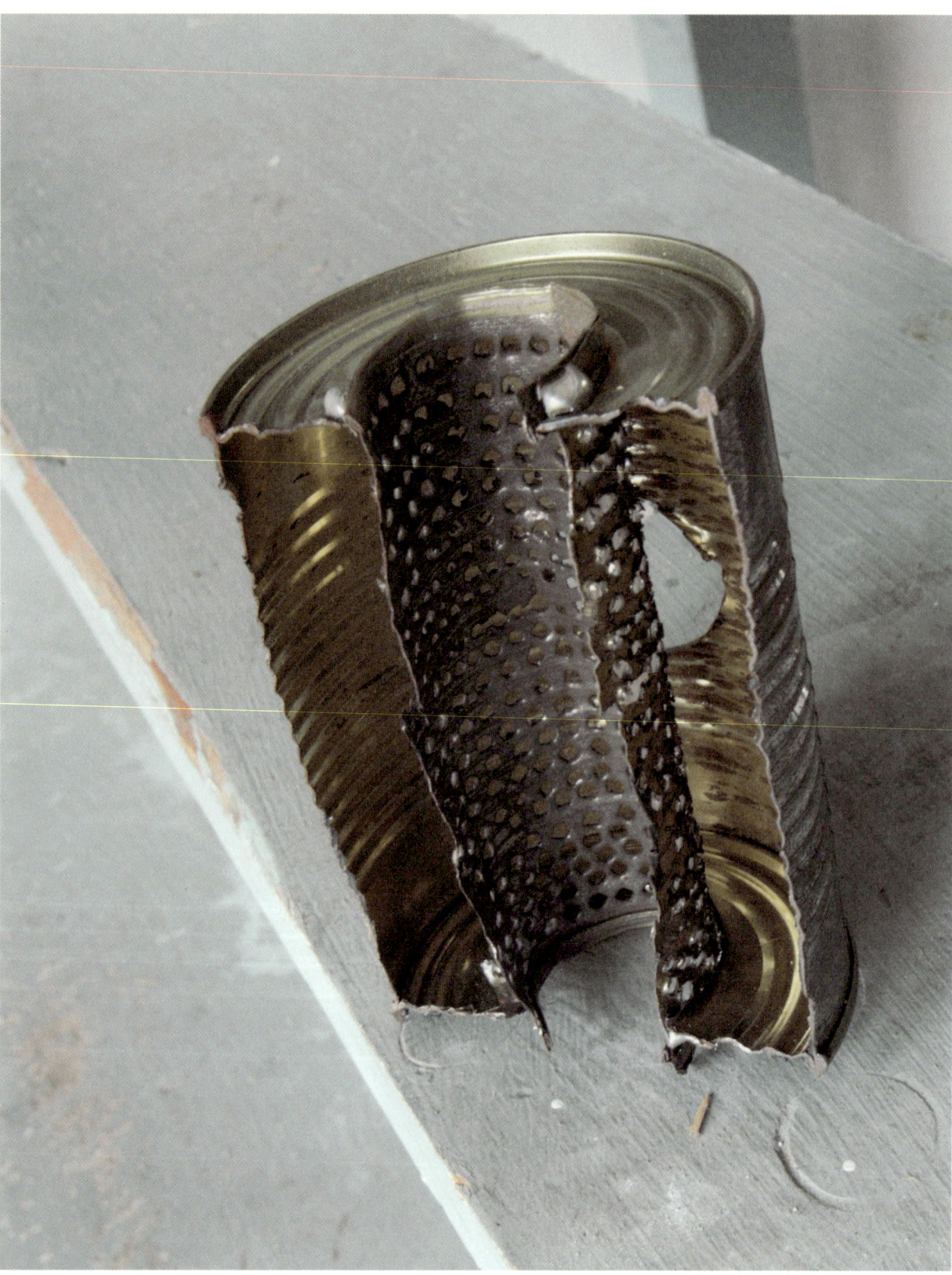

heat distributor

touches immediately on one of the problems of the shell, namely the lack of edibles in front of the entrance. An almost permanent scarcity prevailed at the front of the shell. At the rear, by contrast, the waste matter was piling up. Driven by the shortage at the front and the surfeit at the rear, the shell moved forwards. It let the edible world slip through it. Just as the jet engine of an aeroplane is propelled by taking in as much air as possible at the front and expelling as much exhaust gas as possible at the rear, the shell propelled itself forward, at a more modest speed of course. Bit by bit we evolved into the car-polishing, spin-drying, humping shells we are today.

Symmetry in the human body
My body is too short, which is why my legs are too long. Once a deviation has taken root, it makes its way throughout the rest of the body. When my body was assembled, it seems my limbs were found by rummaging in bins marked 'any old length'. My right arm is too long and so my left arm is too short.
The palm of my right hand has a completely different shape to that of my left. Even my nostrils, although holes aren't really body parts, fail to match. As I see it, this asymmetry, the irregular curvatures, lends my body a certain distinction. Rampant asymmetry can do you favours in your day-to-day life. For example, a guitarist strums with one hand and presses the strings against the neck with the other. So he keeps the fingernails of the pressing hand short and lets those of the strumming hand grow. A favourable asymmetry then. Each of the two hands has learnt a trade of its own. Confusion would reign if the jobs were to be swapped by grasping the guitar the other way round. Even a virtuoso would look helpless in this situation. Although the genetic blueprint is almost identical for limbs in mirror image, so that these can be regarded as identical twins, an asymmetry obtains that evidently has its uses. Usually the right hand lords it over the left. The left hand acts as a feeder. We need feeders too. It's a question of specializing. Once limbs or organs have found their vocation, they have admittedly forfeited their mirror-image condition but have progressed further up the developmental ladder. This means that the human body must have been more symmetrical once. Perhaps it even exhibited vertical symmetry. I mean that fingers looked a lot like toes. Ten fingers, ten toes. Monkeys are able to grasp things with their feet and use their hands when walking. During evolution, hands (on the front legs) evidently came to specialize in grasping and feet (on the back legs) in walking. The

Let go of your ego

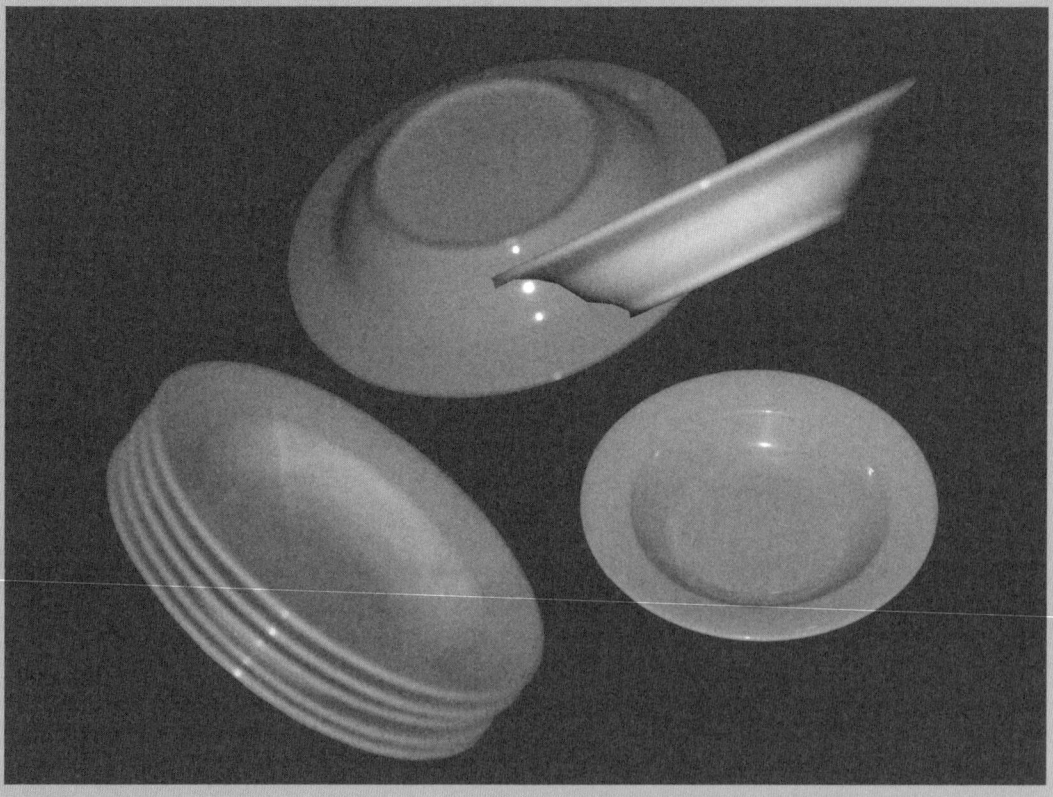

Trekkers on the French motorways can be divided into two types. There are Moroccan families who tie vast quantities of baggage on their car roofs with masses of rope and string.

Towels dangle over the top edge of the doors to ward off the scorching sun. A break taken in a parking area soon reveals four kids on the back seat. They all eat home-made snacks together on the grass.

The larger group of trekkers, however, consists of Dutch families. These hardly need describing. It's familiar ground by now. The way tourist families look has been ridiculed enough already. I would like to add, though, that Dutch families don't ward off the sun with towels but use a sheet of black gauze attached to the side window by a suction pad, which also happens to be the nose of a cat with long whiskers. How cute.

What is it that brings our species to hit the road in the summer? It's obvious why the Moroccans do it. They are off to visit relatives. As for tourists, the reason is more deep-rooted. It has to be. Why would they give up their lives of luxury otherwise? They have to contend with soggy matches, primitive lighting, washrooms fifty metres away, primitive sleeping, primitive eating, primitive everything, bending, lots of bending and getting down on their knees. The urge must be very deep-rooted indeed.

My theory is that we do it to evoke the memory of our nomadic lifestyle. Don't tell me, we've never lived like nomads. But the experiences of our forebears seem embedded in our brains. By playing the nomad, we are remembering the lives of our ancestors. This puts us at a distance from our own lives, everything becomes new again. We are decelerating. We're letting go of our egos, which is what camping is really all about. We imagine ourselves in another life.

This nomad's existence has got me doing the dishes. Armed with a washing-up bowl full of rattling utensils, I set forth daily down the forest path to a tap sticking out of the ground. Eight plastic plates nestle among the mugs, forks and knives. It doesn't matter how you throw those plates in the water; not a day passes when they don't somehow come together amidst the suds. They cuddle up as if they know it's the thing to do.

Adhesion or cohesion is out of the question; they are too far away from each other for that. Whatever, I keep experiencing this miracle. The random movements of the water are bringing the plates together. In so doing they are flouting the laws of probability. Naughty plates that they are. They call from out of the suds that I just don't understand at all. They point out to me the strangeness of life and the futility of the ego, the 'I'. Unasked, they drag god into the equation, the same god the nomads had. Visions of flocks of sheep and tents scurry though my head. Hunting scenes too. They have it in for me, those plates, there in those woods, by that tap.

duties were allocated. With favourable results. The motor skills of the human hand are considerably more advanced than those of a monkey's. The human hand no longer needs to think about walking. What intrigues me most is the symmetry or asymmetry of the organs formed at the openings at the top and bottom of the body, where the limbs converge. Namely the head and the organ of procreation. Did they spring up like cities at the mouth of a river? Did there used to be a greater symmetry between above and below in the distant past? There are times when I feel I can see two halves of a brain down there. I know for sure that the head and the genitals used to be the same organ. Both have hair sprouting freely from them. Both determine the way we act. Together they are the body's directors. The one specializes in spiritual reproduction, the other in physical. Essentially, we consist of two organs: the upper body and the lower body. They live together in symbiosis like Siamese twins. They exchange services and substances, to the satisfaction of both. The foremost service of the lower body is carrying. It carries the upper body from home to work and at the end of the evening carries it to bed. They then go to sleep together, like man and wife. A key duty performed by the upper body is the use of the throat. The lower body is admittedly able to produce sound but this gets no further than the graceless, primitive trouser cough. No subtlety, almost nothing of intonation and at all events no melody. Making noise increases one's chances of survival. Screeching, cheeping, squeaking and growling have their uses. Cries can keep the lineage going. These are mating calls and warning cries. The mating call can be regarded as a belch from the genitals. If genitals had been able to make a noise they would have done it. But it proved more practical to leave that to an organ better suited to it because of its position on the air pipes; the throat. Mating calls bubble up from the loins and erupt in the pharynx. This in contrast to warning cries. These emanate from the top. Brains are capable of forming an image of the surroundings. Pretty much the same things that happen in the real world happen in that mirror world. The future can be simulated this way. Danger is spotted before the fatal occurrence can take place. The result? Fright, constriction of the pectoral muscles and small muscles around the vocal cords. Out comes the warning cry. The way I see it, the mating call gave rise to music and the warning cry to language. Both phenomena are outgrowths comparable with the opulent tail of a peacock or the excessively large crop of a cock pigeon. Early man kept inventing new noises to prevail upon his partner. This turned into singing and the use of

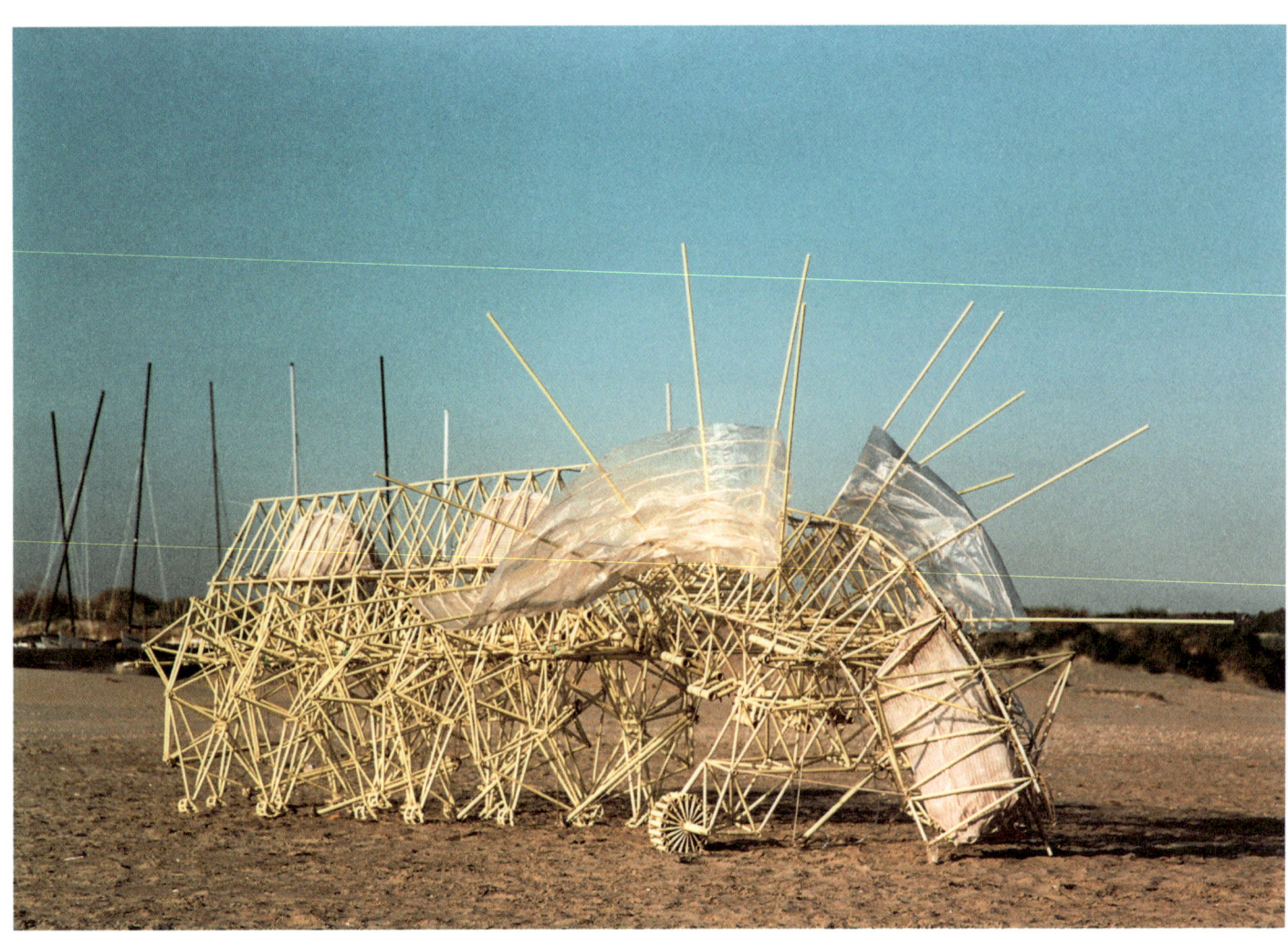
Animaris Percipiere Secundus

struck objects to drive the message home, and ultimately ushered in tonal instruments. These days music has little left in common with its original purpose, to seduce one's partner. It has become a phenomenon in its own right. We see the same thing happening with language. The warning function of language is negligible these days. The most degenerate form of language is literature. Here content is scarcely an issue; what counts is how the words are arranged. Indeed, literature is much less about occurrences than the way these are described. It's the succession of words and the images they evoke that matter. Just as an occurrence can be divided into sub-occurrences, novels can be divided into chapters, sub-chapters, paragraphs, sentences, words, letters. The passage from one word to the next is what brings about the tension and excitement in literature. This tension and excitement is entirely new. It's certainly not useful, rather it's man transcending his basic reflexes. There's nothing wrong with basic reflexes, but there is so much more.

Cells

This must have happened to you as well. When I went to take a look at my old primary school a short while back, it struck me that everything had got smaller. In my memory, the front door had been a gateway much like the Arc de Triomphe but with two doors in it. What a comedown after seeing it again. The door was anything but a triumphal arch. Obviously all kinds of psychological reasons contributed to this, but what is certain is that the reduced size of the door has to do with the increased size of my body. We measure the size of things against the size of our bodies. Something similar happens with our perception of symmetry. We observe deviations in symmetry in objects as a result of deviations in symmetry in our own bodies. A thing can be left or right of our body, in front of or behind it. Four fundamentally different compass points of the body. I would say North in front, South behind, East to your right and West to your left. There are people who have great trouble distinguishing between east and west (right and left). We all have this problem to a degree. I myself remember it from my childhood. You can do it but you have to think about it all the same. Where is your heart, which of the two is your right hand? Again, there are plenty of children who find it difficult to tell the difference between a *b* and a *d*. The confusion between left and right is due to the fact that the left and right halves of our bodies are virtually symmetrical. It's not difficult to discern what is the back and what

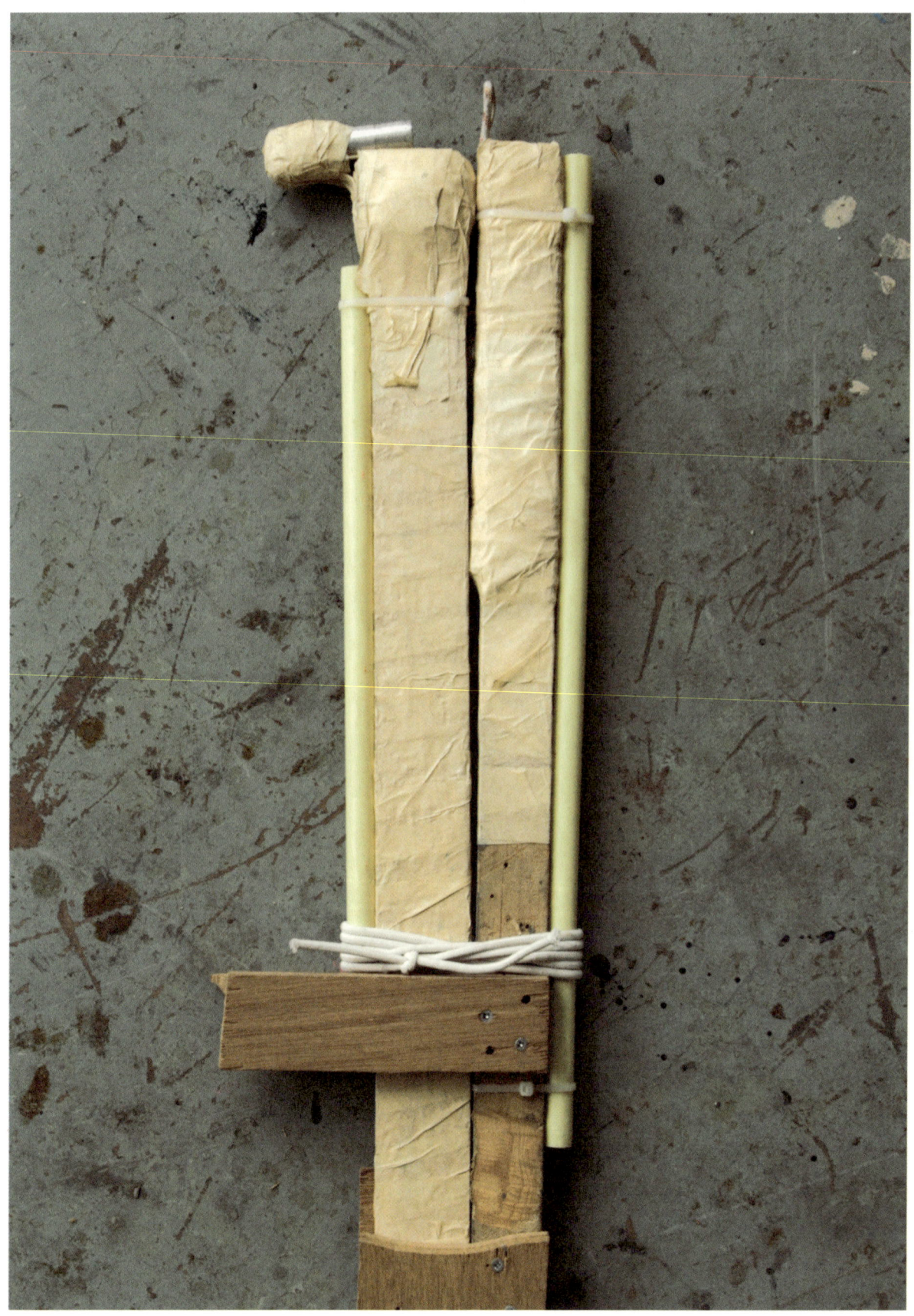

roll-up mould

is the front. The front of one's body is utterly unlike the back. The buttocks are at the back and there are only eyes at the front. There's no way you're going to say something is in front of you when in reality it's behind you.

As I asserted earlier, our bodies also possess a vertical symmetry. Draw a horizontal line through your midriff and an element of vertical symmetry presents itself. We have no difficulty distinguishing above from below. Every one of us can differentiate between a *p* and a *b*. That we have no difficulty here is due to a most persistent memory aid – gravity. Our feet are better at combating gravity, which is why we usually walk on our feet and not on our hands.

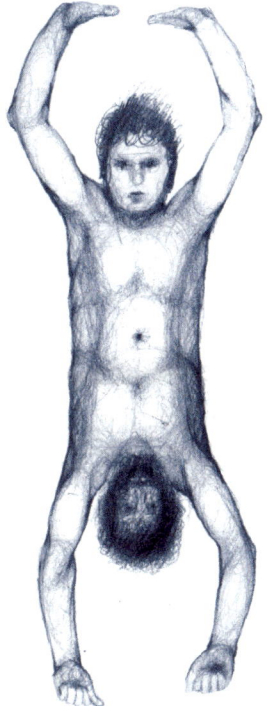

If we could walk as easily on our hands as on our feet, we wouldn't know what was above and what below.

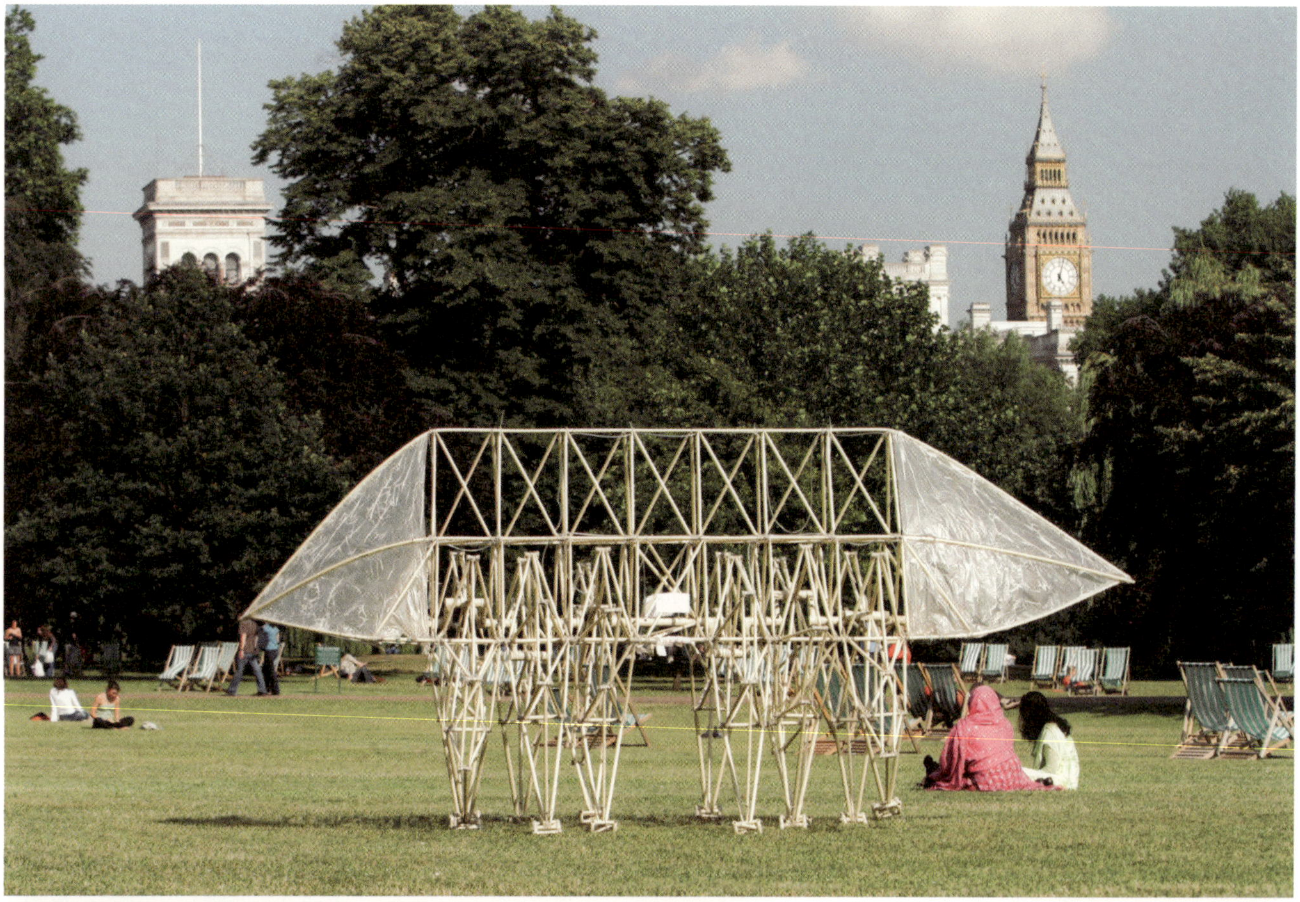

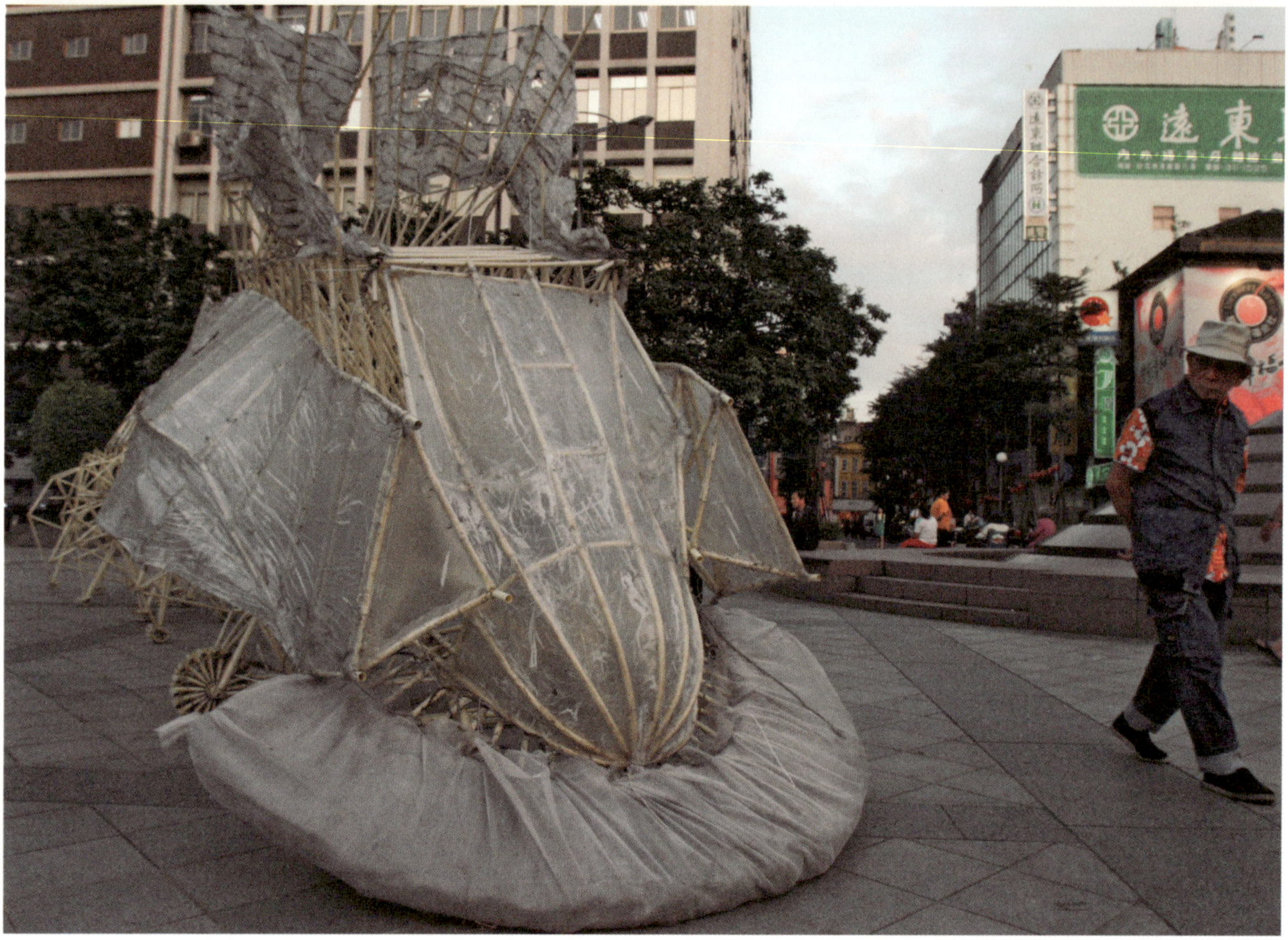

exhibitions in London (2006) and Taipeh (2006)

Thanks to asymmetry, we know what above and below are. This would be far more troublesome if we could walk as easily on our hands as on our feet. What I really mean to say is that differences in direction can only be measured in terms of differences in ourselves, in the human body, the thing where consciousness resides.

And now, the world seen from the perspective of a plastic tube. Such a tube is entirely symmetrical. Like some people, that tube doesn't know the difference between left and right, front and back, above and below. Suppose we were to apply pressure to the tube. In the theoretical case of the tube being perfectly symmetrical, it shouldn't be able to break. In which direction *could* it break? After all, it doesn't know the difference between left and right. So a perfectly symmetrical tube is unable to decide which direction to break in. It is infinitely strong. We know from practice that this isn't the case, however, so we are forced to conclude that tubes are never perfectly symmetrical. Each tube is a bit bent. It's a minimal deviation, but a perfectly straight tube simply doesn't exist. In structures, therefore, the strategy is to get as close as possible to symmetrical tubes. Beach animals, like real forms of life, are assembled from identical building blocks. For simplicity's sake I call these 'cells'. Beach animals' cells are spindly, unlike real cells which are more or less globular. Form follows function. We have skin cells, muscles cells and a whole bunch of others. The cells of beach animals are equally diverse.

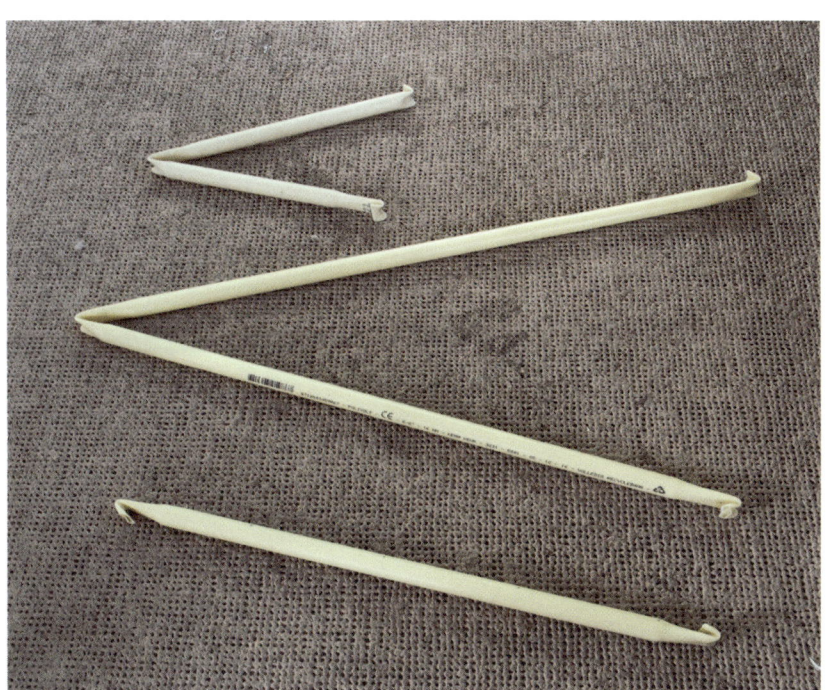

cells

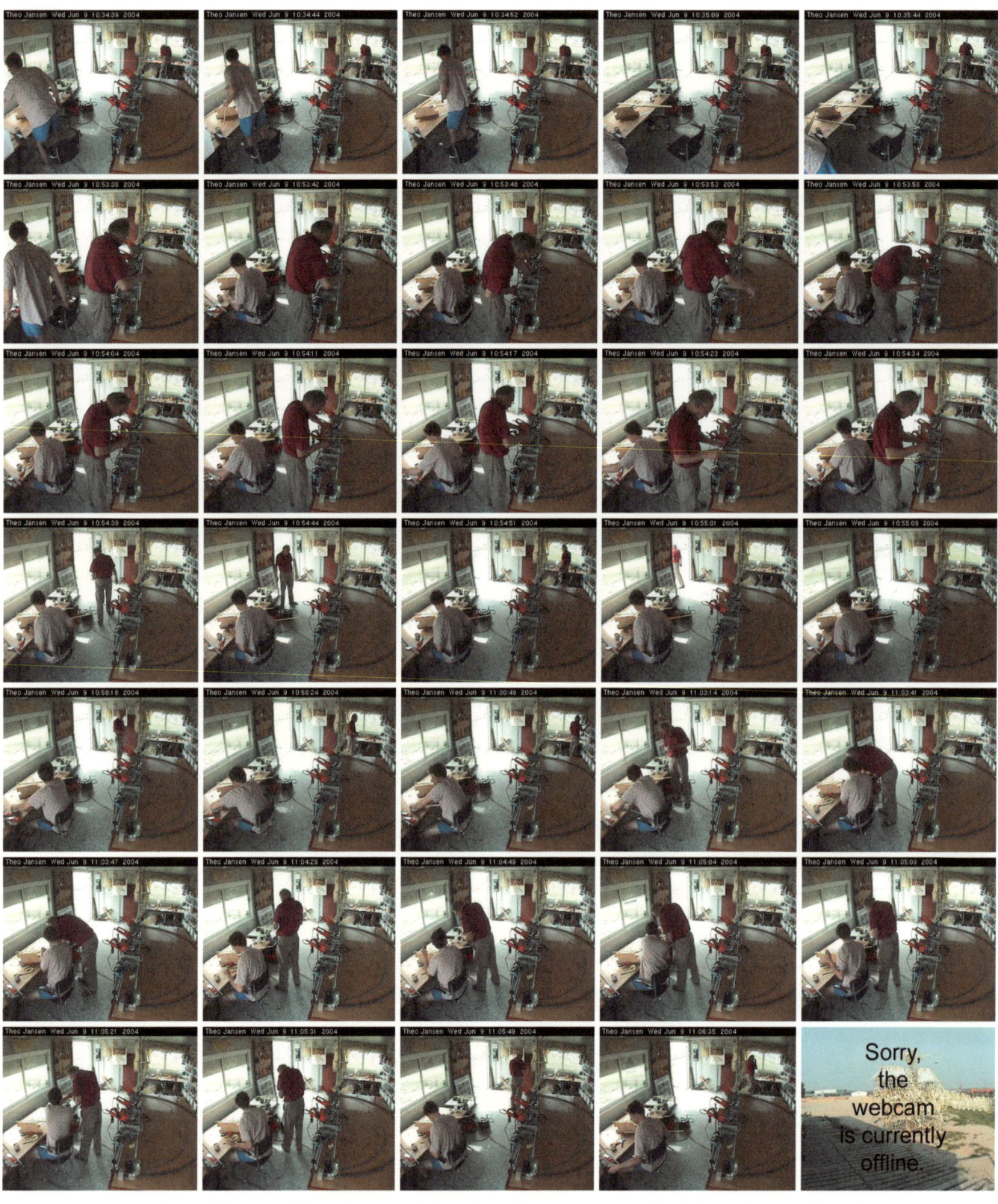

The webcam at www.strandbeest.nl allows you to follow what goes on in the workplace. Here you see Loek van der Klis helping Theo. Loek is a photographer, but a few years back he was Theo's regular assistant and often accompanied the beach animals on their forays into the world. 'The first time I observed Theo's battle with the elements – it was on the beach at Scheveningen – a light went on in my brain. The doors of the sea container were opened. Theo crept underneath what to me then was just a tangle of tubes, hitched it onto his back and headed out to the beach. Once the thing was in place the wind caught it and it began to move. Theo went with it. So did I, holding the camera. Then something broke and Theo said "Hold on to this, would you?" That's how it all started.'

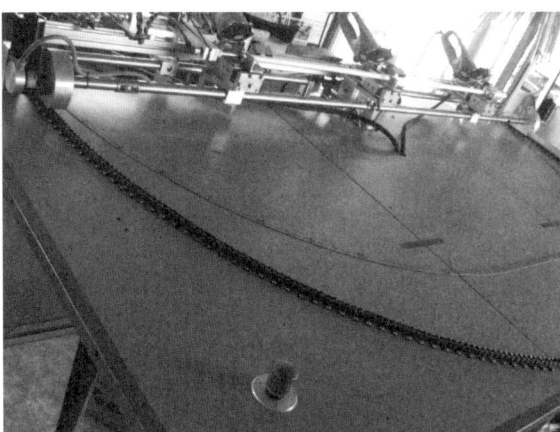 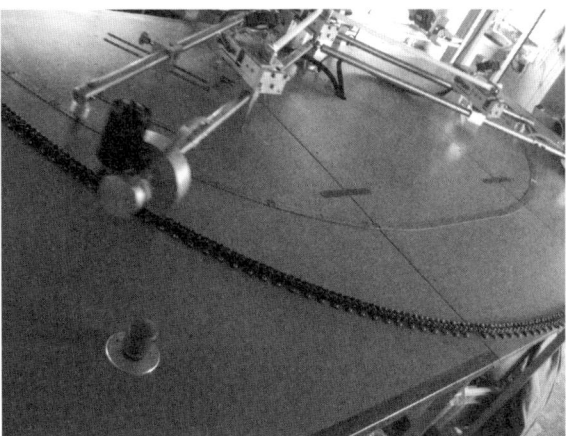

current cell machine

The illustration on page 121 shows cells that are more or less V-shaped and one with an I-shape. Rigid structures can be built with I's and V's. Take the Eiffel Tower. It is so rigid that it's still standing tall after more than a century, all thanks to the letters V and I.

The cell machine does the same as the wooden mould from the Calidum. The difference is that the distances between the small metal tubes can be varied. In addition, the plastic tubes are not fixed in place with cable ties but wedged between two metal clamps. To make a V-shape, a long tube is slid into the machine from the side. It is then heated at three points simultaneously with hot-air guns. Next, the heated tube is clamped in place and bent cleanly into a V. An important detail is that the tube is stretched slightly while it is being bent. This is to keep the shape of its extremities absolutely symmetrical. Stretching the tube in its heated state causes all transverse forces to fall away. This is comparable to the system devised by the architect Gaudí to build the Sagrada Familia in Barcelona. He constructed the models upside-down using chains and sacks of pellets. The arches then took on such a form that all transverse forces (forces perpendicular to the arch) fell away. Later, when the arches were copied in stone in an upright position, it was as if they didn't know which way the forces in them were to act. The same thing happens with the nodes at the extremities of a beach animal cell. These are heated until malleable and then stretched. This gives them their supremely symmetrical shape. When pressure is applied it is as if they too don't know how the forces in them are to act. A tube never breaks at the ends, always in the middle. A new cell machine was constructed in 2004, not by me but

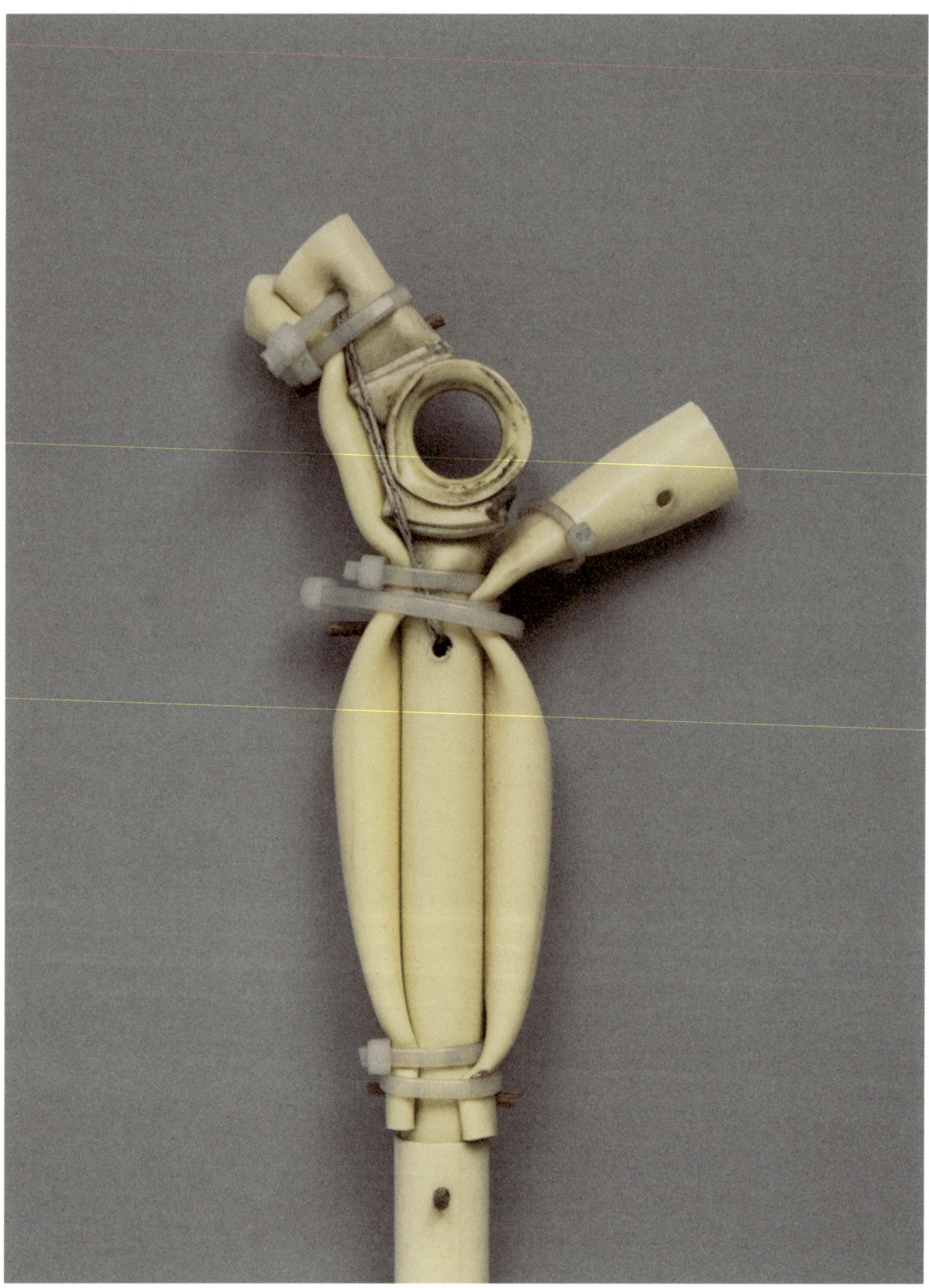

vertebra of Animaris Sparug

by Job Kneppers. The new machine proved to be far more reliable and accurate. This is important, as the cells have to be identical if the resulting shape is to be consistent.

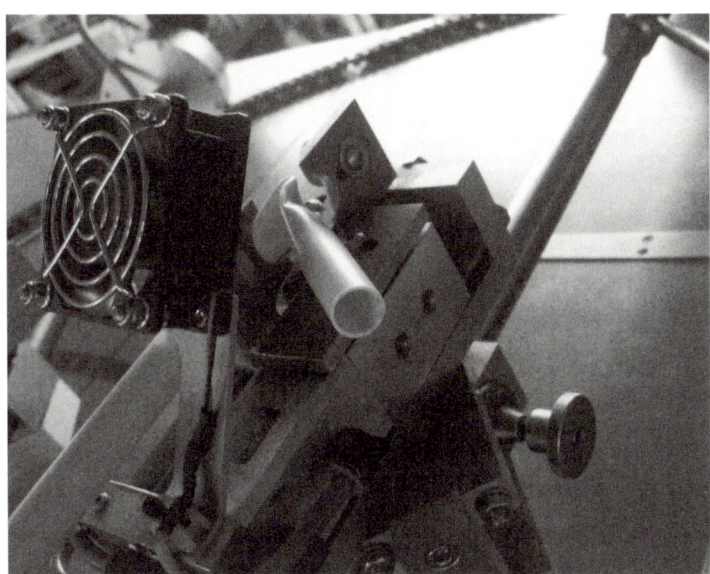

pinching mechanism

At the extremities of the cells are hooklike protuberances of bent tube into which another tube fits exactly. The cells are tied to other tubes with Dacron strings using a constrictor hitch. This is an excellent knot for permanent binding.

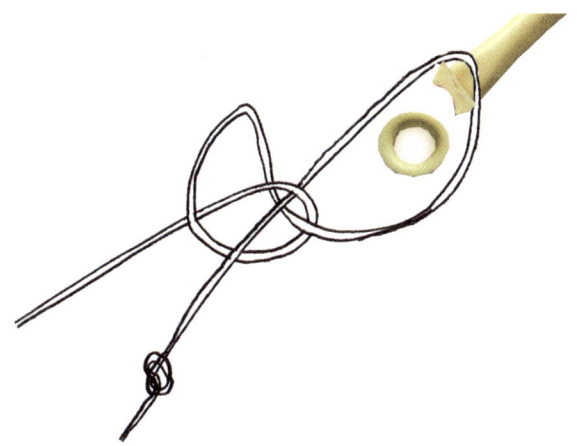

constrictor hitch

Cable ties can also be used for lashing the cells together, though they do deteriorate in time. A year or two is about their limit.

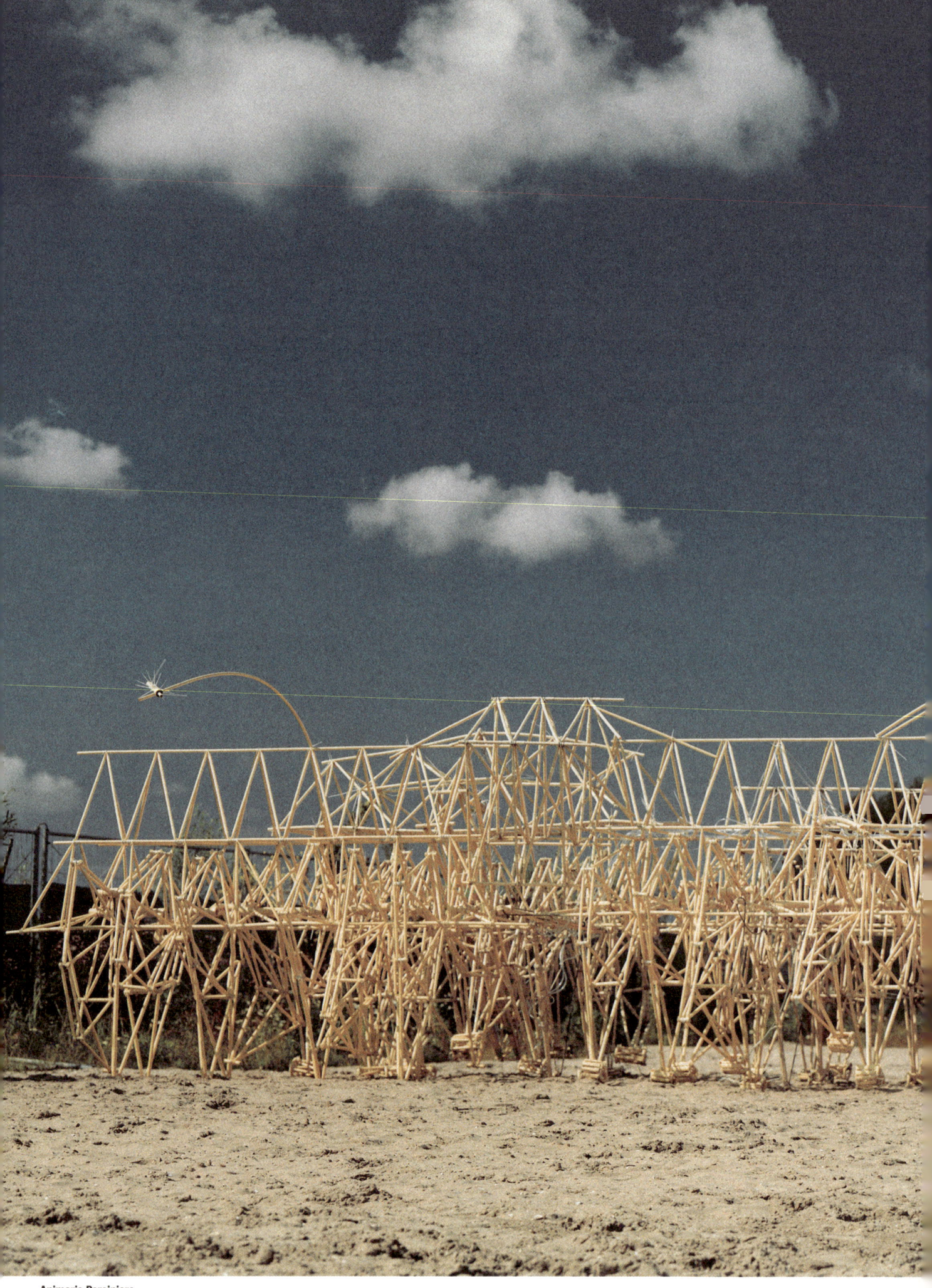

Animaris Percipiere

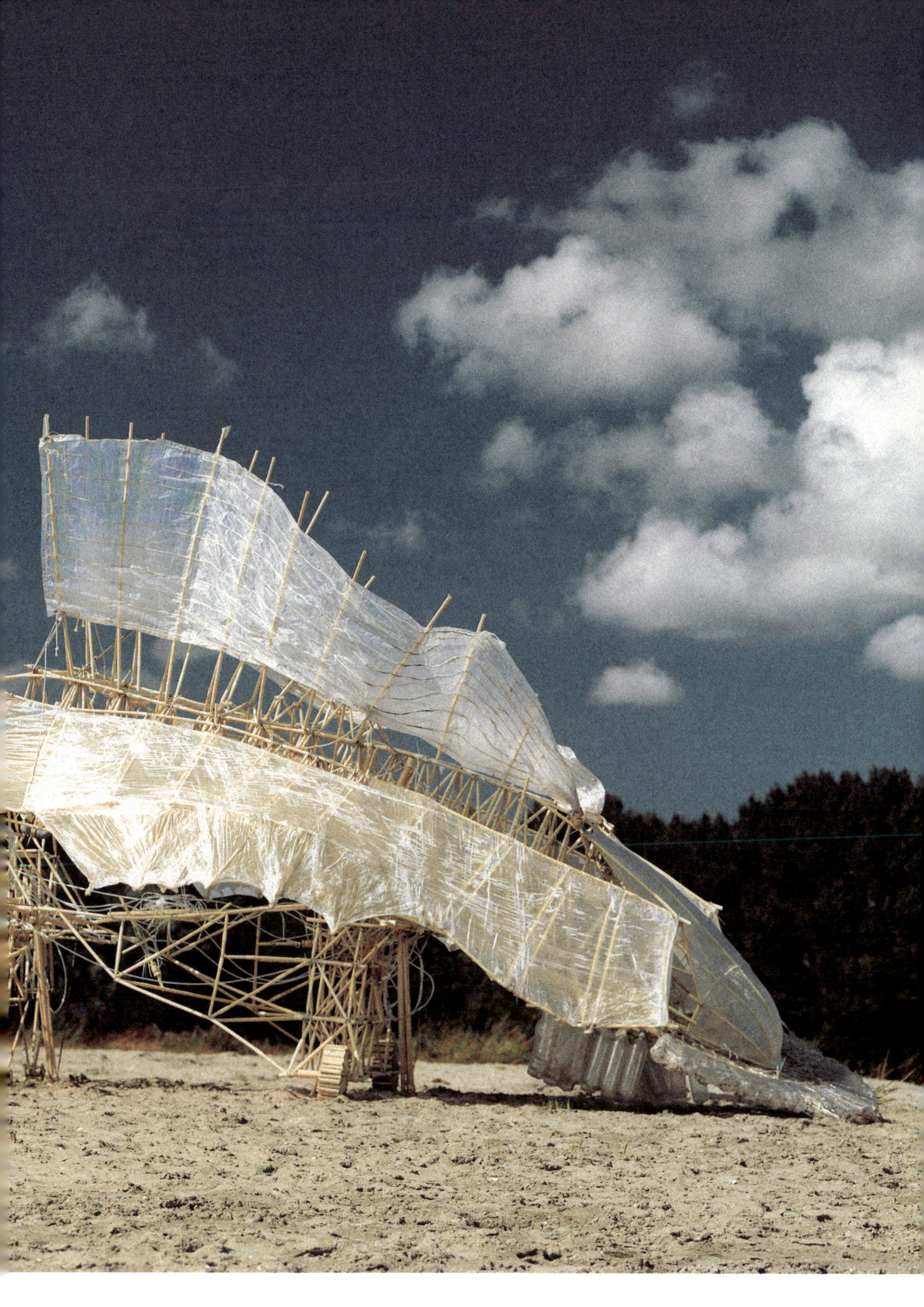

Football

This time we are going to talk about football. Most scientists seem to thoroughly detest the game. But I know that when the World Cup or the European Championship comes round, their love of the ball is greater than many of them would care to admit. Something scientists are bound to appreciate is the experiment whereby a chip is mounted at the centre of the ball so as to register to the nearest millimetre the ball's position with respect to the network of chalk lines. It can then be established, with a far greater accuracy than that of our own observational capacities, whether or not the ball is out.

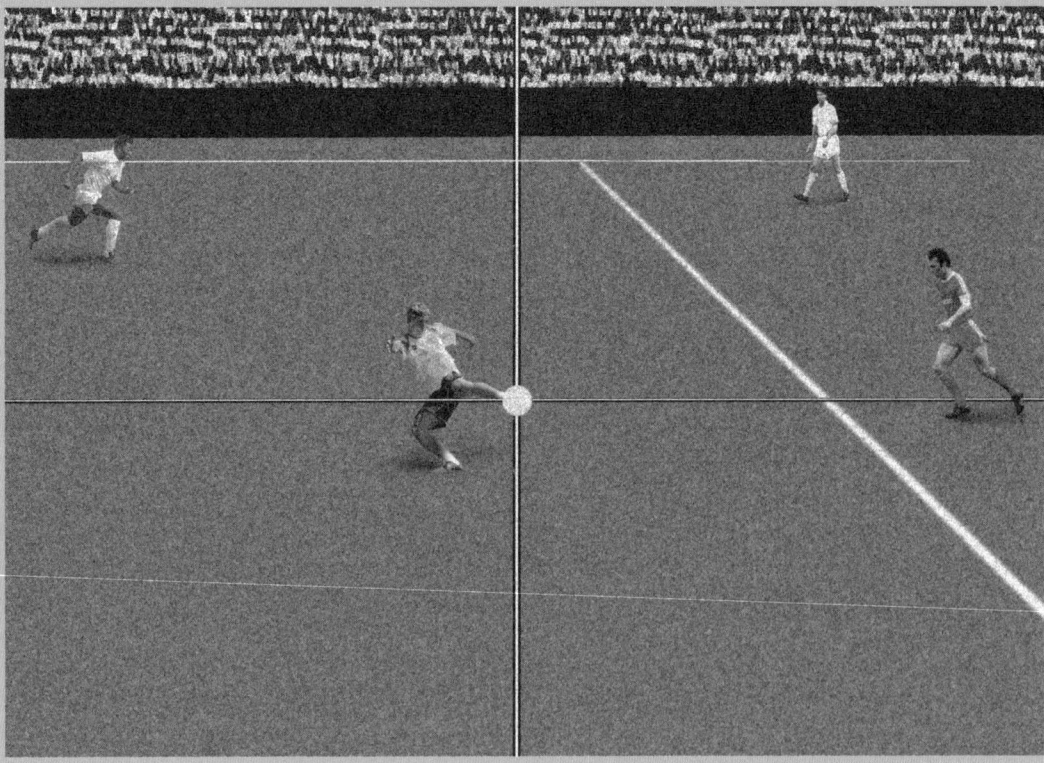

A brilliant discovery! At the same time it offers the possibility to realize a long-cherished idea of mine that would radically change the way television registers football matches. A TV camera alongside the pitch can use the chip to stay focused on the ball so that this will always be at the centre of the screen as a static white dot. You could, in a manner of speaking, attach a little white sticker to the exact centre of the screen where the football is fixed and where the players keep coming to kick it. When the ball is kicked, the entire TV set, so to speak, is kicked along with it. This will get the viewer more into the spirit of the game. Another advantage is that you never lose sight of the ball.

With this new approach comes a complete transformation of the game of football. At the moment a player kicks the ball, the ball doesn't move; instead, the pitch is pushed backwards by the player's other foot, the one standing on the ground. This brings the goal of the opposing team closer. The ball is just a launching pad for the action. The art of football therefore is a question of manipulating the pitch instead of the ball. The pitch must be moved in such a way that the net of the opposing team's goal catches the ball, much like catching minnows in a stream.

All accepted mechanics are thrown overboard. The ball behaves as though it possesses infinite mass. Footballers by contrast become as light as feathers. A keeper who catches the ball stands still in an instant, when his eighty kilos would lead you to expect a period of deceleration. The entire pitch seems to have relinquished its mass when it bounces against the ball. All that grass, all those tonnes of earth are as light as a wad of cotton wool, all thanks to one tiny chip.

There will come a time when a camera will be mounted in the ball for that optimum ball feeling. There must be a way (perhaps using a gyroscope) of preventing the camera from following the rolling movement of the ball. Just imagine the tension as you wait to see which head or foot the ball will land on next.

There will also come a time when tiny cameras will be taped to the players' foreheads. Viewers at home could then 'surf' from player to player. To *really* get into the spirit of the game.

Symmetry in Animaris Geneticus

The legs of beach animals are identical. Since the cranks of the crankshaft are rotated away from each other, every leg is in a different phase of the locomotion at a given time. The major advantage of all legs being identical of course is that they are easy to produce en masse. The question is whether those legs would still be identical if we were to reproduce and select as in the case of Animaris Geneticus. I think not. Errors would creep in when replacing the tubes. They might not get pushed properly into their sockets, increasing the distance from joint to joint. They might be measured wrongly, cut off inaccurately, badly reinserted. In most cases this error would have negative repercussions for the quality of locomotion. Yet in some cases it would be favourable. For example, I know that beach animals are poorly equipped to deal with side winds. In the beginning especially, there were occasions when those side winds were the cause of a leg breaking. And once one leg is broken, the rest follow suit. Now I can imagine that when a tube has become longer, as shown in the drawing, the leg will stand at a slight angle.

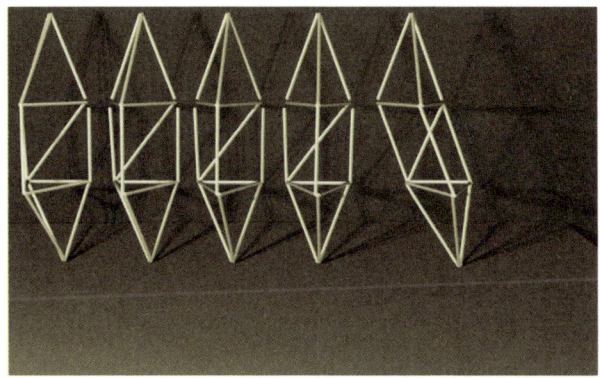

the leg at right braces itself

It's as if it were bracing itself against winds from the side. There is a good chance that this animal will survive the walking races longer than the others. The braced leg will be carried forward into subsequent generations. This mutation can also occur in one of the other side legs. And if you wait long enough all four side legs will contract it. This works differently in the human body. I can't imagine a mutation occurring in the right hand without the same thing happening to the left. Gene mutation in biological animals seems to be organized centrally.

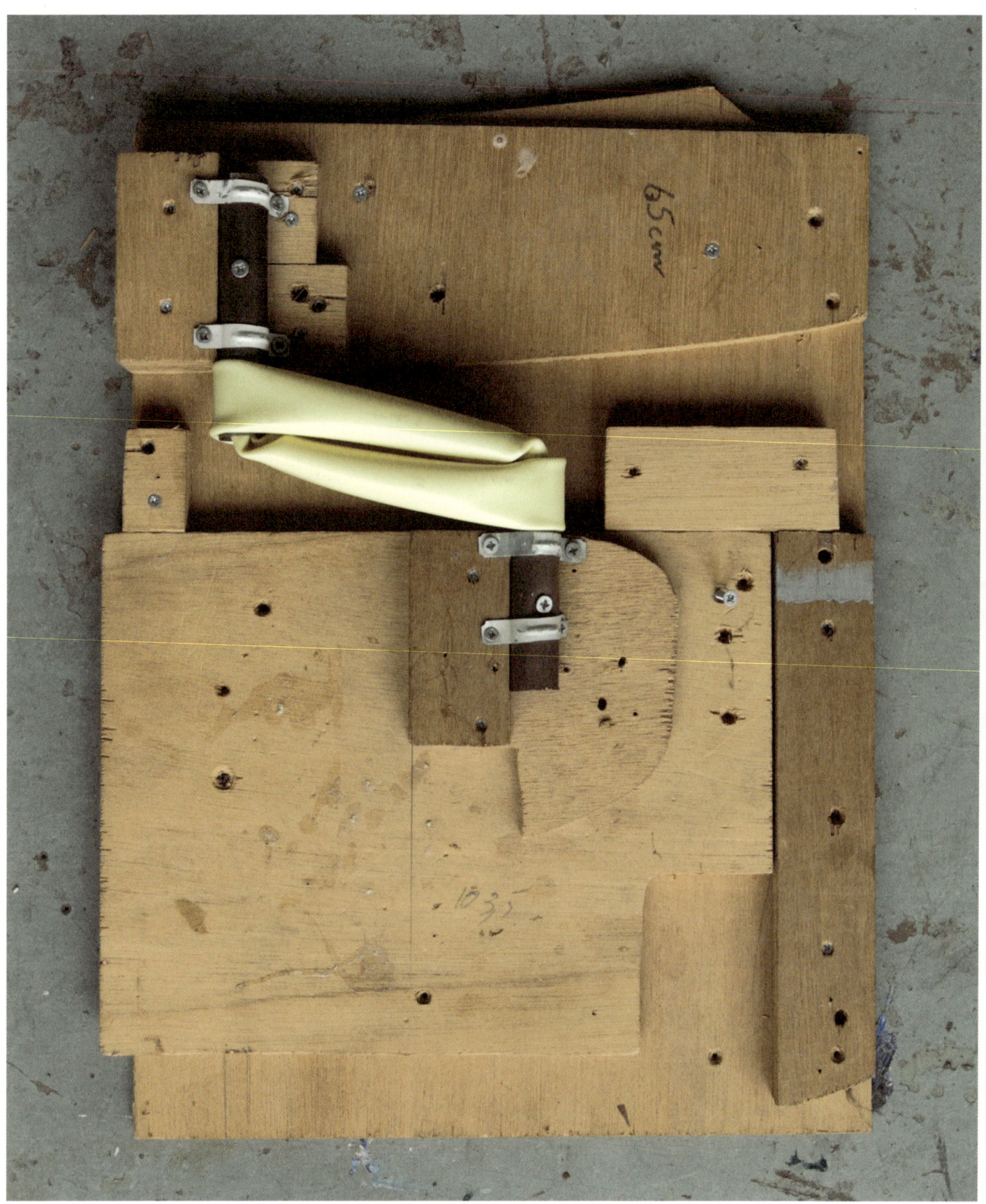

shaft mould

The code of things
The similarity between a stick insect and a stick is purely visual. The two taste quite different. The insect is much softer and juicier and vigorously resists being eaten. But the most important difference is in the genes. A stick insect has genes utterly unlike those of a real stick. By this I mean that totally different genetic codes are apparently able to produce similar forms, regular sticks and stick insects.

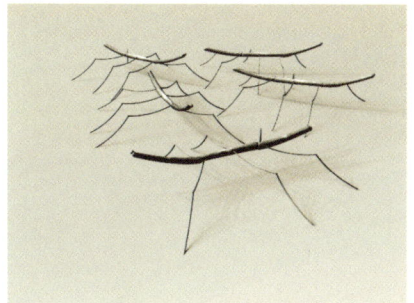

stick insects of zeros and ones

This can be seen in the above illustration. The stick insects you see there are made not using genes but with the zeros and ones in the computer. There are programs with which you can make virtual three-dimensional objects. Architects often use such programs to construct in the computer the likeness of a yet-to-be-built house so that the client can get an impression of it. But it's not just houses that can be constructed with zeros and ones in a computer; so can animals. It had been my hobby for a time to build beach animals in the computer. They were virtual animals.

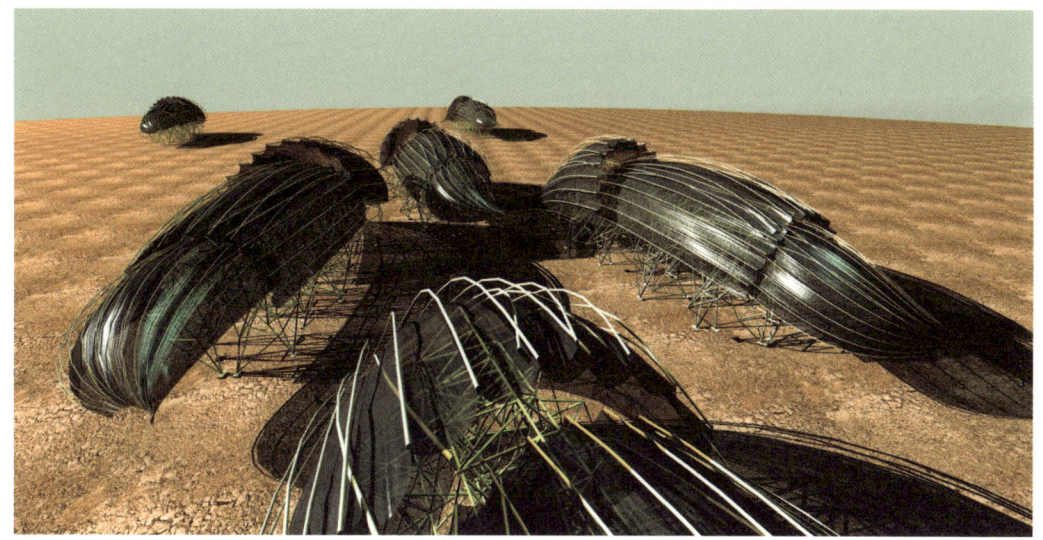

Animaris Gryllothalpa (woodlouse animal)

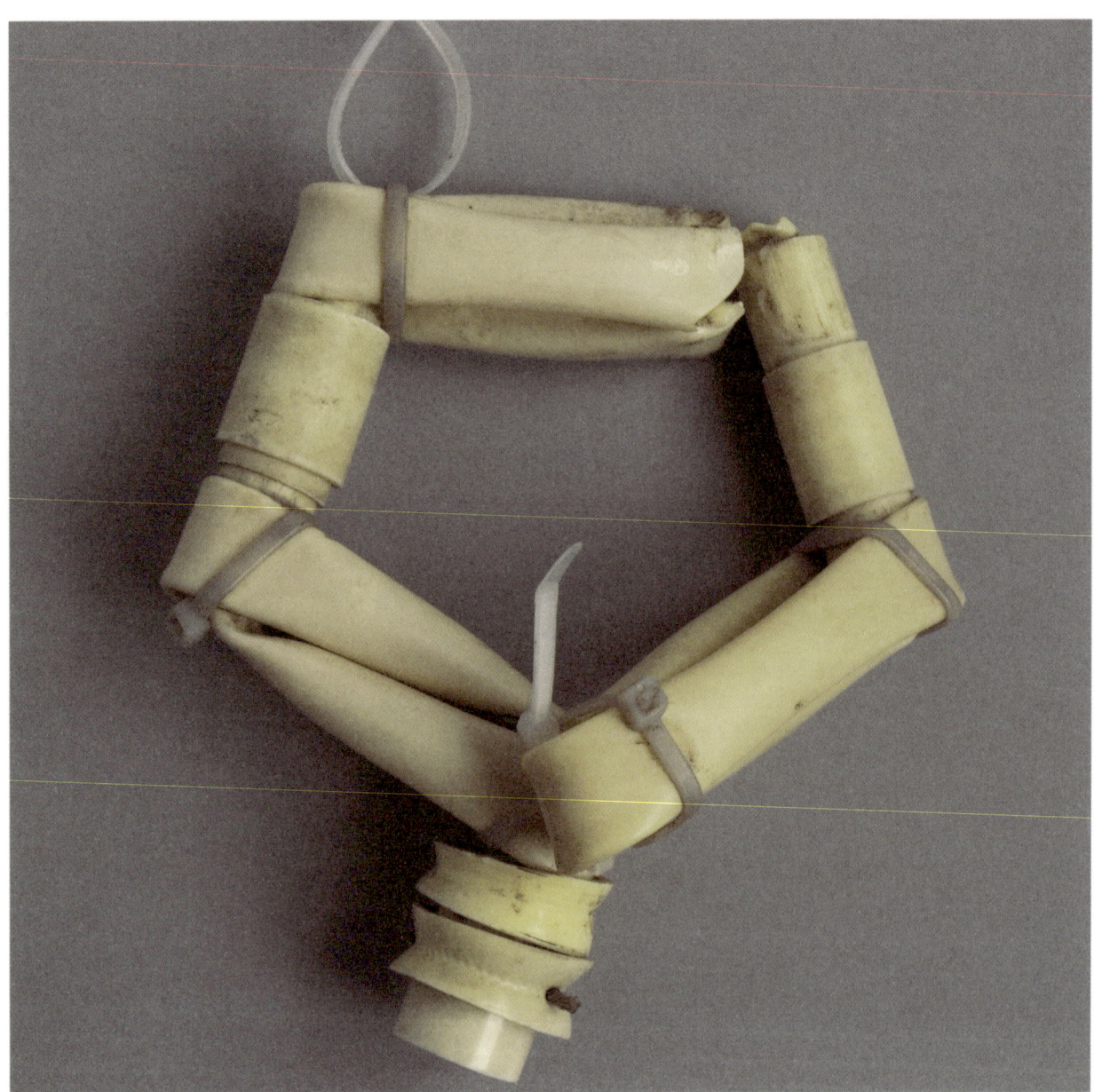

Animaris Currens Ventosa, two crankshafts

But you can also make models of biological animals such as stick insects. You might compare that code of zeros and ones with a genetic code. After all, it is an animal you're making. The stick insects in the illustration all have the same code. There are five of them, so they must be quintuplets. Now I just happened to make stick insects, but it could have chairs or tea cups instead. In other words, not only animals and plants have codes, everything can be hallmarked with a code, a number.

It could have been the title of this book: *The Code of Things*. A dictionary that gives the code as well as the meaning: breadboard 010000011110..., chest of drawers 1000101000... It seems to me that such a book already exists in our heads. Objects we see are coded and stored. It's the only way to get all that visual information to fit. Sticks and stick insects clearly have altogether different genetic codes, but in our heads the two codes are stored close together. Nor is it just visual information; the mechanical world is coded and stored there too. We can carry out mechanical experiments. Take the matter of aiming the arrow in a bow. We are calculating the arrow's path in our mind. In our mind's eye we see that arrow in flight. Is it hitting the prey or not? That depends on the tension of the drawn bowstring, the direction and the wind. The hunter tries out different directions. In the end he launches the arrow in the direction corresponding to the event in his head where the prey was hit. He's conducting experiments in his mind, the way an engineer simulates a bridge in a computer. In fact we're all just trying things out in our minds. Without consequences. Such trials are scouts in uncharted territory. And that territory is the future.

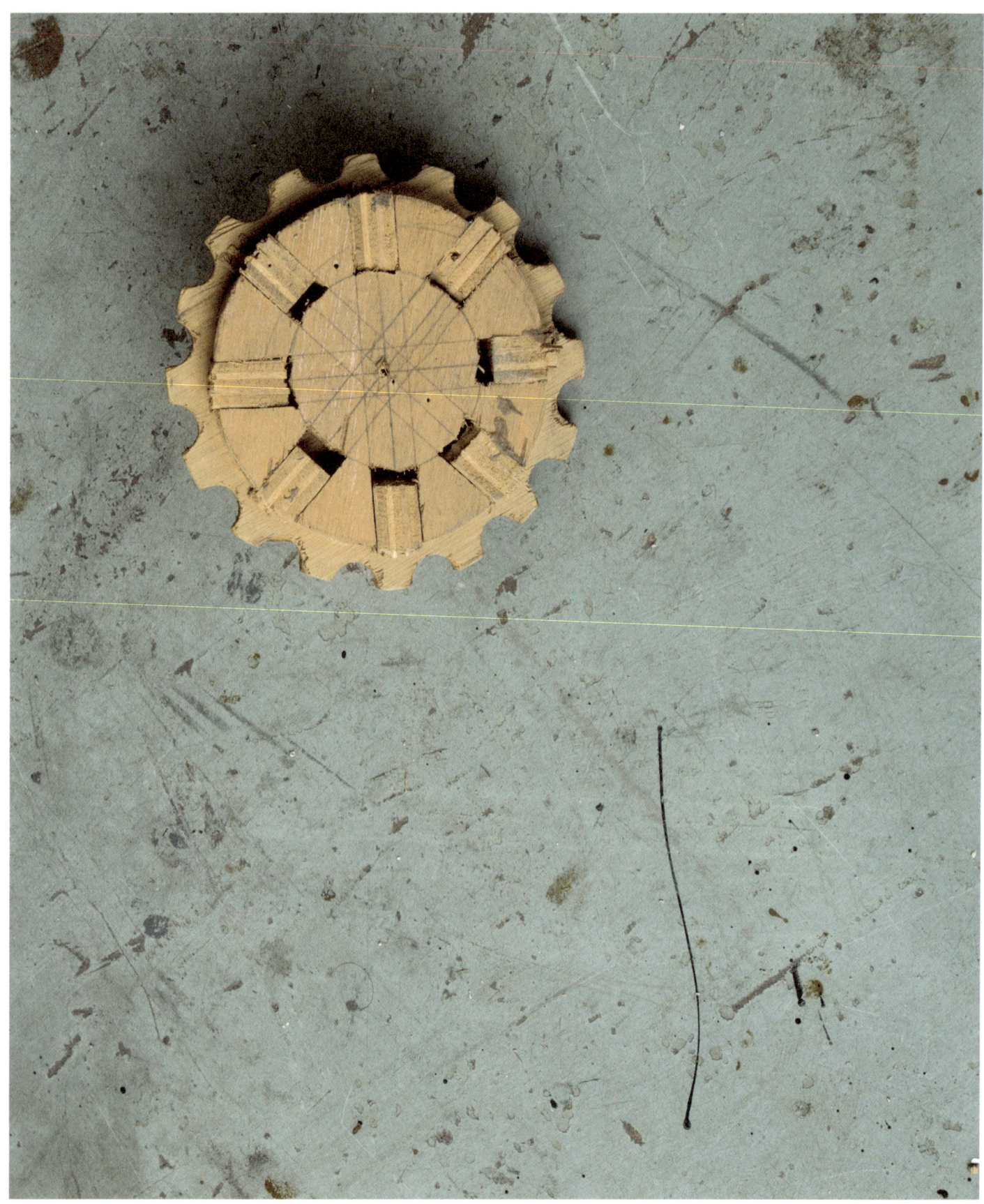

cogwheel mould

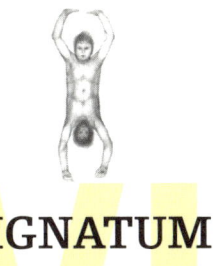

VI
LIGNATUM
the wood period (1997-2001)

The Lignatum was a period of infidelity. I cheated on my plastic tubes. I was still smitten by them but had been seduced by wood as a building material. It grew into an affair. For at least two years I had two relationships going with two different materials, plastic and wood. From tests I had done with Animari it transpired that the length of the legs was a key element in preventing the joints from rubbing. The taller the animal (the longer the legs), the lighter its gait. Now you can't increase the size of a beach animal of electrical tubing ad infinitum. Large beach animals lose their rigidity. If I was to make large animals, I had to start thinking about other methods of construction and other materials. Gradually I found myself drawn to wood; it was pallets that gave me the idea of sandwich construction. Translating plastic tubing into pallet wood gave the resulting beach animals an appearance utterly unlike that of the Animari.

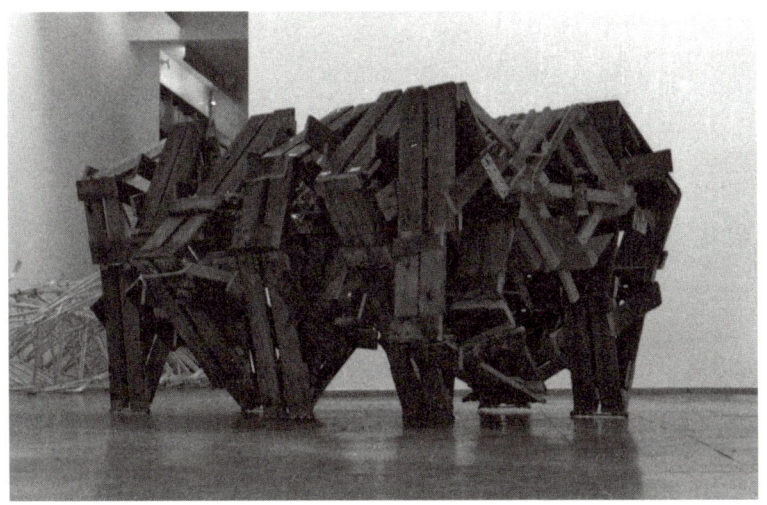

Animaris Rhinoceros Lignatus (wooden rhinoceros animal)

Pallets are cheap transport structures found lying around the street. They are tough, and rigid in all directions. They can be hitched together in freely articulated assemblies using hinges. To these sandwich constructions I gave the name Rhinocerae. One pallet beach animal was actually built. Christened Rhinoceros Lignatus, it weighed a cool 250 kilograms.

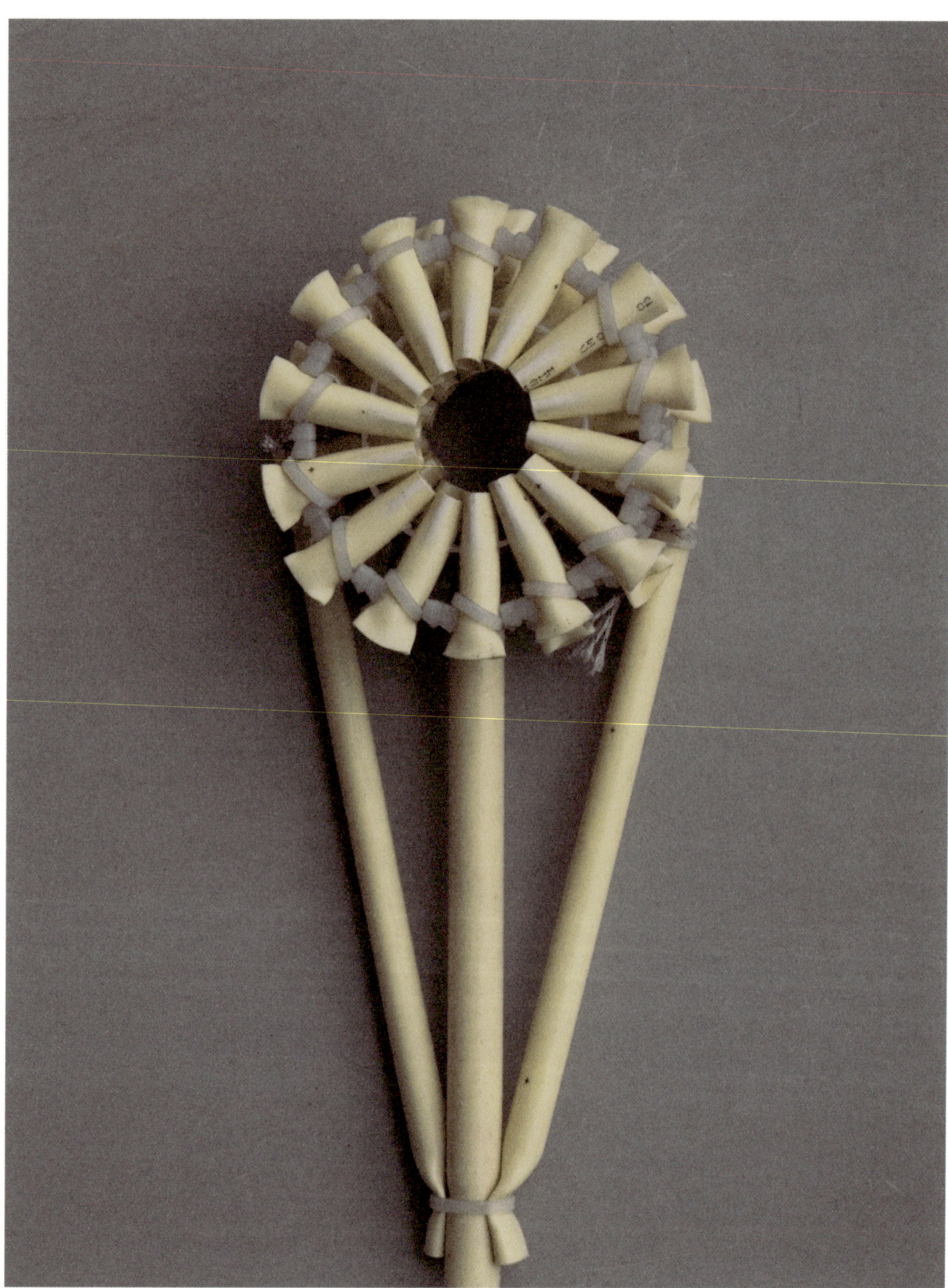

vertebra of Animaris Circodentis Primus

Pallets

My experience of house removals has taught me that heavy boxes need not be a problem. You just have to limit what you carry. And walking back and forth one more time never hurt anyone. In fact, walking is good for relaxing the back. Particularly if you whistle on the return journey. Then straighten the back and lift. The difficulty is in raising the box off the ground. First you have to get your fingers underneath the thing. The middle finger was not designed to be a crowbar and yet this is what it has to be now. You have to lever the box with your fingers to prize it loose from the ground.

This same problem reared its head when the forklift truck emerged on the scene. They tried to solve it by ramming the fork under the pile of boxes at speed, but that cost too many boxes and caused too much mess. Not long after, they introduced a special accessory, the pallet. Since then there are dozens of different types of pallet on the market: small lightweight ones, thick ones for heavy loads, pallets of wood and pallets of steel.

With the explosive rise of mechanical lifting the number of pallets rocketed. There are now millions scattered across the world. And because there are so many, they cost next to nothing. Two euros for a second-hand pallet. Treat them with creosote and you can leave them out of doors for years. But of course you don't do that. Leave them outside? You have to use them for something, whether it's floating houses, adventure playgrounds or flexible fencing. Give them a new lease of life!

Now I must point out to you the mechanical properties peculiar to the pallet. It is assembled in a sandwich construction, a layer of beams set between two layers of planks, rather like cocktail sausages between two slices of bread. This is what makes the pallet rigid in all directions; no torque, nothing. It's a rigid rectangle. This can't be said of a sheet of hardboard, which flaps every which way. It was because of its mechanical properties that I entered into a relationship with the pallet. If you add two hinges you can make something like a door. Six such doors give you a deformable hexagon.

You need twelve of these hexagons for a roving assemblage like Rhinoceros Tabulae. They are mounted on either side of a body which consists not of hexagons but of squares. Each hexagon is joined to the body at certain points using small beams. These also attach other points of the hexagons to a crankshaft located inside the body. The rotations of the crankshaft are converted into an ambulatory movement in the pallets via the small connecting beams. It's a minor miracle –

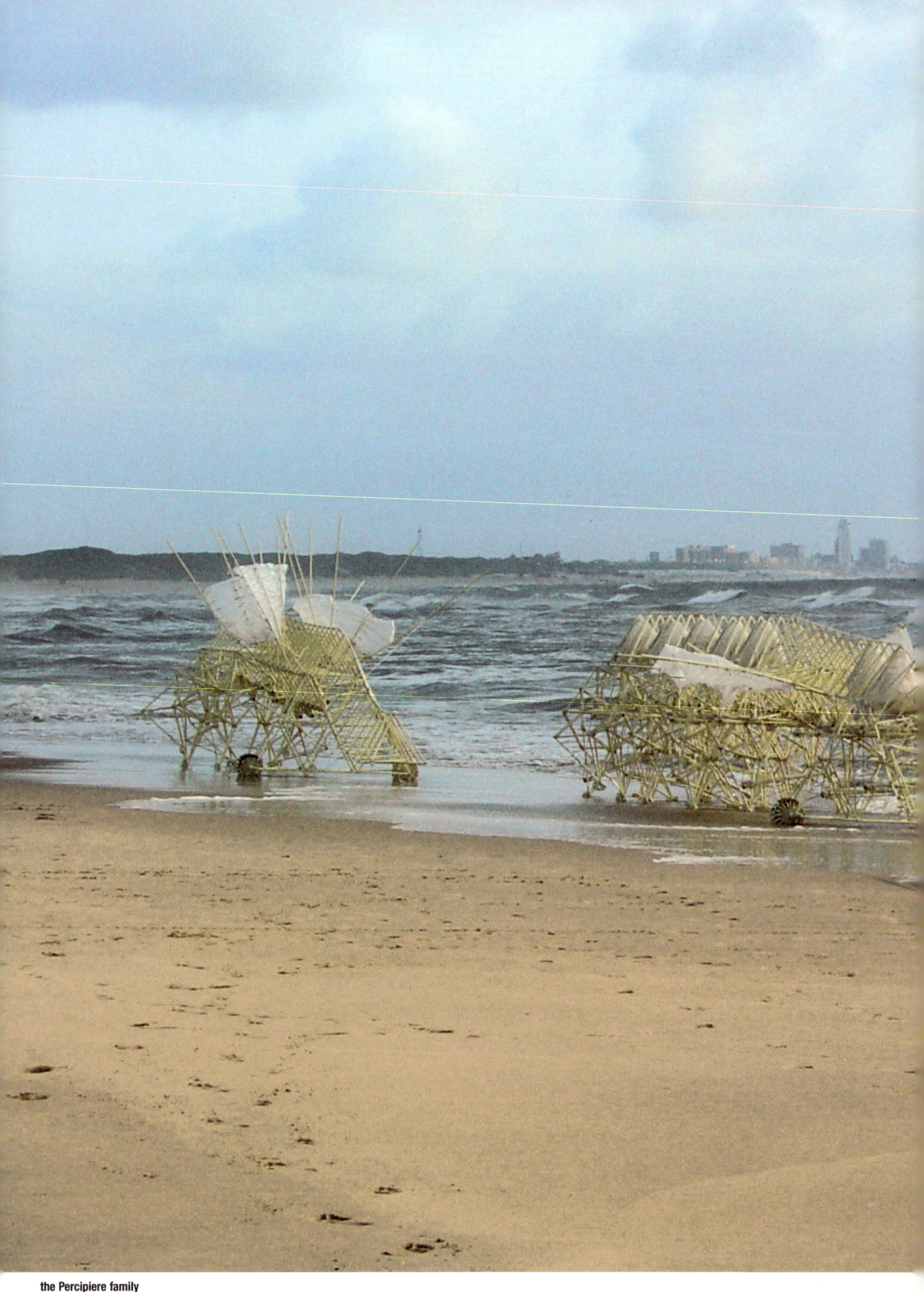
the Percipiere family

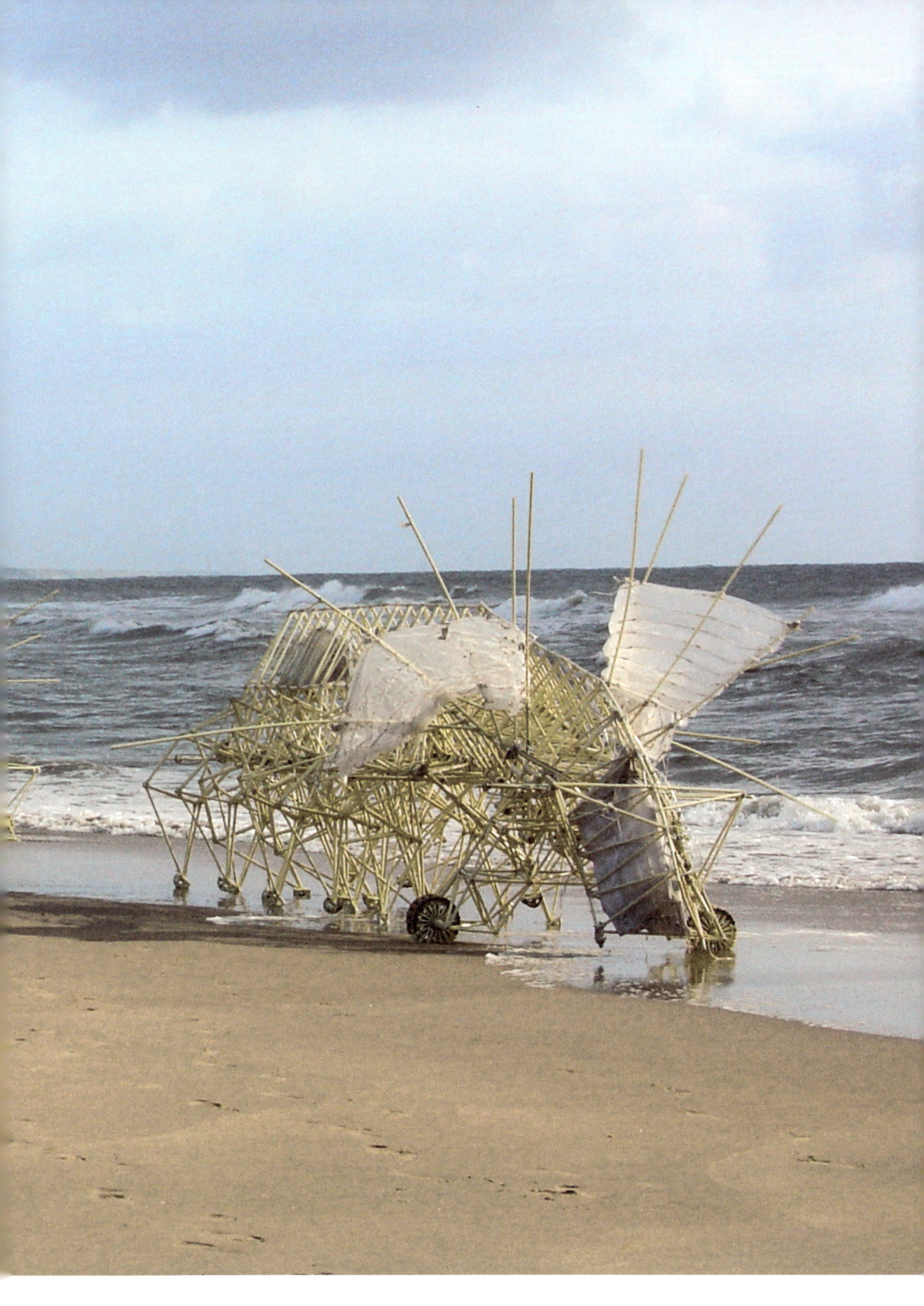

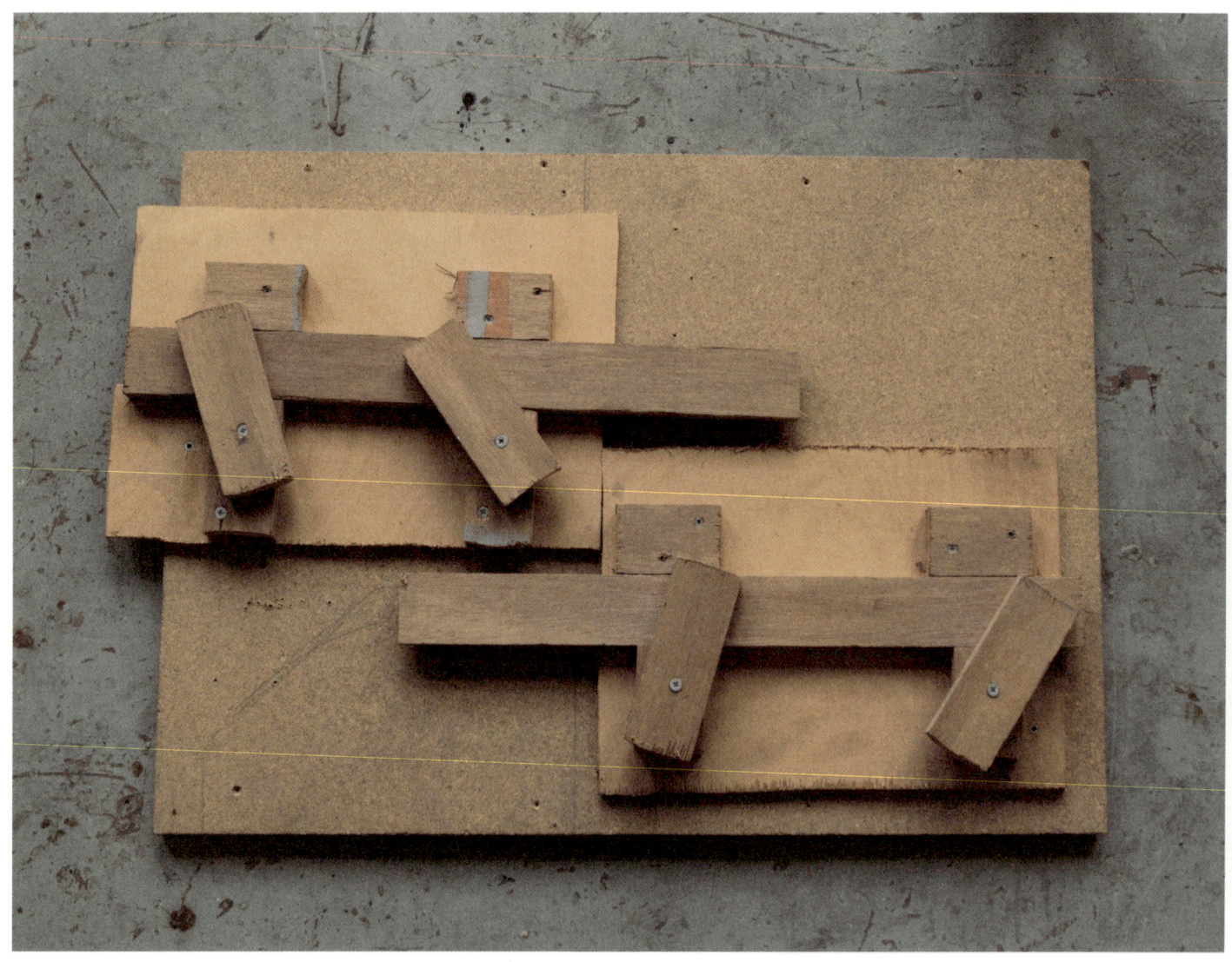

double crush mould

the animal walks without lurching. The freedom of movement enjoyed by the hexagons and the distances between the points of attachment interact in such a way that the structure moves.

In September 2004, Animaris Rhinoceros Transport was ready. It weighed 3.2 tonnes and was 4.7 metres tall. But it wasn't assembled from pallets. Steel proved more suitable after all. The frame is of powder coated steel and the skin of polyester. It has a gritty look to it. And you can sit in it. Just as a car is a horse for passengers and a plane a bird for passengers, Animaris Rhinoceros Transport is an Animaris for passengers. Steel is an obvious choice of material for vehicles. You don't make a car from protein because this is what a horse is made of. So I didn't need to make Rhinoceros from plastic tubing.

One man was enough to get Rhinoceros moving. Tests were carried out on the runway of the former Valkenburg airfield northwest of Leiden. Rhinoceros walked crosswind like a dream. In fact, it walked too fast and the joints gave way.

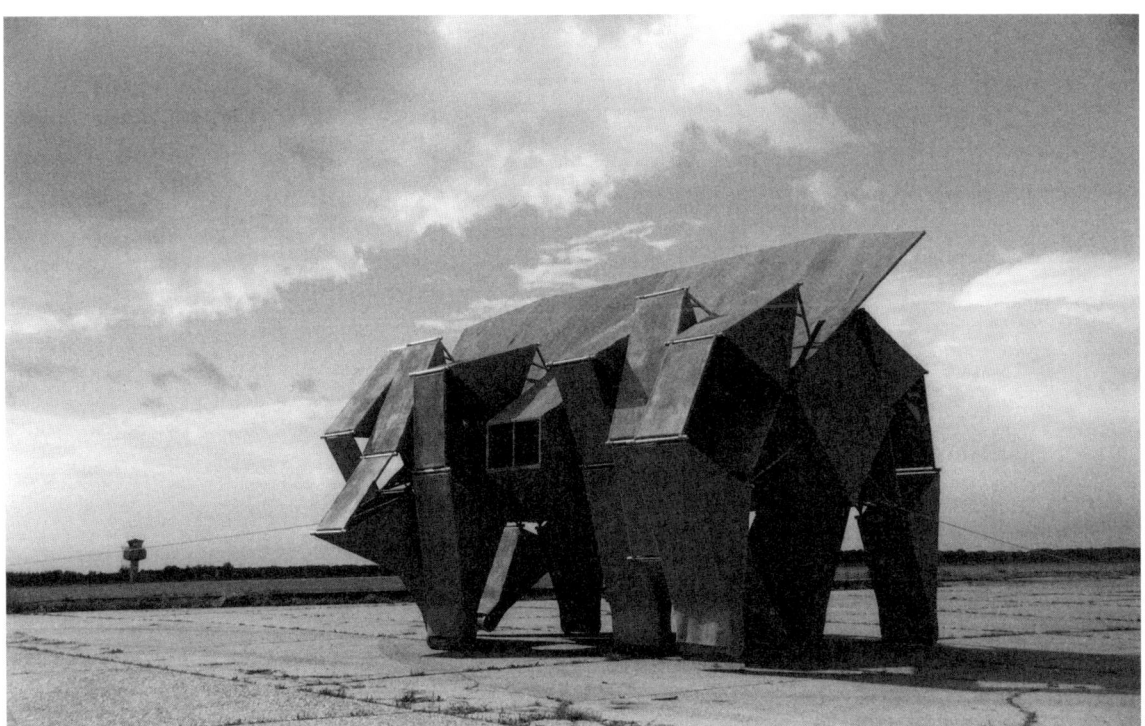

Animaris Rhinoceros Transport on the runway at Valkenburg airfield

The necessary repairs were made. On September 18th 2004 it lumbered through the streets of Amsterdam, pulled by local children. It now stands in the water

Seam

It's seldom that I address a particular target group. I don't like this practice. But there are times when you have no choice in the matter. So this is specially for the executive boards of Toshiba, Nokia, Alcatel and Ericson.

Long ago I wrote a piece that prescribed just the thing for the mobile phones of the future. The idea was there for the taking. They could have it, absolutely cost-free. Did they want to sell mobile phones? Of course! Did they listen? Did they hell. Now we're the ones to suffer.

Last week it was my turn when I went to buy an even smaller mobile than the one I already had. A real improvement, twenty tunes, eighty games and a doorstep of an instruction manual. But it had such tiny keys! You needed women's fingers to operate them. I could only work the thing using my fingernails.

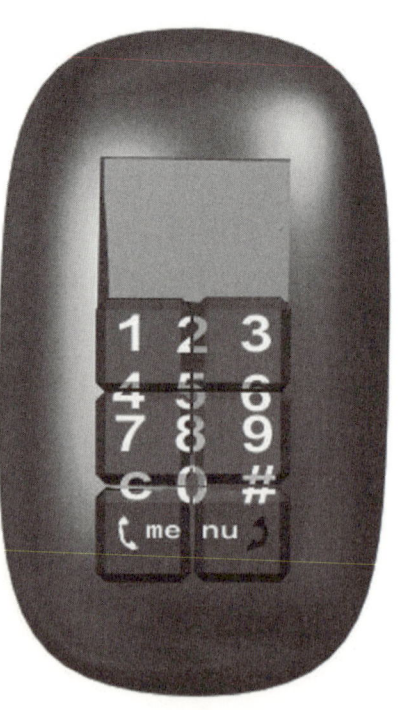

About seven years ago I described in this column a keyboard for fat fingers. Even then it was clear that computers and mobiles were to be slimmed down. In those days mobiles were called GSMs and were big and black. Everything became smaller, but of course fingers stay the same size. Keys are a restriction on small appliances.

Normally there are sixteen little keys on a mobile phone. That should be brought back to six, just six nice fat keys close together.

The idea is that the seam between two keys should provide information too; you then press two keys simultaneously. Let's begin with the four uppermost keys. These together replace the number keys 1 to 9 and are given the values 1, 3, 7 and 9. Now let's say that the seam between two keys gives you their mean value.

For example, we press the seam between the 1 and the 3. In other words we press the 1 as well as the 3. This would confuse a regular mobile: does my master (or mistress) mean the 1 or the 3? The fat-keys mobile knows that we mean 2, that is the mean of 1 and 3. Between the 1 and the 7 is the 4 and between the 3 and the 9 the 6. There is also a number at the intersection of two seams, the 5. If we press all four keys, we then get the mean value of all four: $(1 + 3 + 7 + 9) : 4 = 5$.

The two bottom keys are for making and breaking the connection. These two keys are bordered by three seams and an intersection (of seams). These are for the hashes, the ats and the resets.

At last we have entered the era of tiny mobiles. People everywhere, from Vietnam to Brazil, are grappling with the darn things. You can feel the irritation. There are even those who want the large mobiles back. It's a tight fit in the trouser pocket but at least you can punch those numbers and letters easily.

We may be a clever bunch, but whenever I see people wrestling with their mobiles in the street, in buses and trains and at the market, that cleverness takes on a ridiculous edge.

near Lambertus Zijlplein in the Geuzenveld district in the west of that city. Its walking days are over. You must have heard of that spring festival in Pamplona where bulls are let loose in the streets. I suggested making a tradition of allowing Rhinoceros to descend from its pedestal once a year and trundle through the streets of Amsterdam. Nothing has come of it yet.

The photograph below on the left shows the cardboard scale model used to secure the commission from Geuzenveld District Council. I let it amble over the meeting room table, across the papers, under the noses of the officials. The animal shown below on the right I made in a chalet down south in Zeeuws Vlaanderen where I was holidaying one Christmas with a bunch of people. The idea was to put a card model of it on the market but that hasn't materialized either.

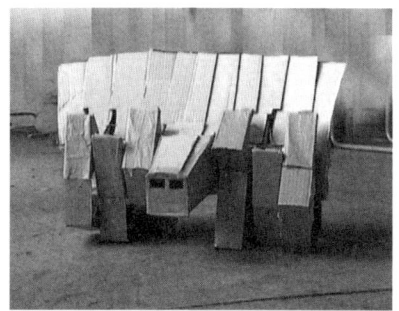
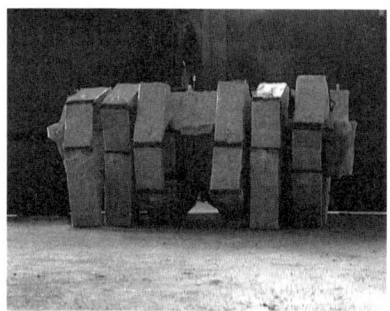

Animaris Spissa Carta (cardboard animal) and Animaris Spissa Gluton (masking tape animal)

There's almost no limit to the size an Animaris Rhinoceros Transport can attain. You could easily make one fifteen metres tall. Then it would be heavier, but would catch more wind. I could imagine a mode of transport for travelling in the tundra. It would have lots of compartments, as you might have to wait for ages for the appropriate wind. This has to be strong enough and be blowing in the right direction. There would be a library on board to kill time, an observation tower for detecting deer and of course a cosy sitting room for playing cards in the evening. Until the right wind rises. Then it's anchors aweigh!

Wind, wind, wind, wind
If you repeat the word 'wind' often enough, it will start blowing. Try it some time. The negligible amount of wind produced by one's mouth picks up by degrees to become a breeze and in some cases even a full-fledged storm. An atmosphere filled with hundreds of words can't avoid the meaning of those words. You'll see. It starts blowing sooner or later.

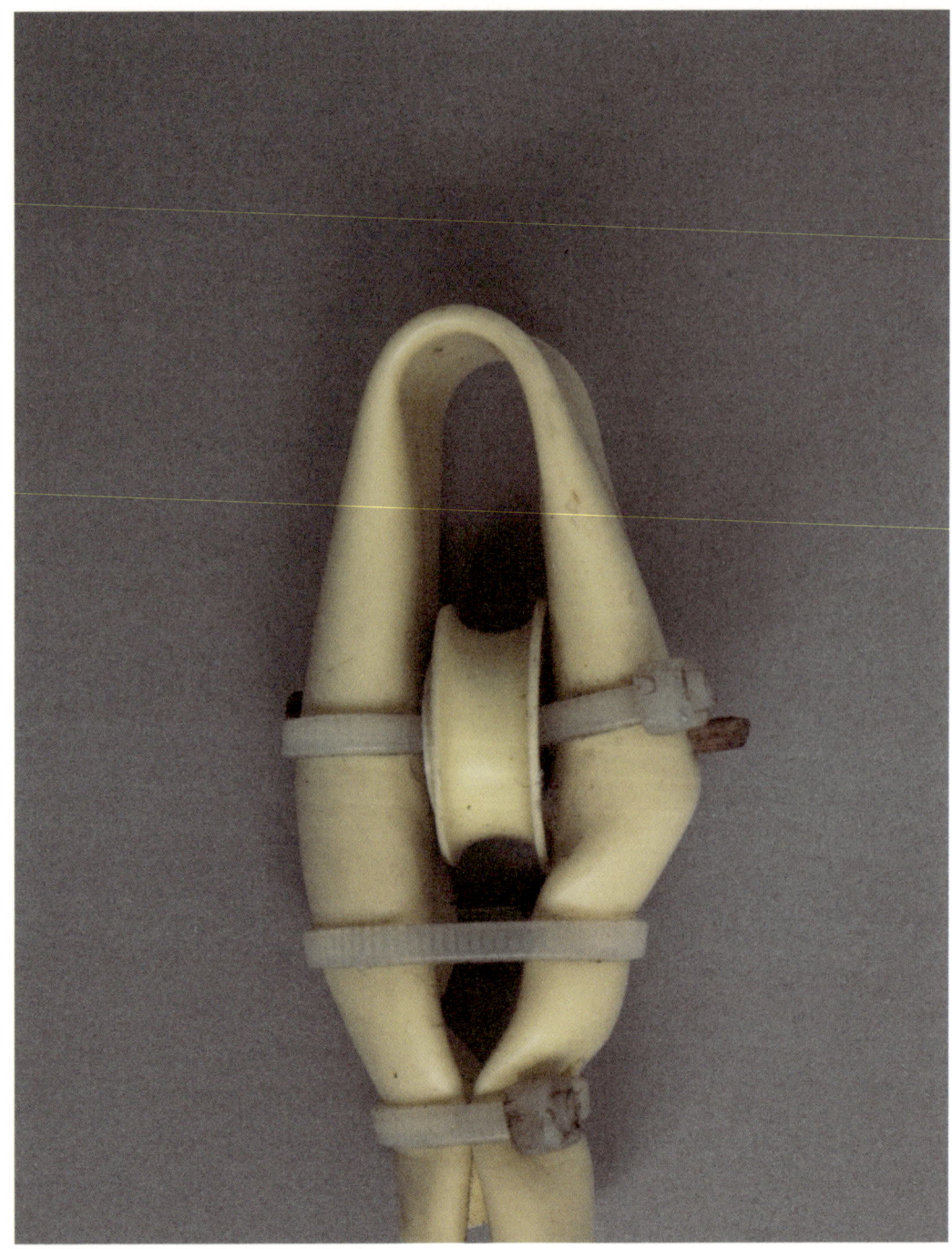

Animaris Percipiere Primus, end of reverse muscle

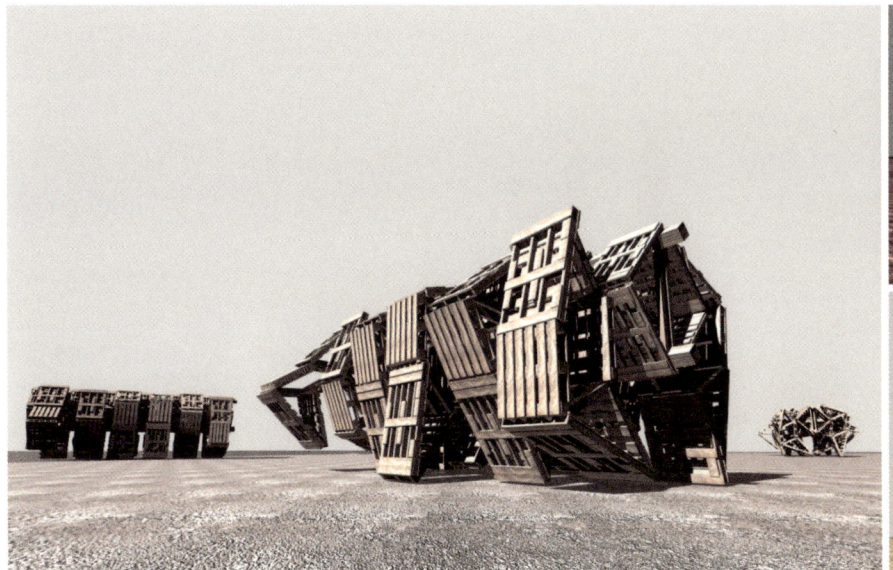
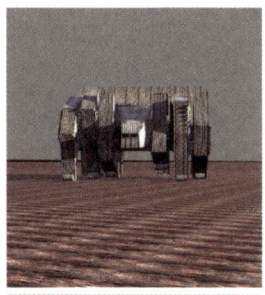
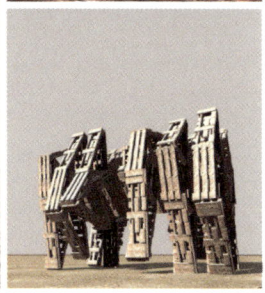

Animaris Rhinoceros Vulgaris (regular rhinoceros animal), Animaris Rhinoceros Transport Primus (first transporting rhinoceros animal) and Animaris Rhinoceros Tabulae (plank rhinoceros animal).

Transport
Anything can be transported but the most important is transportation of the human body. Bodies can be lugged from one place to the other, seemingly for no real purpose. Back and forth, back again and forth again. Long ago, everyone had to walk. Forever walking, pacing, trudging, tramping, strolling, ambling, toddling. To the shed, to Amsterdam, to Rome, always by Shanks's pony. But even in those days, so very long ago, we humans harboured the inclination to get others to do the dirty work rather than do it ourselves. And so it was with this walking business, and that was where the horse came in. To my knowledge, the horse was the first form of servomechanism. Just as power steering assists in turning wheels and brakes can be servo-assisted, the horse does the same for our legs. It's true that a horse's leg movements are not literally the same as ours, yet I have been given to understand that a rider does dictate the rhythm of the horse's gait. This he does by means of his posture. Apparently, he uses a rhythmic up-and-down movement of his hips to let the horse know the pace it is to maintain. This automatically sets the speed the rider desires for the ground he is to cover. This servo-assistance extends to the hands, too. Pull on the bit in the horse's mouth and the horse will pull the wagon. All hand movements are followed by the horse and ultimately transferred to the wagon with greater power.
We used the horse as a seat, as a chair that moves. With a bit of practice, the horse moves to where you want it to move. Giddyup, you bag of bones.
It would then be logical to suppose that to fly one can use a bird as a seat, but

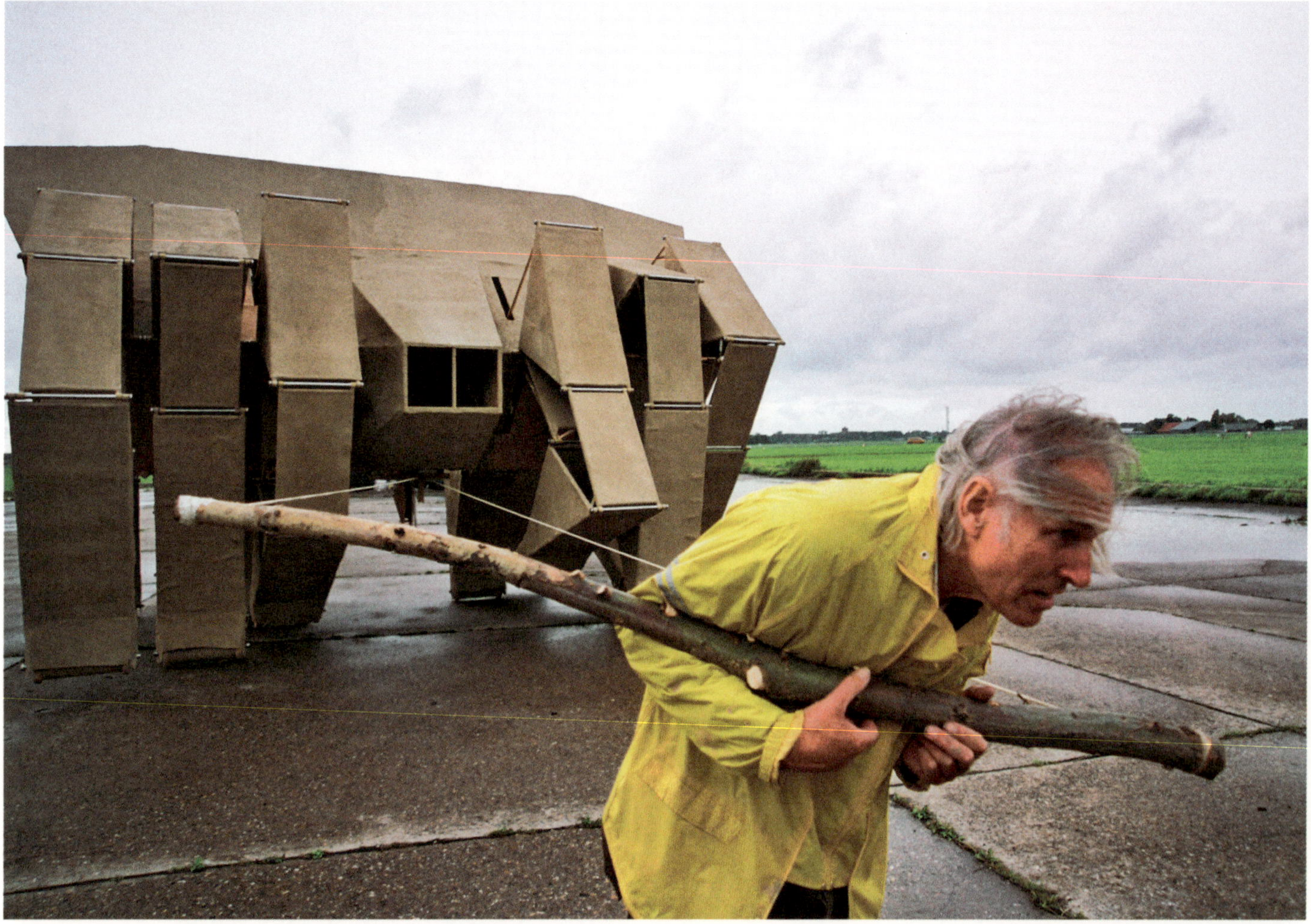

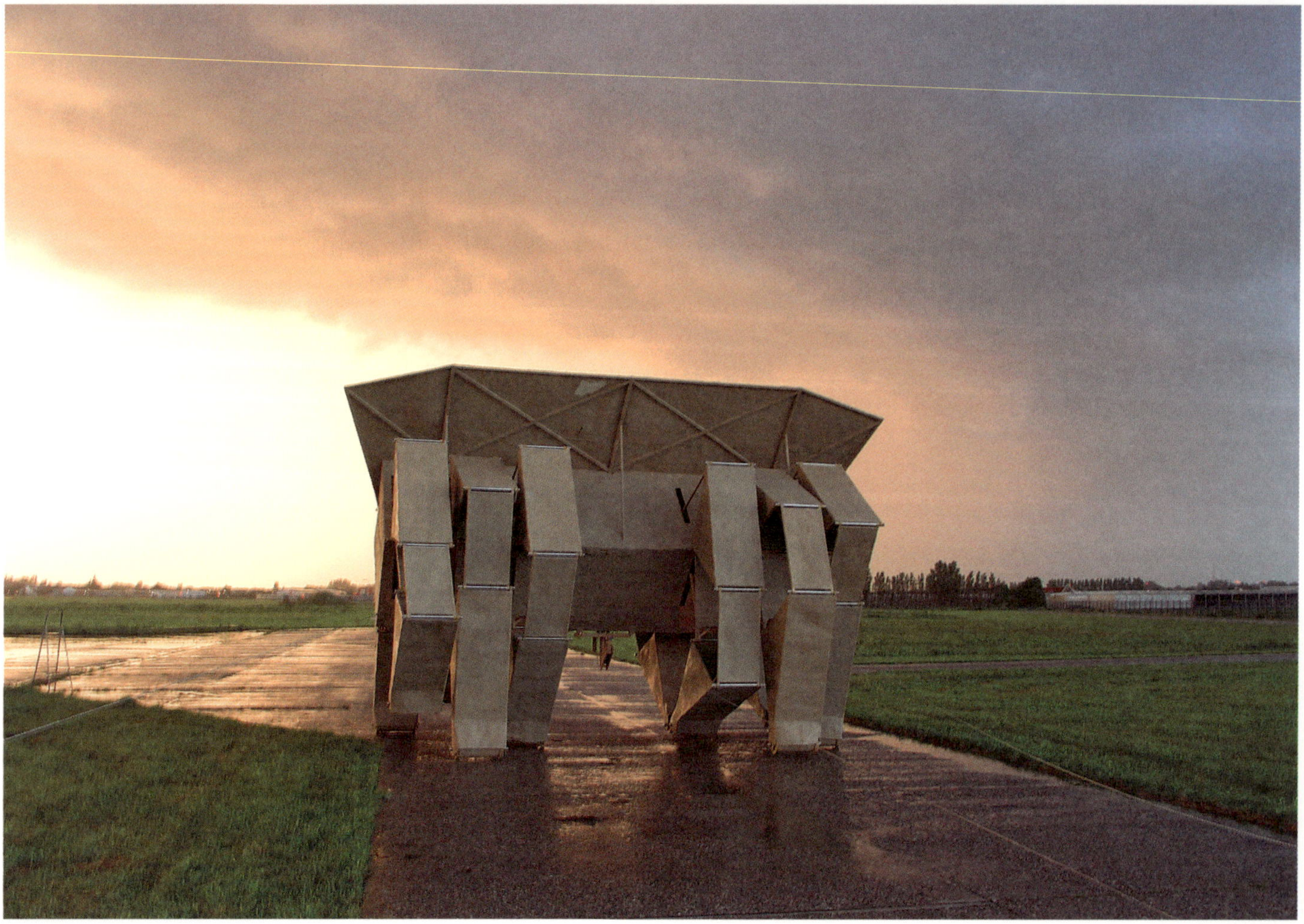

walking tests at Valkenburg airfield

this is easier said than done. We are too heavy for birds. We keep waiting for our fingernails to turn into flight feathers, but it never happens. This is why there are large imitation birds of aluminium. We take our seats in the belly of a motorbird. You might even say we are donning our birdsuits. Once enfolded in aluminium, we have the capacity to fly.

Walking

Walking is fun, but still I prefer cycling. The major advantage cycling has over walking is that you can stay sitting down. This is impossible when walking. You can't sit and walk at the same time, but you can simultaneously sit and cycle. When you walk, your legs are constantly engaged in lifting your body and putting it down somewhere else. It is handier to dump the body on a cart and push it. The wheel has made our lives a lot easier for five thousand years already. Its history began with the trunks of trees. Massive stones were moved using rolling sections of tree trunk. The disadvantage here was that the trunks needed to be continually collected at the rear of the transportation process and brought to the front. The wheel made that redundant at a stroke. Now the sections of tree trunk were fixed to the load permanently. Think of the Flintstones' car. The sections of trunk were automatically brought along with it. That's what the wheel is, an automatic tree-trunk bringer-alonger. A wonderful invention indeed.

Wheels are quite useless on the beach, however. It's fun to have a quick cycle there, of course, but it's not long before you're walking the bike instead. What you need on the beach is a cart on legs, a carriage that walks. Just like the axle of the wheel, the hips of the beach animal describe a horizontal straight line. And that in fact is what the beach animal is. You can push it like a cart and use it to move a load. This really is a case of reinventing the wheel, the wheel for soft surfaces. What began thousands of years ago with tree trunks and switched to the wheel thousands of years later, has now, thousands of years later still, developed into the wheel for soft surfaces – Animaris. The invention of legs is in fact an improvement on the wheel, a new means of transporting heavy loads.

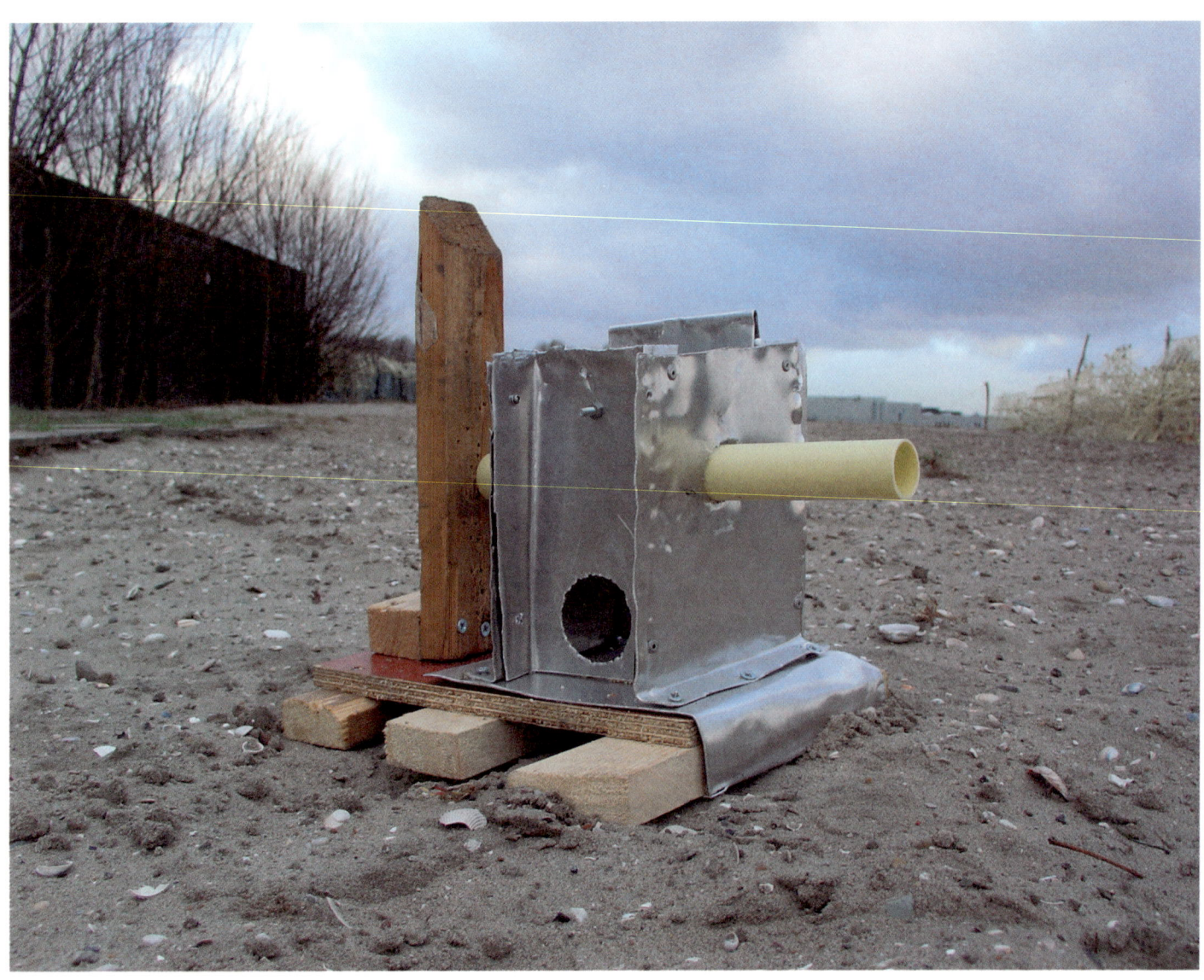

kiln

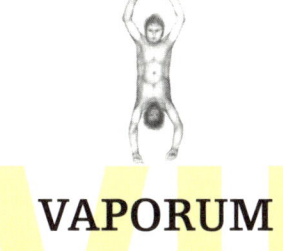

VAPORUM
the pneumatic period (2001-2006)

Muscles

In the Vaporum, Animaris progressed from *being moved* to *moving itself*. In all the preceding periods the moving had been done for it, by the wind no less. There's a clear distinction between moving and being moved. There are only two types of object capable of moving across the earth's surface of their own volition: animals and motorized vehicles. Plastic rubbish bags, chairs and cups of coffee can all move but not of their own accord. There are objects whose shape, physical properties and/or size make them likely candidates for travel. An empty plastic bag lying in the street is liable to take a walk, or rather take to the air. It weighs next to nothing and has a large surface area, two qualities that make it susceptible to the wind.

A euro coin is slightly heavier and if I leave it at home on the table, nothing happens. It just keeps lying there. But if I put it on the pavement in a shopping centre, before long that coin sets off on a zigzag journey across Europe. Its shape and physical properties make that coin susceptible to moving about. The beach animals from the earlier periods were susceptible to the wind because of their build and their wings. But being dependent on the wind can lead to dangerous situations. One is if the wind drops at the time of an incoming tide. Too much wind is dangerous too, of course. In the event of a storm it's more sensible for a beach animal to stay where it is. Its moving legs are no match for such lateral forces. The animal can better face downwind and keep still, like a stick insect, until the danger passes. Animaris Percipiere Rectus was capable of anchoring itself to the beach. On its nose it had a hammer that drove a thick tube into the ground. If the nose is secured, so is the rest of the animal. Should the wind turn, the animal turns with it and keeps facing downwind so that the wind has less effect on it. Animaris Percipiere Rectus survived a force eight gale on the beach at IJmuiden.

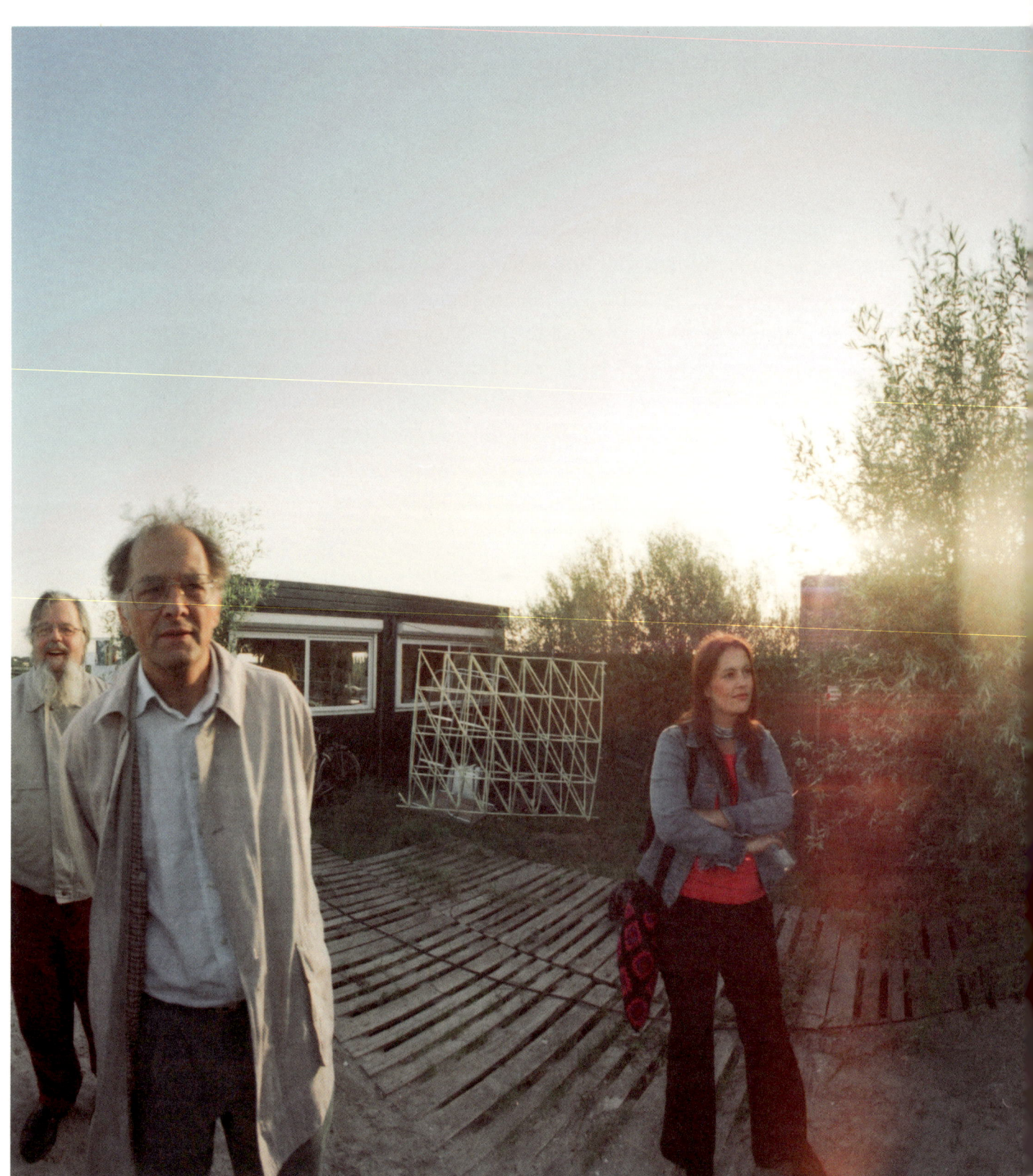

a summer evening at the workplace

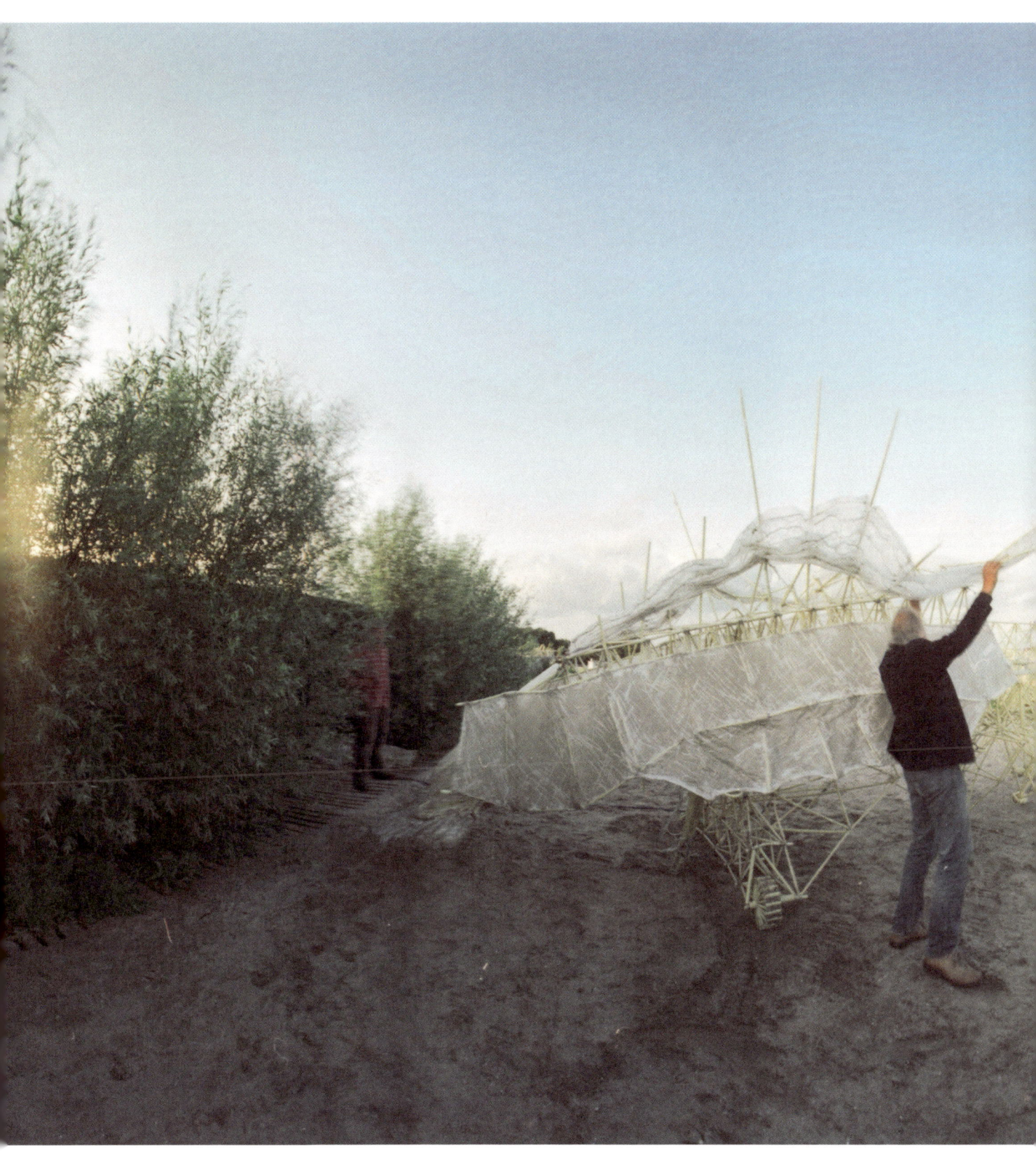

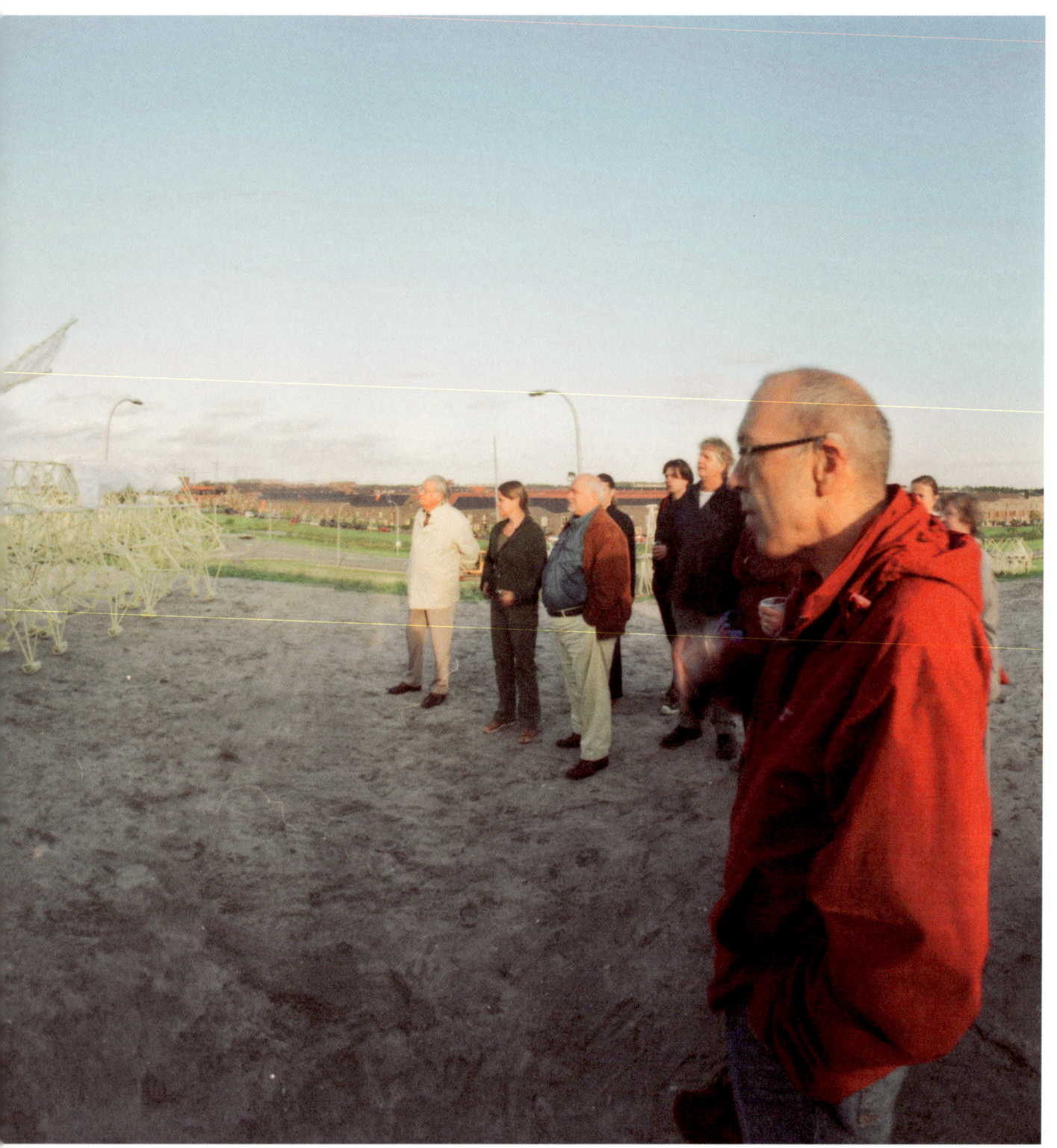

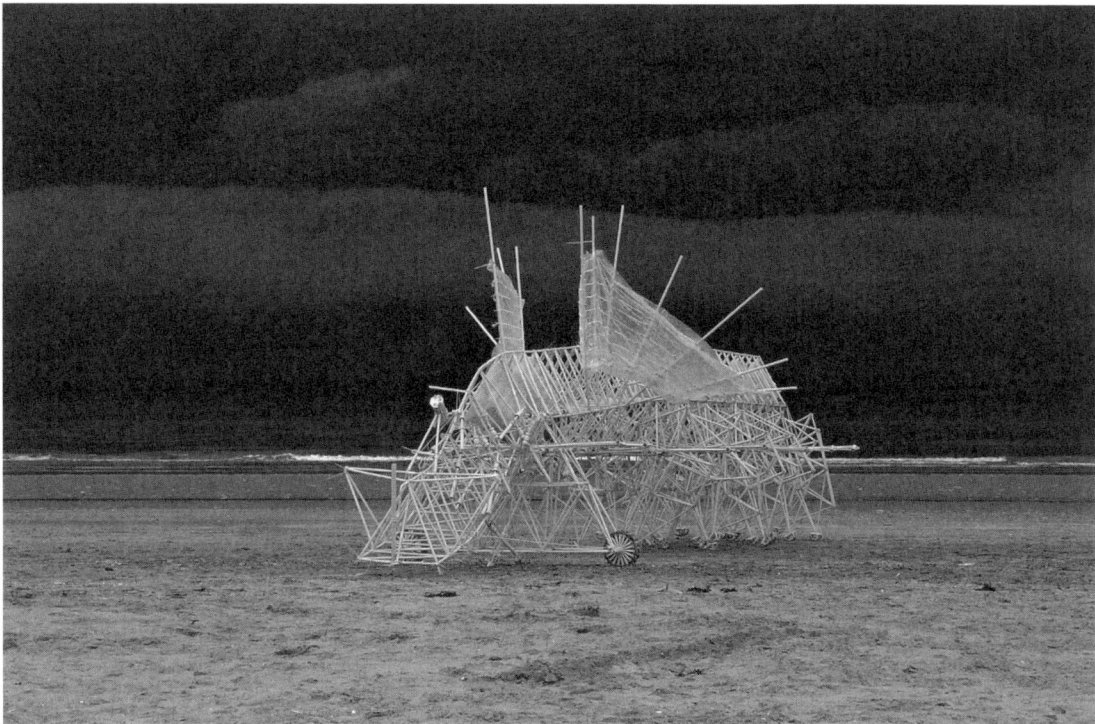

Animaris Percipiere Rectus (obedient animal with sensory organs)

Universal anatomy

The anatomy of self-propelling objects is subject to universal laws. Those of cars and animals are similar in many ways. First of all, there's the stomach. Stomachs are like petrol tanks. It takes energy to move around, and that energy must accompany its user. All self-propelling objects carry provisions with them in a sealed-off space, a stomach or a tank. Self-propelling beach animals like Animaris Percipiere have a stomach too. This consists of recycled plastic bottles containing air that can be pumped up to a high pressure by the wind.

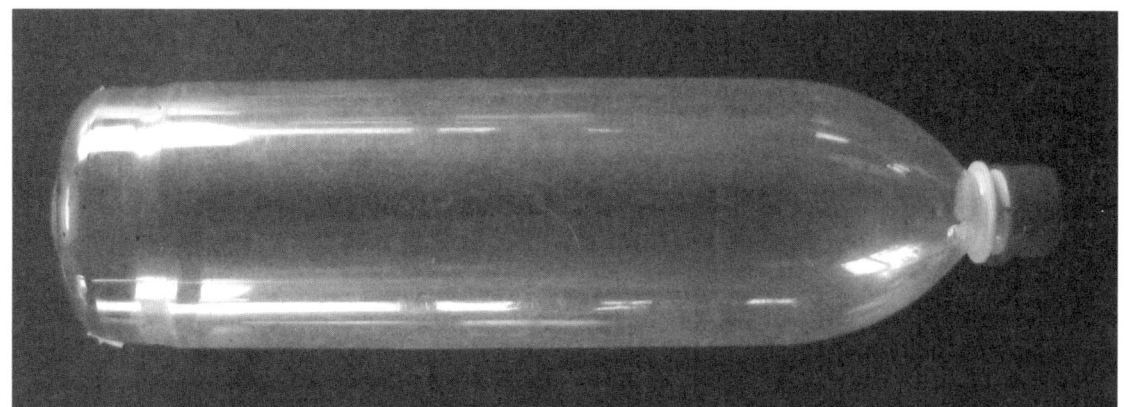

recycled plastic bottle

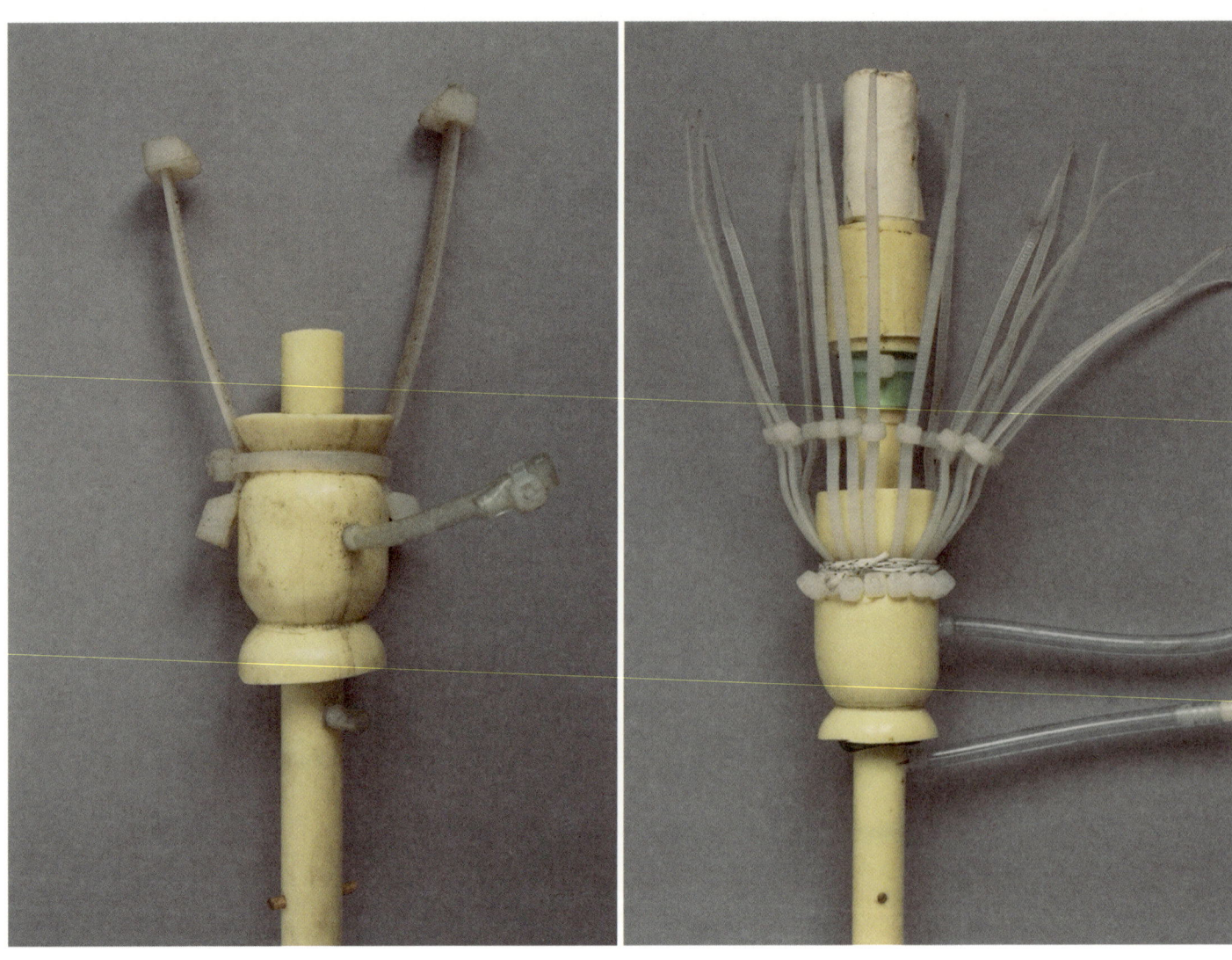

end of nerve cell of hammer (Animaris Rectus), end of nerve cell (Animaris Vaporis)

This is done using a variety of bicycle pump, needless to say of plastic tubing. Several of these pumps are driven by wings up at the front of the animal that flap in the breeze. It takes a few hours, but then the bottles are full. They contain a supply of potential wind. Take off the cap and the wind will emerge from the bottle at high speed. The trick is to get that untamed wind under control and use it to move the animal. For this, muscles are required. Every self-propelling object has muscles. The word needs to be taken in a broad sense. A muscle is an object that can become longer or shorter to order. The muscles in your arms become shorter when instructed to do so. They are contracting muscles. They pull on the bone so that you can lift things. But there are also muscles that push. These get longer when instructed to do so. Take the muscles of a mechanical excavator. Hydraulic cylinders pull on the excavator's arms. They are muscles too. There are muscles like these in the engine of a car. There the system of pistons and cylinders gets longer when instructed to by a spark from the sparking plug. Beach animals have pushing muscles which get longer when told to do so. These consist of a tube containing another that is able to move in and out. There is a rubber ring on the end of the inner tube so that this acts as a piston. When the air runs from the bottles through a small pipe in the tube it pushes the piston outwards and the muscle lengthens. The beach animal's muscle can best be likened to a bone that gets longer. And herein lies the advantage of the expanding muscle over the contracting. We and all other vertebrates need a bone and a muscle to be able to move. But the expanding muscle in the beach animal is itself a bone; muscle and bone are integrated. This means a considerable reduction in weight. Beach animals have better muscles than ours. They are lighter and stronger. The muscles of Animaris Excelsus weigh two kilograms and can exert a force of 100 times that weight.

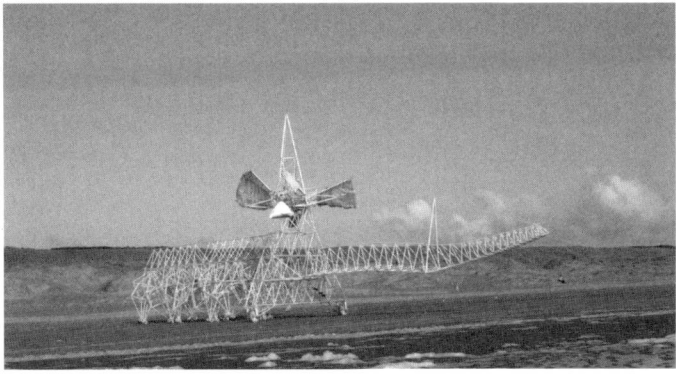

Animaris Excelsus

Out among the clouds, what a treat that was

What became of the time when we crossed the ocean by airship? Well, what *did* become of that time? It drives me to despair. You should know that an airship really is a ship, with all the attributes of a ship: a propeller, stem and stern, foredeck, afterdeck, ropes, lots of ropes, walls you could climb in even when travelling among the clouds. Now and again the ship's engines were turned off and passengers could make their way outside onto the balcony. Just imagine yourself up among the cumulus clouds, heading towards New York, looking out across the sea. The wind in your hair, the contact with the cold outdoor air; that must have been a truly *oceanic* experience. You tell me, what became of that time? I can't find anything of it in a flight on a jet airliner. You peer through a tiny window, a hole, a peephole with a double layer of plastic that reflects everything and anything.

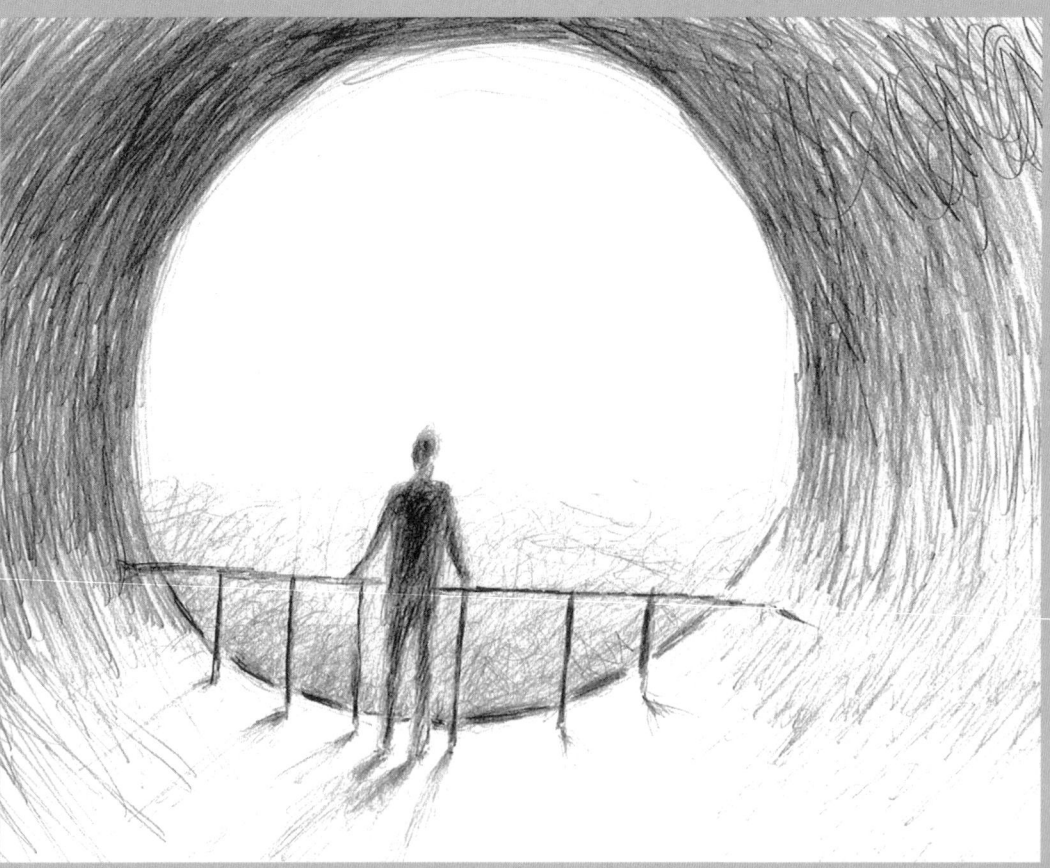

The curvature of the plastic distorts the edges of your view out. There are greasy patches on it. You're folded almost double in your chair. A muggy atmosphere of sweat permeates the cabin.

And now it's time you told me where those times have gone. You sit there bored witless, slumped in your seat looking up at the stewardess who has just brought you some peanuts. I'm inclined to give you a prod, grab you by the hair and pull you away from the film you've been watching for the past hour and point you towards the window. 'Look outside, damn you. We're suspended ten kilometres above the sea. It cost hundreds of lives to get air travel up to this standard!'

Ten kilometres up - that's quite something. Just imagine if we were to be able to go outside at this height, onto the balcony. That wouldn't just be an oceanic experience, but a *cosmic* experience. I know: the plane's moving too fast, the air's too rarefied and it's too cold. Three good reasons not to go outside. It should be possible all the same.

Let's suppose we slice the nose off the plane. And behind that we make a space that is closed off at the rear but open at the front. The air gets in here but doesn't go anywhere else. We can enter and leave the space through an airlock. That front space will probably have its share of turbulence, but you won't be blown out. On average the air is still, just as it is in the cabin. Air pressure is normal due to the pressure being applied at the front. The air's composition is more or less the same, with just a little extra ozone. We can breathe freely.

Not just that, the air at the front will be warmed by the rush of outside air. Think of the heat shield of a space capsule. So it is possible, that cosmic experience. But you don't *have* to go outside, you know. Stay glued to the video screen if you prefer.

Muscles can open taps to activate other muscles that open other taps, and so on. This creates control centres that can be compared to brains.

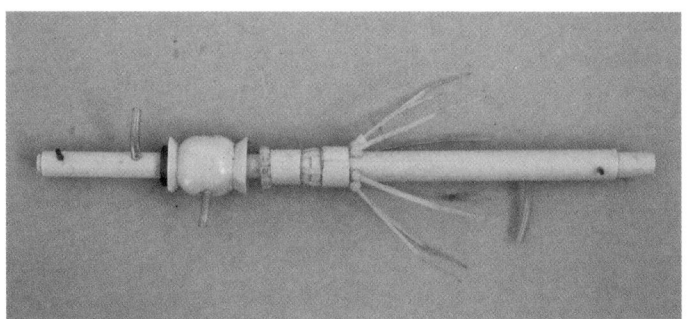

old nerve cell (2003)

The left-hand part of the nerve is a tap which is attached to the pressurized air in the bottles by means of the flexible air hose on the left. A capsule has been slid over the tube, concealing a small hole made in the tube at this point. This capsule (the 'toffee') is made from a thicker tube (38 mm) which has been heated and narrowed near its extremities. Two pieces of garden hose seal off the gap between the thin plastic tube and the toffee. A piston reaches into the tube from the right-hand side. When the tube is at left, the tap is in the 'off' position. At any other time, the air flows from the bottles through the small hole inside the toffee and exits the nerve through the middle air hose. The piston is operated by the muscle (pump) at right. As soon as air enters the hose on the right, the tap is turned off.

I would go on to make sensory organs – water feelers, storm feelers, calm feelers, soft-sand feelers, wind-direction feelers – that work on compressed air. I dreamt about the consequences of this invention for nights on end.

Self-locomotion
Long ago, animals were still just living objects. There were sponges, for example, or one-celled forms. These objects had a hard time of it, and died in droves. This was because they could scarcely move and had a fixed shape. To change that shape, there was nothing for it but to reproduce. Generations came and went before there was any visible difference. The cactus is another example. It has

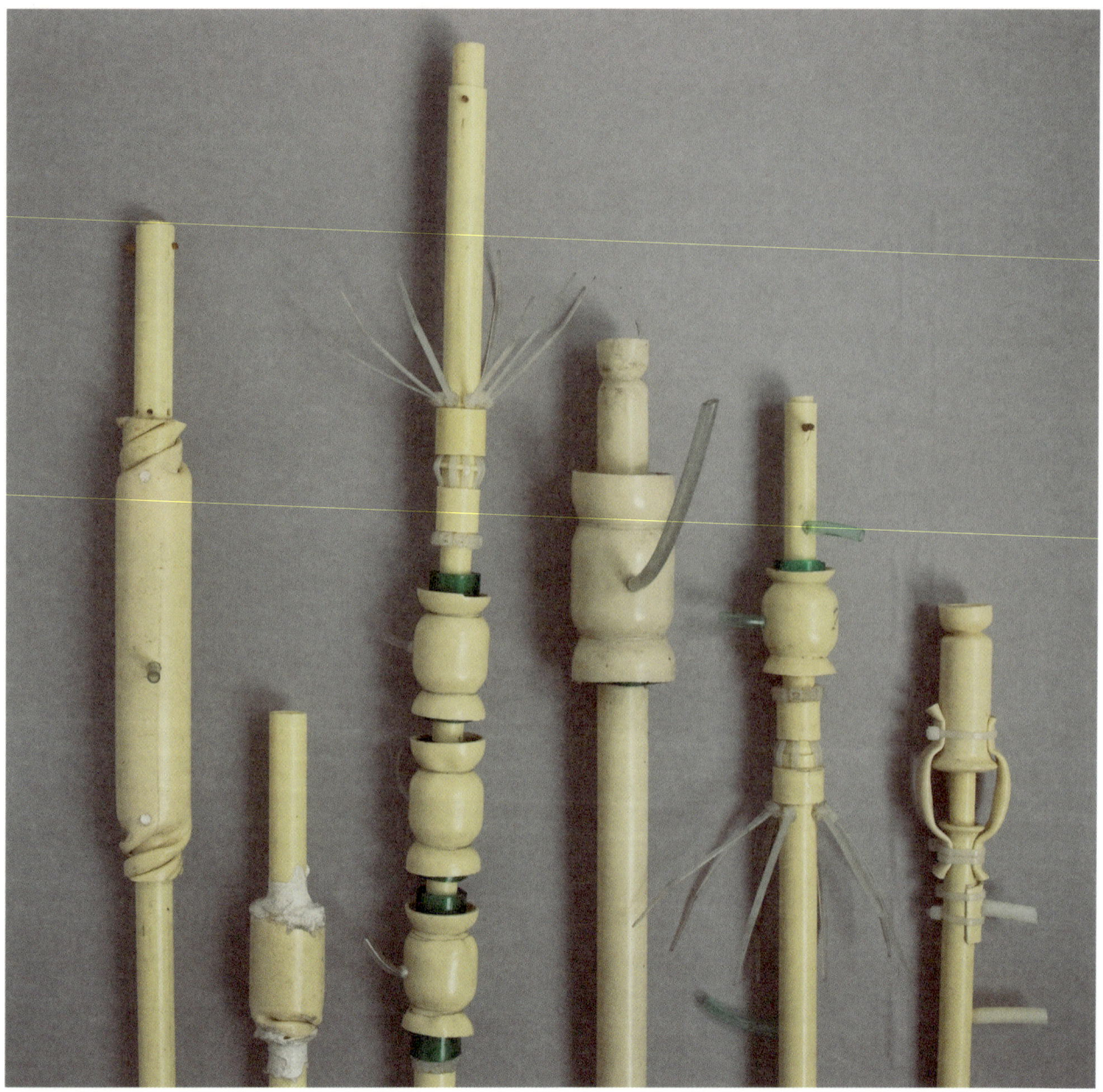

evolution of nerve cells

acquired a round but prickly form, though heaven knows how many generations that took.

Then along came the amoeba. This unicellular creature could change its shape during its own lifetime without needed to reproduce. The amoeba can do in a matter of seconds what other living objects took millions of years to do. Talk about time saving. Armed with the potential of a creature to change shape at will, evolution really got into its stride. From that moment on, everything developed a thousand times faster.

If a rabbit finds itself in danger, it changes its shape in such a way that the earth shifts beneath it and places the danger at a distance. We call that walking. To walk is to continually change shape with the result that you move forward.

A line creature from the Pregluton couldn't walk. In fact it couldn't actually move at all, it just drifted across the screen. Its development can be compared with that of a cactus. It took a good forty generations to be able to roll itself up. The hedgehog took a lot less longer than that. It could curl up in a fraction of a second from day one. A hedgehog lives several years. A line creature lives several minutes. It's muscles that make all the difference. The arrival of muscles in biological nature lengthened the lives of animals. It must have caused a revolution. The arrival of muscles in beach-animal nature certainly did. Everything changed. Now I can make animals that are almost real. They are dreams but at the same time I can see a chance of succeeding. I can release animals on the beach without giving them a second look. They can swim. They lie on their backs when a gale blows up. They pull other animals along. It'll come, mark my words, it'll come!

Vermiculi

To give my animals a greater chance of surviving on the beach, I have worked on ways of getting them to move of their own accord, just like their zoological counterparts, rather than being moved. Although I don't wish to use 'real' animals as an example, I can't ignore the fact that biological nature had invented self-propulsion first. I would now like to take you through the different modes of self-propulsion, beginning with rolling. My children didn't want to crawl when they were babies, they rolled instead. Before going where they wanted to go, they estimated, lying on the ground, the distance and the direction of their objective. They got in position and rolled very fast, in a horizontal pirouette, to their destination. This rolling motion was driven by a repeated twisting movement. Twist-

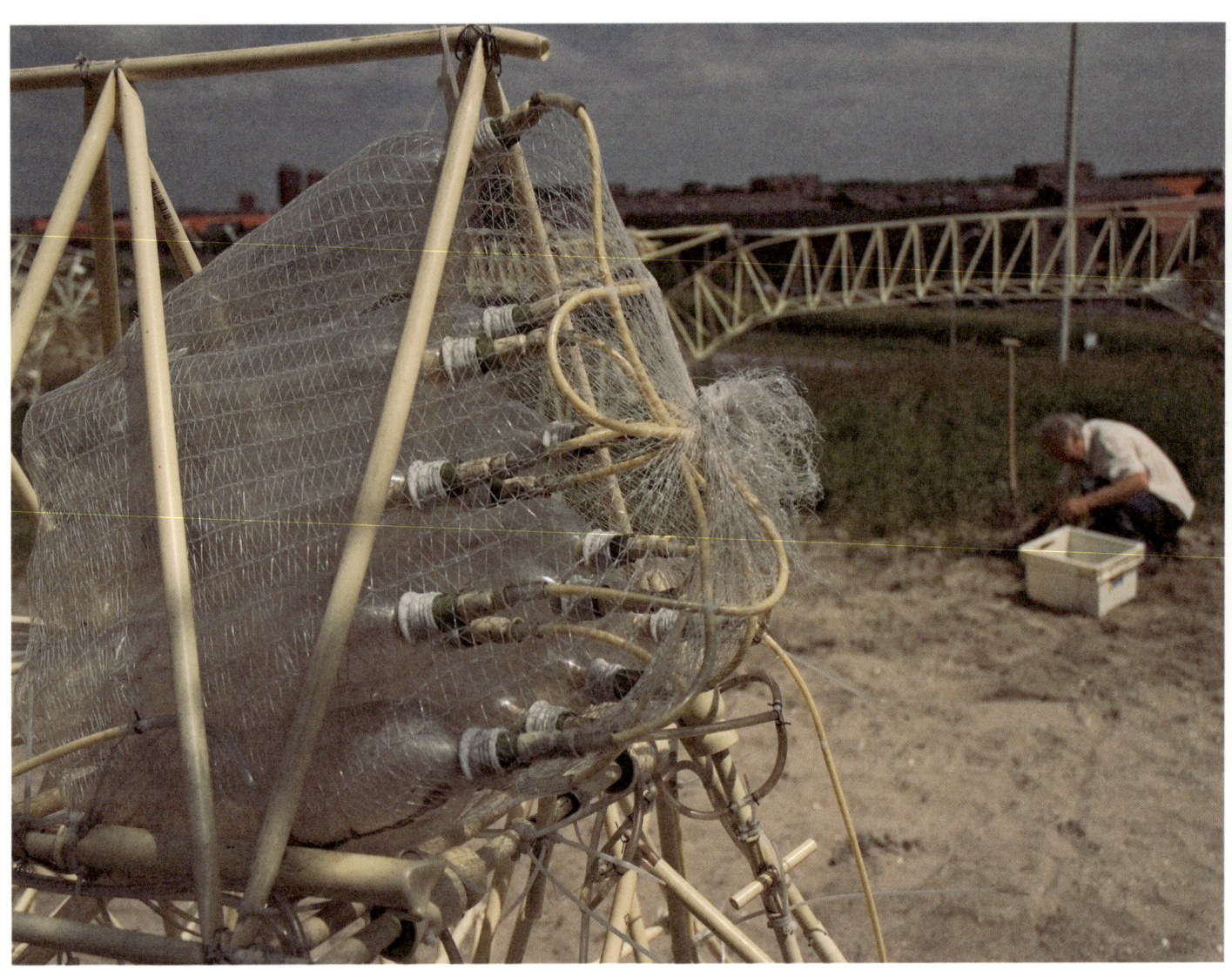
Nets surround the plastic bottles as a safety measure: they might explode. They did once.

ing means changing shape. A caterpillar changes shape, so does a worm, even a walking human changes shape. This obtains for every means of proceeding forward: change shape in such a way as to bring about a change of place. Repeated changes in shape result in repeated changes of place. All these changes of place together give a total movement forward from A to B. Besides rolling there is also undulating. Eels and snakes undulate horizontally, seals and caterpillars vertically.

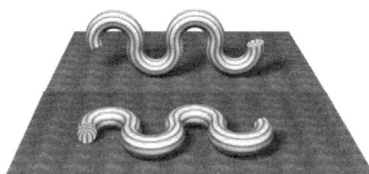

vertical and horizontal waves

The advantage of horizontal undulation is that it can make the change from land to water. An eel can propel itself forward on land and in the water using the same undulating motion of its body. The sine curve begins at the head, moves back through the body and disappears at the tail. In seals and caterpillars, which undulate vertically, the direction of the undulation is reversed. It begins at the tail and ends at the head. If seals and caterpillars were to continue this undulation in water, they would move backwards. We haven't discussed longitudinal waves yet. These are peristaltic contractions in the body, parallel to the forward movement. The animal in question crawls by means of a fringe of fine hairs on the underside.

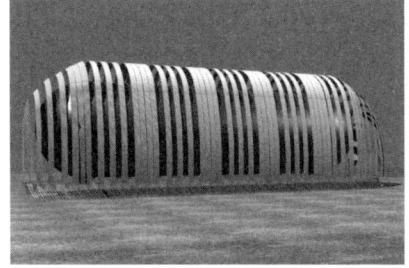

longitudinal wave

The entire animal shifts to the left, seemingly without changing its shape. This locomotion is based on the same principles as that of an ear of corn. An ear of corn caught in a coat sleeve is shifted by the fortuitous movements of the arm. The fringe of hairs on the ear propels it forward in the direction of the neck. As

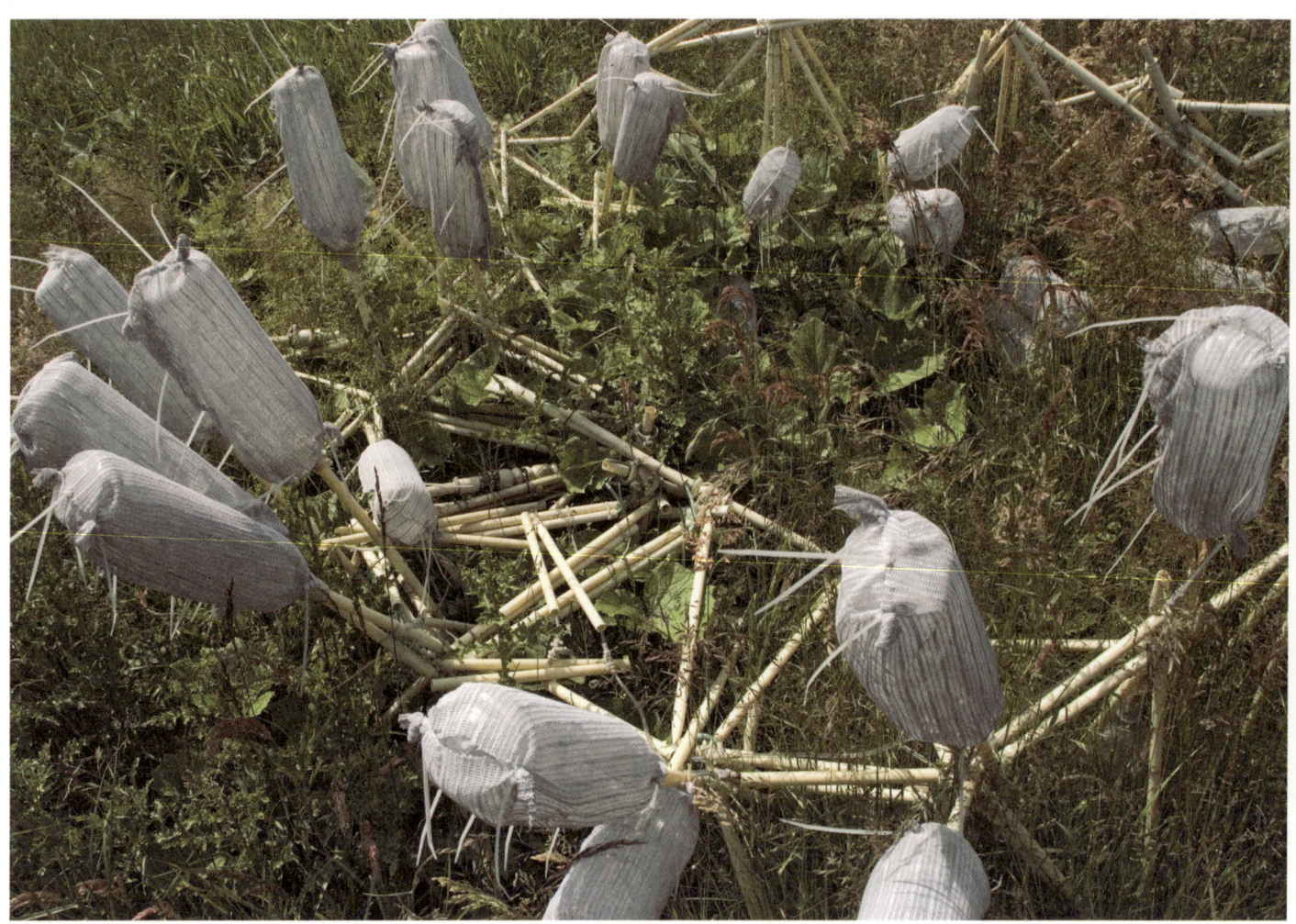

Animaris Vermiculus

already mentioned, we animals began as shells with an entrance and an exit, featureless objects without arms or legs. What amazes me is that such shells can move in such a variety of ways. There are times when primitive forms of locomotion assert themselves elsewhere. You don't need much imagination to see an undulating caterpillar in a galloping horse or a rabbit on the run.
The Vaporum began when I made the Vermiculi. Vermiculi are worms and caterpillars. And boy, do they twist.

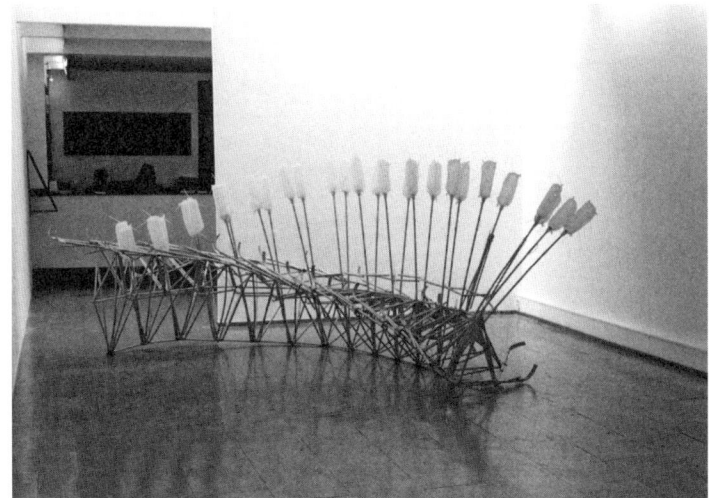

Animaris Vermiculus (worm animal)

Peristalsis of the muscles in beach animals created itself, so to speak. There were special twist muscles. A twist muscle consists of a cylinder and a piston like any other muscle. What makes it special is the little hole at the end of the cylinder. When the muscle expands, the piston passes the hole.

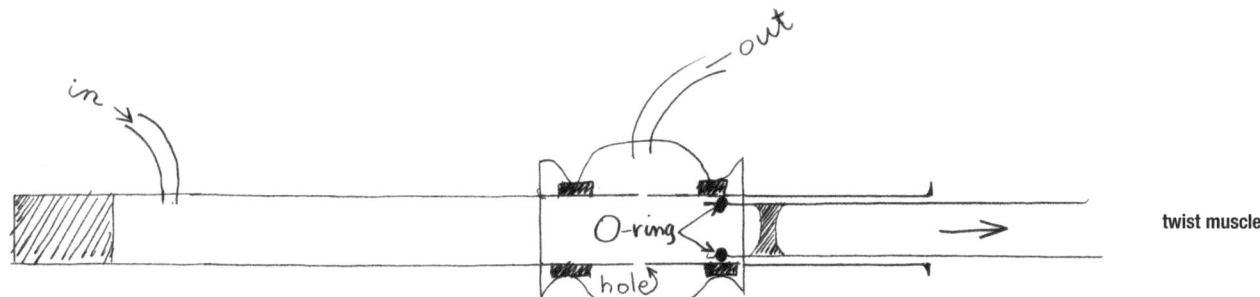

twist muscle

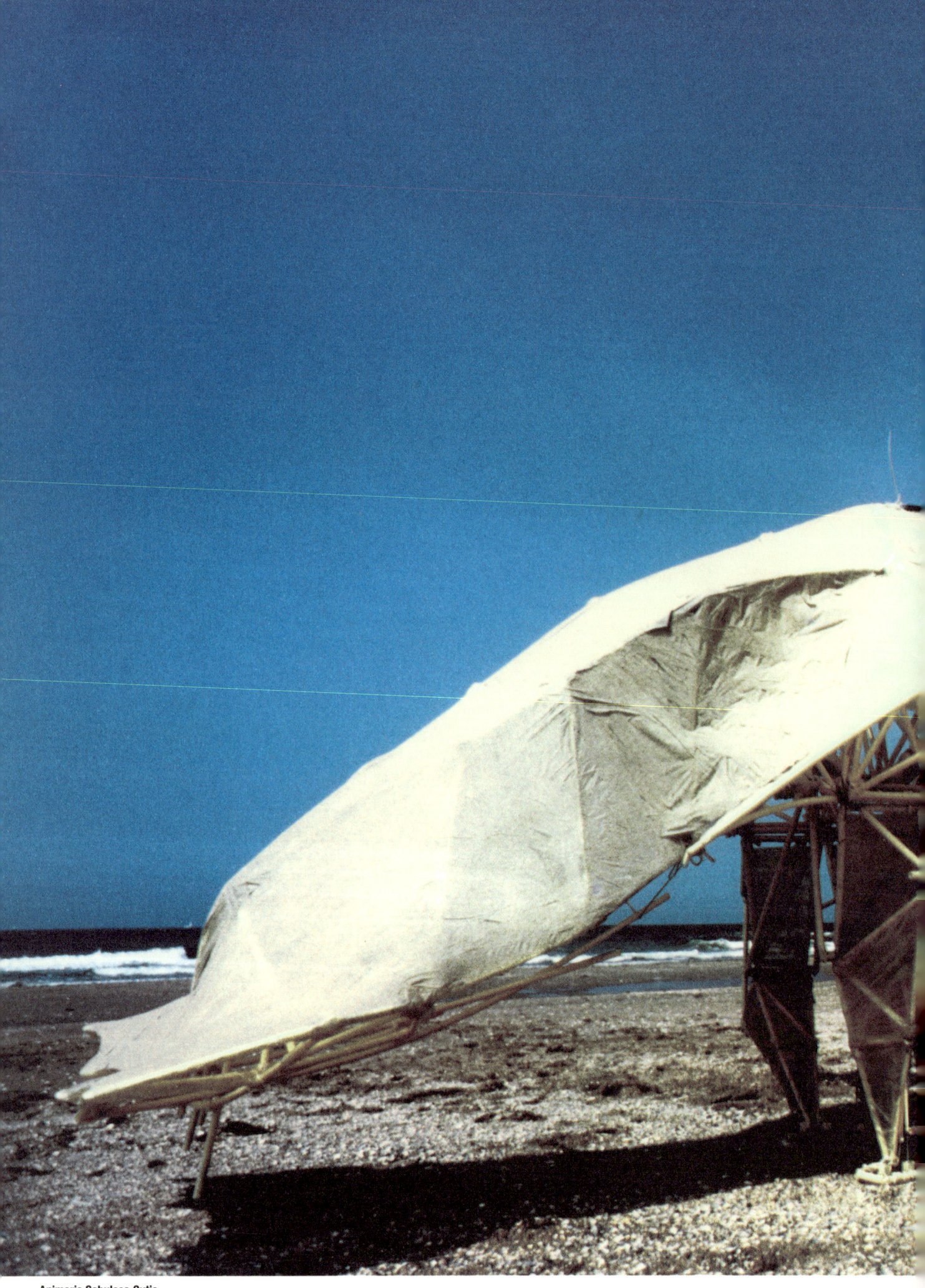
Animaris Sabulosa Cutis

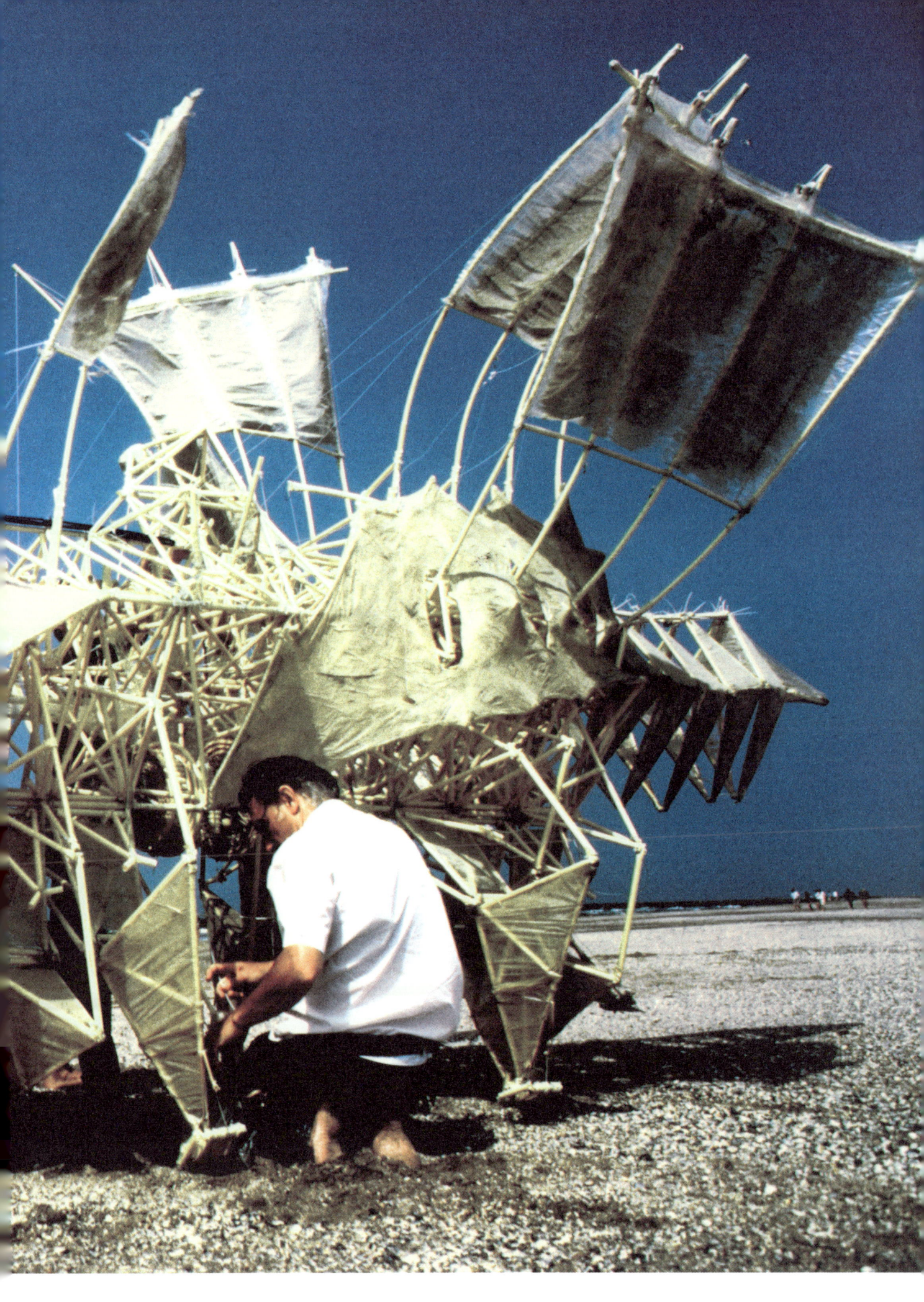

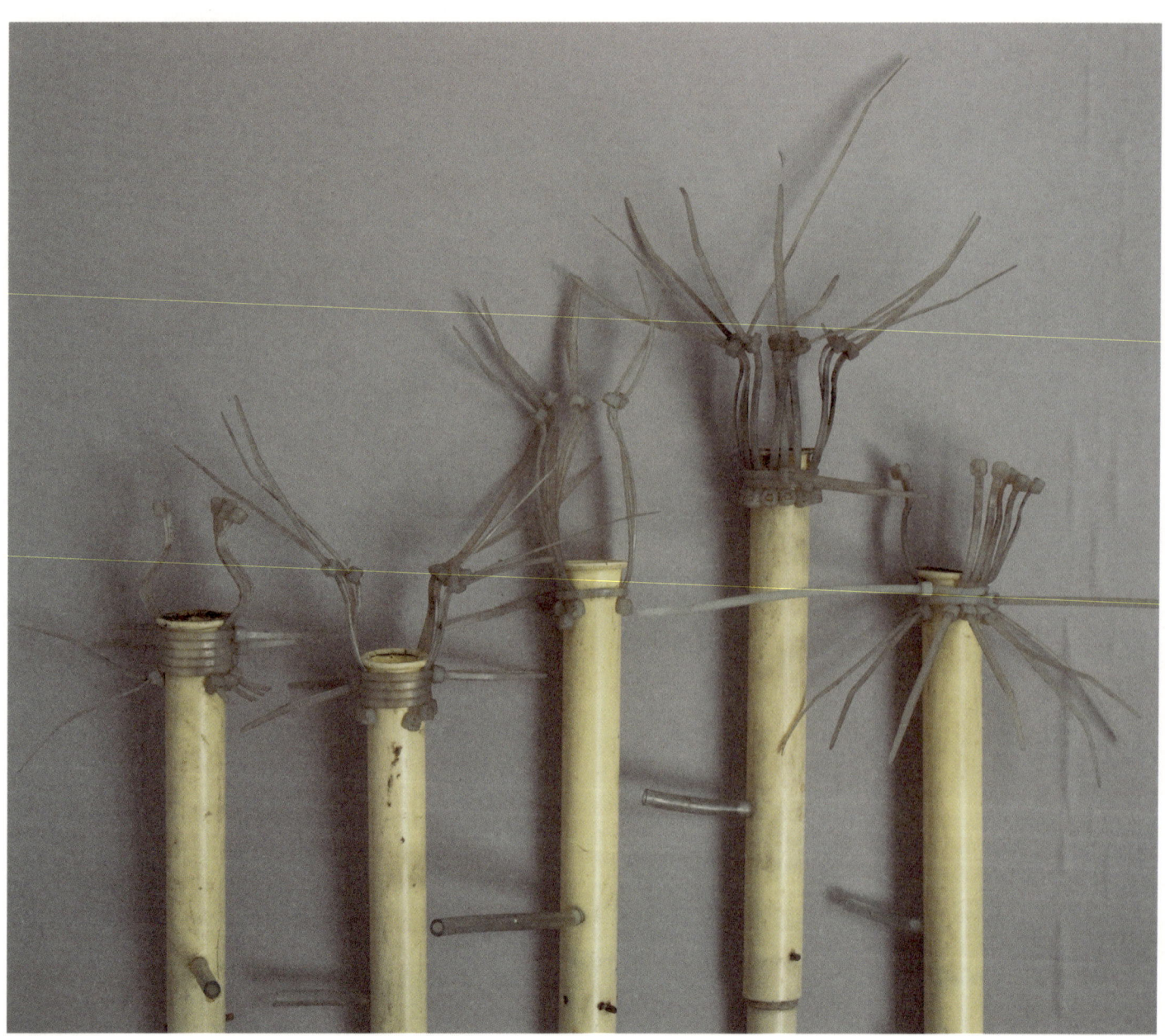

input nerve cells

Air entering the hole is taken up in the 'toffee' enfolding the tube from where it moves on to the following muscle which is then activated. This process is self-repeating. Muscle 1 activates muscle 2 (at bottom left) which in turn activates muscle 3 and so on. This effect sends a wave through the animal and brings about a peristaltic twist. The numbered twist muscles at bottom left are connected to the numbered muscles at top right. Muscles identified by the same number extend at the same time.

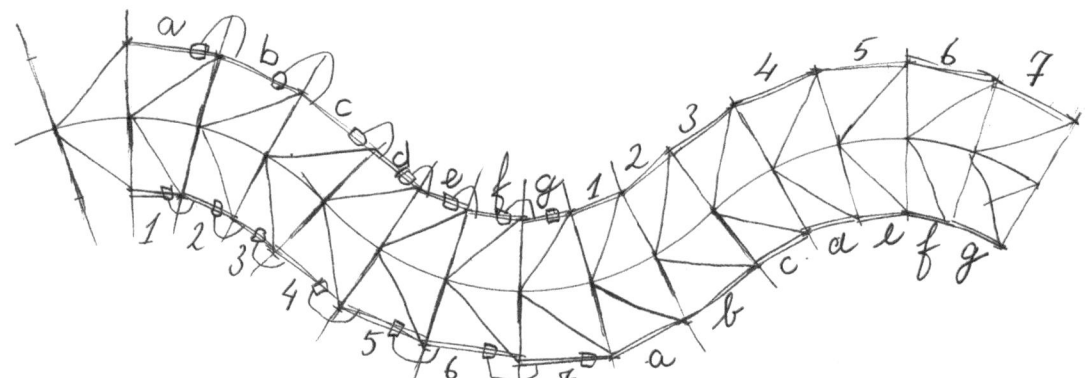

action of Animaris Vermiculus

Once all the numbered muscles are extended, the 'counter-muscles' lettered *a* to *g* on the other side of the worm are activated one by one. This sends a new twist through the animal, producing a curve in the opposite direction.

The two segments, left and right, execute an oppositional twist. This is because the muscles of the first segment are hitched to those on the opposite side of the second segment. Together the two segments present a complete sine curve.
Later there were worms with four segments that presented two complete sine curves. There was also a caterpillar, Animaris Rugosus Peristhaltis or ripple-beast. This made a vertical twisting movement.

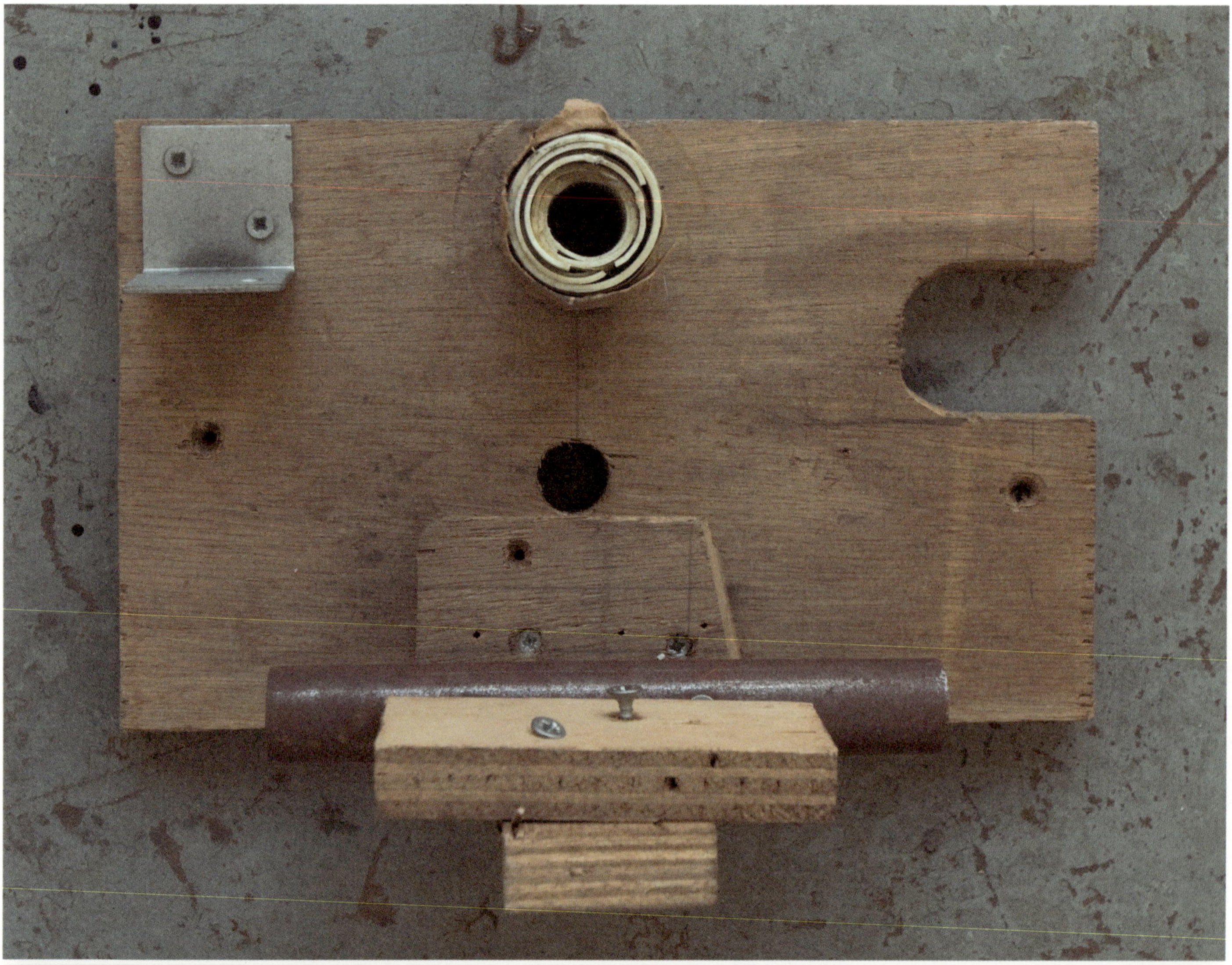

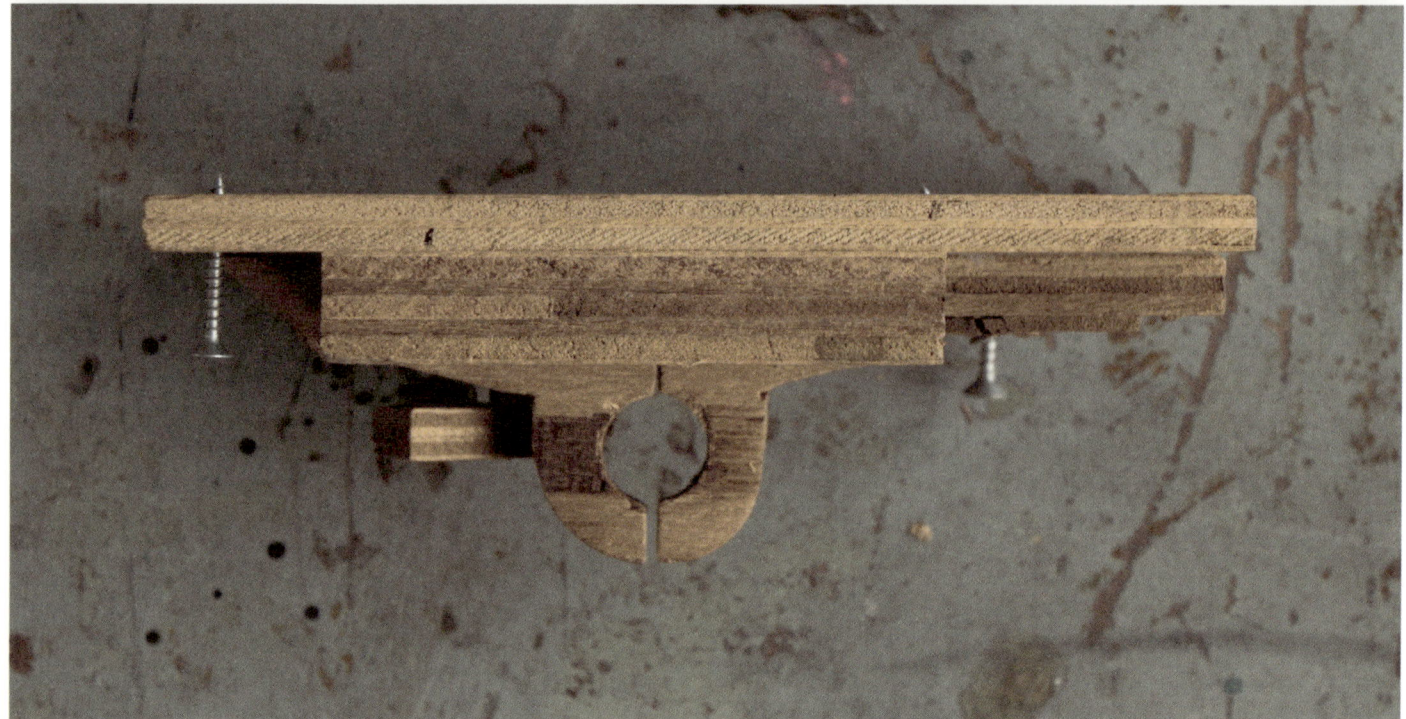

**saddle mould,
tightening mould**

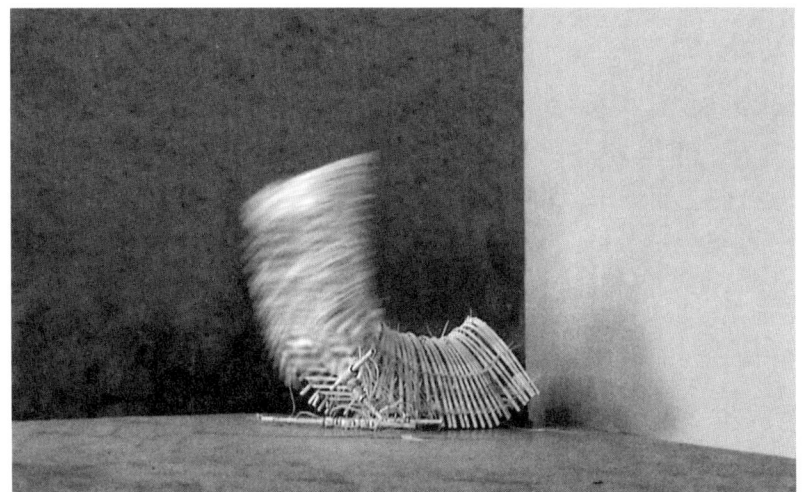

Animaris Rugosus Peristhaltis

None of the worms or caterpillars moved from place to place. They twisted and turned but stayed where they were. The family of worms reached a dead end in 2003. Worms have the advantage of being scarcely affected by the wind. They are low-lying and stable. Another possible advantage is that they could move from land to water, sticking to the same twisting motion. The recycled plastic bottles could act as floats. They could skim across the surface like water snakes. These water worms would undoubtedly possess cinematic qualities. But what do I stand to gain from this? Is this what I want? A question I often get asked is why don't I make flying animals. It can't be that difficult. Plastic bottles are light and would make excellent energy storage tanks in the air. On the ground, the wings would beat in the wind and pump the plastic bottles full. Once the pressure is high enough, the pumps function as muscles and the animal takes off. Airborne plastic bottles, that would be quite something. But what would it contribute, apart from looking good up there?

My concern is not to make animals that look good. My concern is survival, period. No frills, no fripperies: survival! Nothing more, nothing less. I only have perhaps thirty years left. In thirty years' time I'll be ninety. No time to lose, no swimming worms, no flying birds. We already have those anyway. What we haven't had yet is the roll worm. Of all the life-forms I am engaged on, the roll worm would seem to have the best chances of survival. To my knowledge, there is no other animal that moves the way the roll worm moves. If anything is to survive after my death, it's the roll worm. It hasn't been made yet.

Urrggghh!

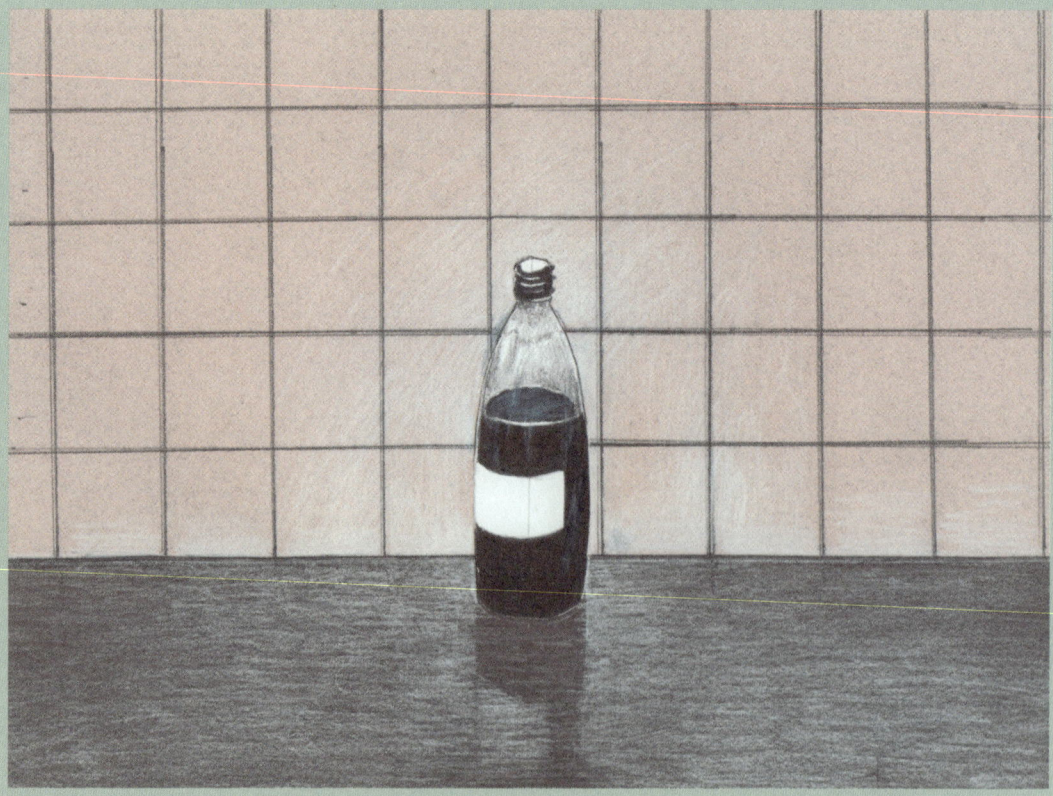

I've never complained about her salad dressing. There isn't a papilla on my tongue that would dare to question the combination of ingredients: a little pepper, a little garlic, olive oil; and vinegar, the most expensive, the tastiest. So why should anything be wrong with the salad? I mean, it all adds up. She's just nipped out to the toilet. The wine vinegar - that oh so expensive, exclusive red liquid ripened in oak casks - is still there on the kitchen worktop. I'll just give it a taste, a quick swallow .... just a small one. Urrggghh! .....Good grief, what ... urrggghh! ... acetic acid molecules scatter throughout my length and breadth. I'm tingling all over, even my brain. It conjures up memories. Where am I? In bed. I'm lying in bed. It's been a heavy night. It feels like there's a steel band wrapped tightly round my head. What a night. B. is lying next to me. Her face is swollen, white, clammy. Yesterday's make-up has run, her mouth hangs half open. She was amazing on the dance floor. I'm no dancer, so I hoofed around a bit, my heart throbbing, a stabbing sensation in my spleen, but I got what I wanted: B. in bed. Where she promptly fell asleep. Alcohol, Dutch courage, the great screwer-upper of the young and enamoured. But where's that whiff coming from? A strange bed, that must be it. A squatters' mattress slept on by half Amsterdam. I'll just check; no it's not, it isn't the mattress. Surely not .... it's, it's .... B. Urrggghh! Good grief, girl, you stink to high heaven. What an unedifying vinegary smell. What were our chances now? None, as it turned out. But it's only now I realize what wine vinegar actually is. It's putrid wine, wine gone sour, wine puked up. How on earth have I always managed to shovel in the salad flavoured with that muck without noticing? So here in the kitchen I arrive at a startling theory. Taste has nothing to do with the chemical composition of the victuals. We taste a memory. Most of these memories have to do with things that didn't even happen to us. They are relics of experiences our ancestors had. A chicken that has been grilled on wood is tastier than one out of a microwave. Why? It's the evenings our ancestors enjoyed round the campfire that make the chicken taste good to us. Lettuce is green and doesn't taste of anything in itself. This has to do with the fact that lettuce leaves were originally used as cutlery. Armed with a lettuce leaf, one could pick up sticky stuff like feta and olives without getting one's fingers greasy. To make life easier, one ate the cutlery as well. As lettuce is rife with vitamin C, this habit added years to one's life. Which is why the lettuce eaters had more descendants. It was only much later that those descendants came to appreciate the cutlery as food. But then always combined with other food. The fact that I like the taste of lettuce with wine vinegar has to do with memories of sultry evenings on a Greek island. Not with B. but with my beloved L., specialist in dressings and fervent advocate of a particular brand of wine vinegar. Can't beat that dressing.

It's so wonderfully primitive. No wings, no wear and tear to speak of. And it lies on the ground, so the wind has no effect on it. It thrives on patience. the patience to move. For the roll worm only moves a few seconds a year. And that's it. Which is why it wears so well. It rolls in the sequence described below.

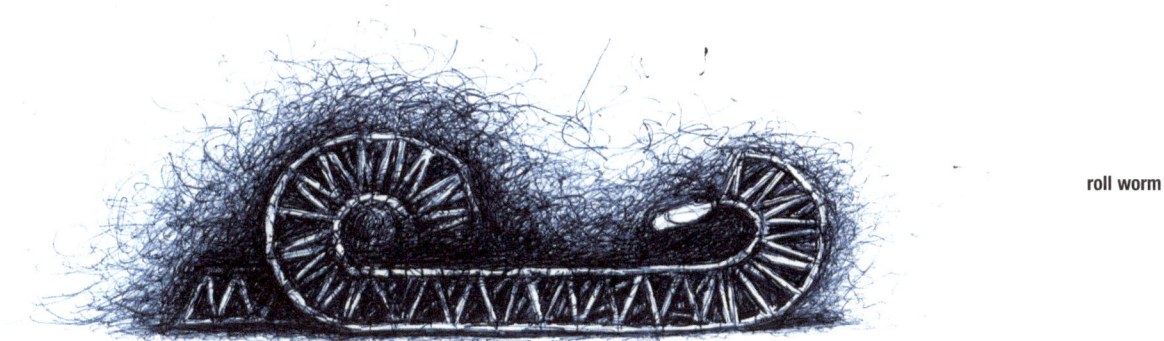

roll worm

It costs the roll worm almost no energy to move. All it needs is a plastic bottle. This wobbles to and fro in the wind. On one side of the neck is a little pump, a very simple affair. Each wobbling movement causes the pump to send a little air

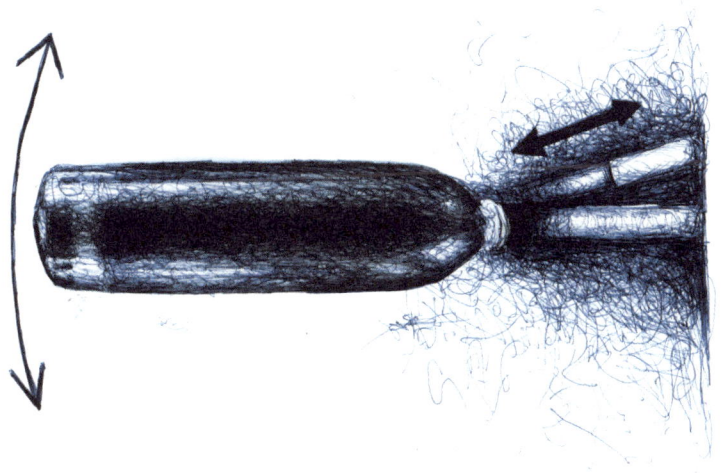

wobble bottle

into the bottle. After a year the pressure is high enough for the worm to move. This movement only lasts a few seconds and carries it one or two metres further. After that, it has to wait another year. The position of the tail indicates the pressure level. The muscles feed directly off the pressure in the bottle. When the pressure reaches a certain level the tail curls up slightly. One day the pressure

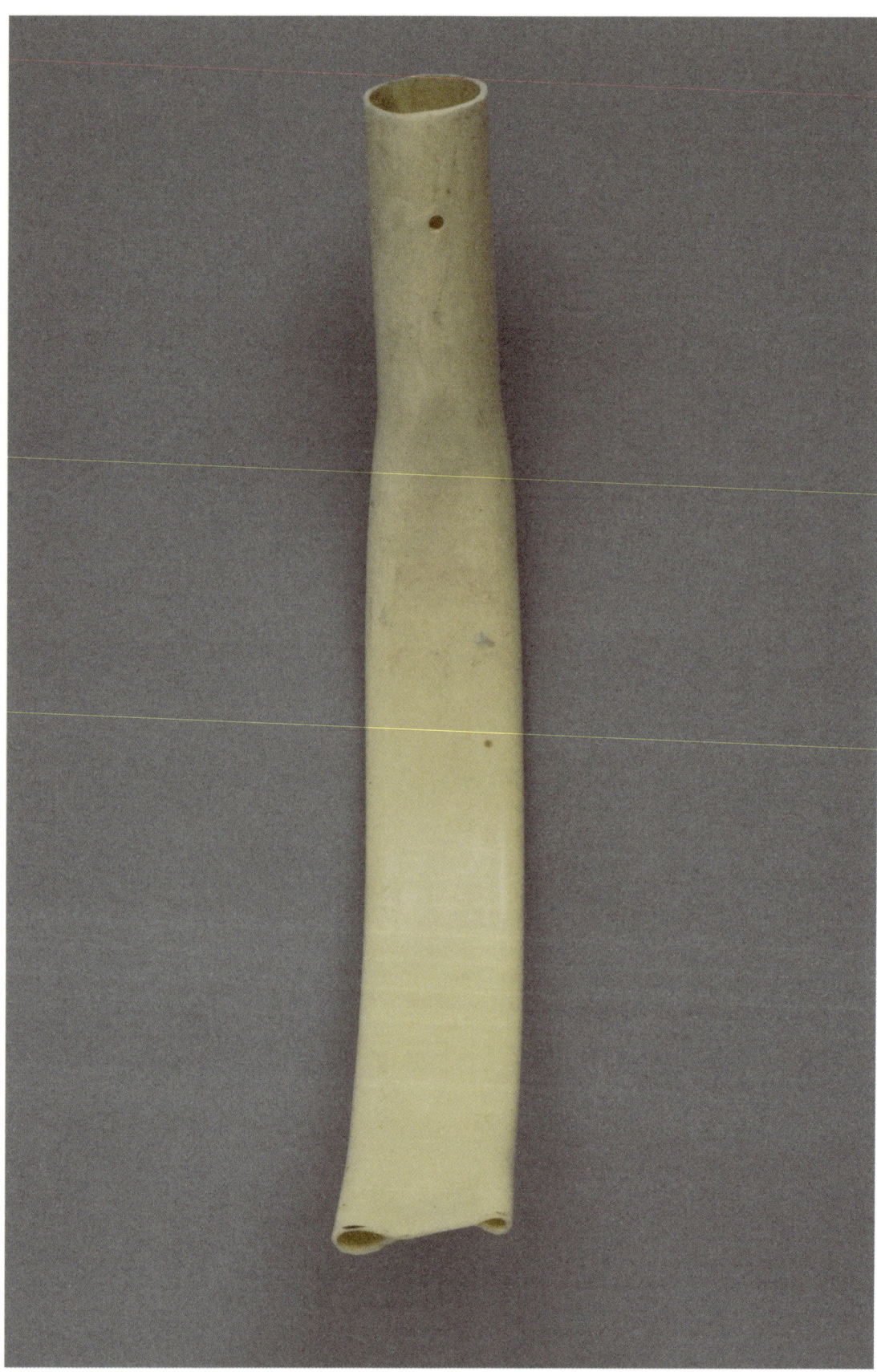

Animaris Vermiculus, sliding piece

will be high enough to get the tail curling in earnest and the worm will start rolling. A gust of wind may help. I can imagine the national press waiting for days on the beach to record the rare movement made by the roll worm. They might even give it a sneaky push.

Gerrit van Bakel (1943-1984) has been a major source of inspiration for the roll worm and for my work in general. He made machines that ran on the forces of nature, such as the London Machine (1979-80) which was propelled along by changes in the level of humidity in the air. It gets its name from the London fog. Other machines of his include the Rain-cart, which is driven by the weight of rainwater. He also made a cart that ran on expanding ice. In 1982 he took part in the Documenta in Kassel and achieved world fame. His Tarim Machine (1979-82) was 7.2 metres long and travelled 3 centimetres a year. The machine got its energy from differences in temperature between day and night. It would take some 36,000,000 years to cross the Tarim Basin in the far west of China. Cossacks, Kyrgyz, Uyghur and Uzbeks would care for its upkeep. Thirty-six million years. As a tribute to van Bakel I included one of his pieces in a major retrospective of my work in the Rotterdam Kunsthal in 2002.

Sealant and O rings
The Vaporum was full of hissing and squishing. Everything leaked. Occasionally an explosion rent the air. Once, a piston shot out of a tube past my head and through the closed window of the shed where I worked every day. Making life is fraught with dangers. The problem of leaks was ever present. A piston in a tube should fit tightly. This was solved for the first three years with sealant. Sealant is used in day-to-day life for filling in cracks in sanitary ware. What I did was slide a thin tube into a thick tube and then fill the result with sealant. After three weeks' wait, the sealant had hardened to produce a rubber piston which fitted exactly into the tube in which it was to move.

sealant piston

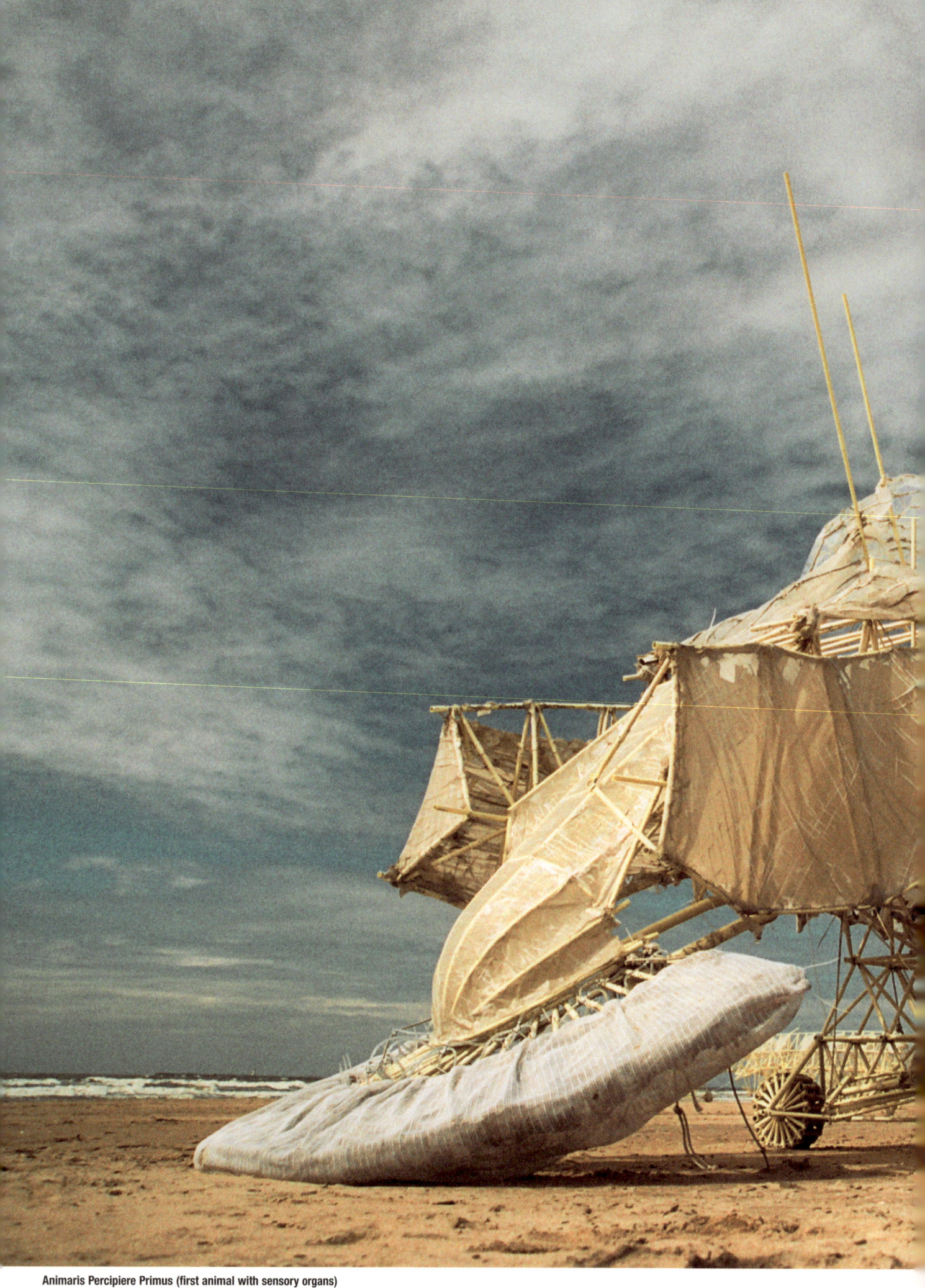

Animaris Percipiere Primus (first animal with sensory organs)

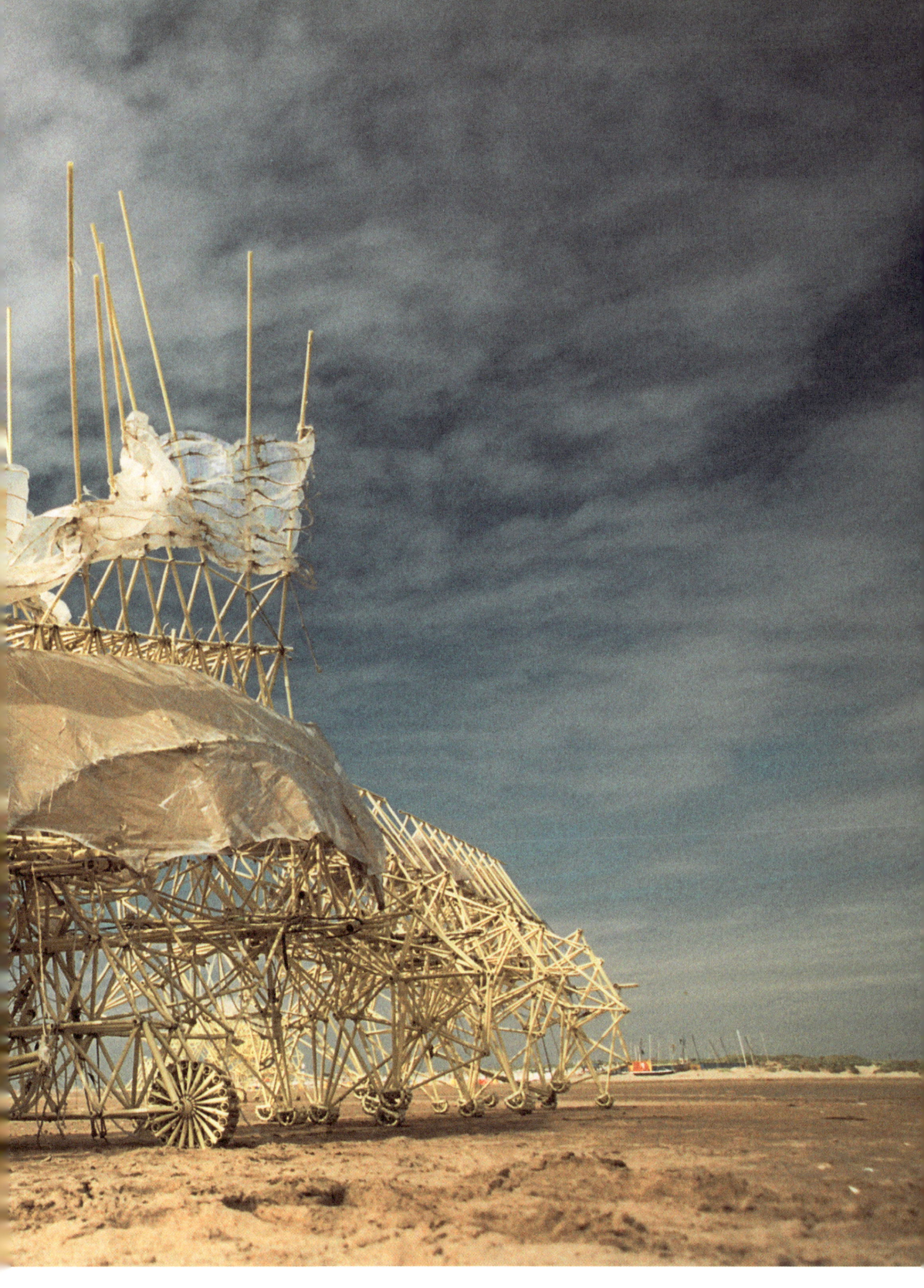

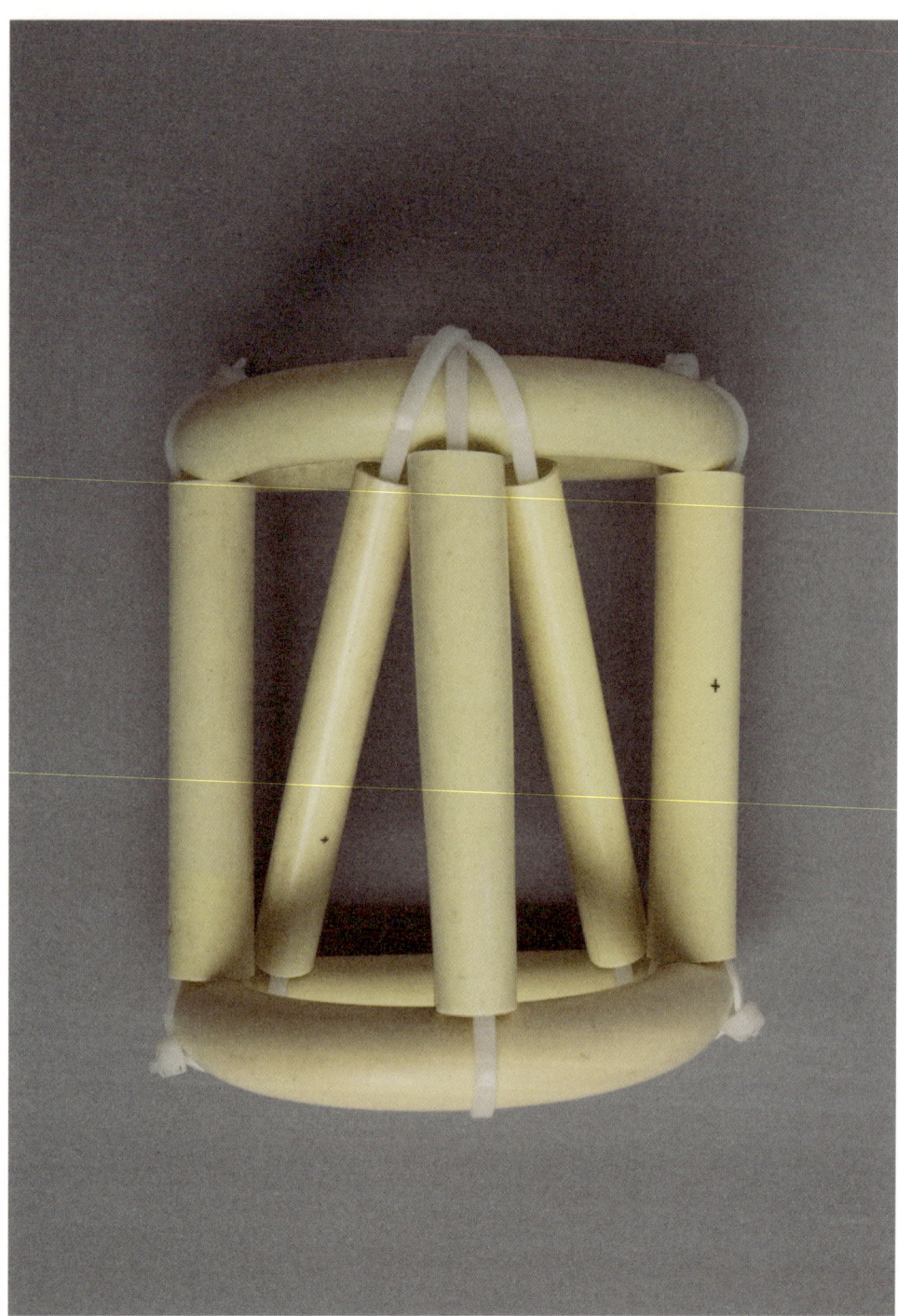

foot of Animaris Percipiere Secundus

It sounds ideal but in practice it was less successful. There were often air bubbles in the sealant, the pistons were frail and getting them to move through the tube was more arduous than I had expected. It was a real struggle to get the pistons to work. Already at the start of the Vaporum, acquaintances were advising me to try O rings. I had always brushed that advice aside, idiot that I was. Finally it was Job Kneppers, the one who had vastly improved the cell machine, all praise to him, who took the law into his own hands and made a piston with an O ring himself. It was only then, after three years, that O rings proved to be the solution to all problems with the pistons. They cost thirty cents apiece.

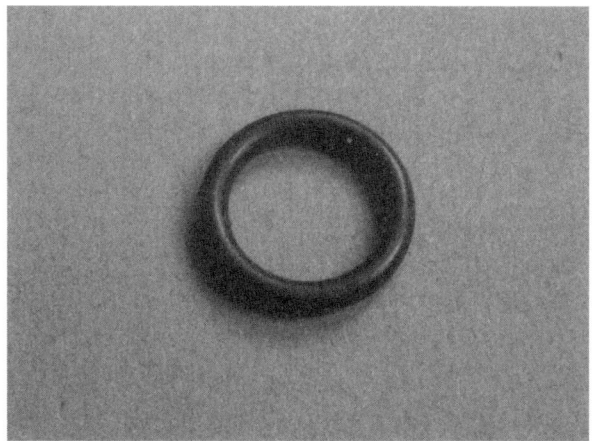

O ring

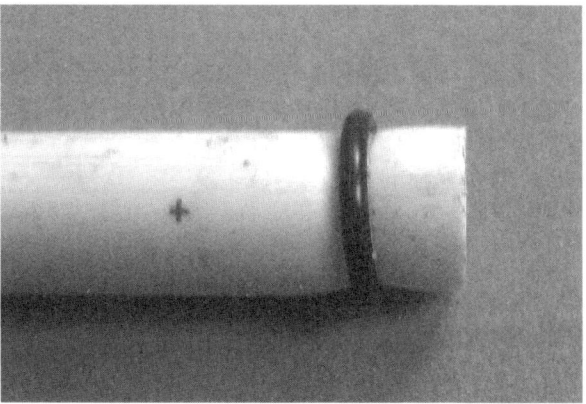

piston with O ring

I grease the pistons with a mixture of ball-bearing grease, Sanex shower gel (with Dermoprotector) and a little water. Those pistons run really smoothly and don't leak.

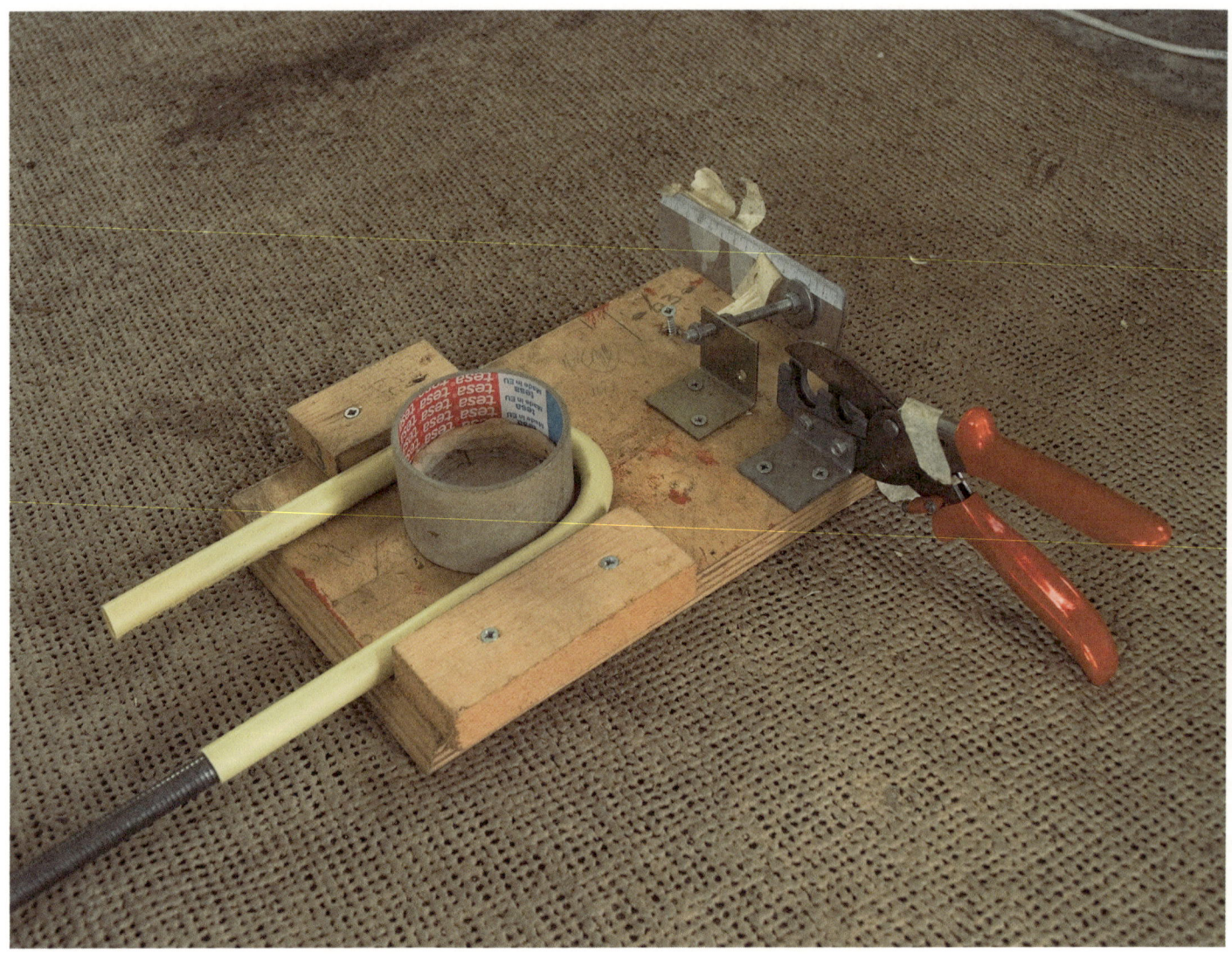

leg mould

New nerve cells

The year 2005 saw the emergence of a new type of nerve cell. It's smaller, lighter and easier to produce. It isn't as small as the chips in a computer yet, but the two evolutions run parallel. Everything is being made smaller so there is more room for brains. The performance of the new nerve cell derives from the properties of an air hose. When the hose is bent tight, air is unable to pass through.

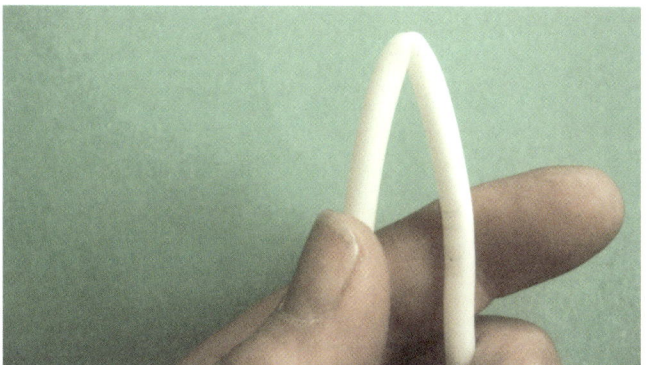

air hose bent tight

In the photos below of the nerve you can see that the air hose is fastened to the tube with cable ties. In the situation at left the hose is a bit bent but the air can still get through. When a piston shifts the tube upwards, the air hose is bent tight. It shows that hoses can be opened or shut using air pressure.

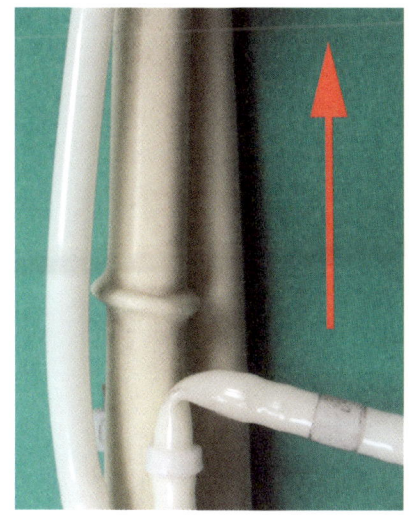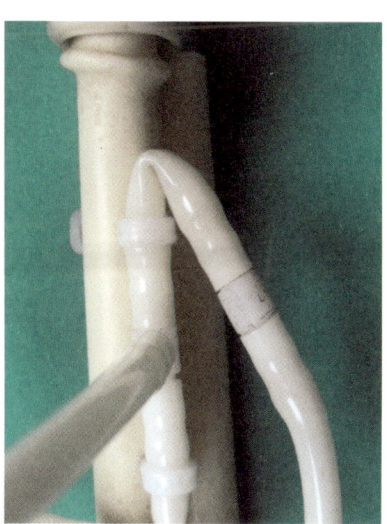

barely open and shut

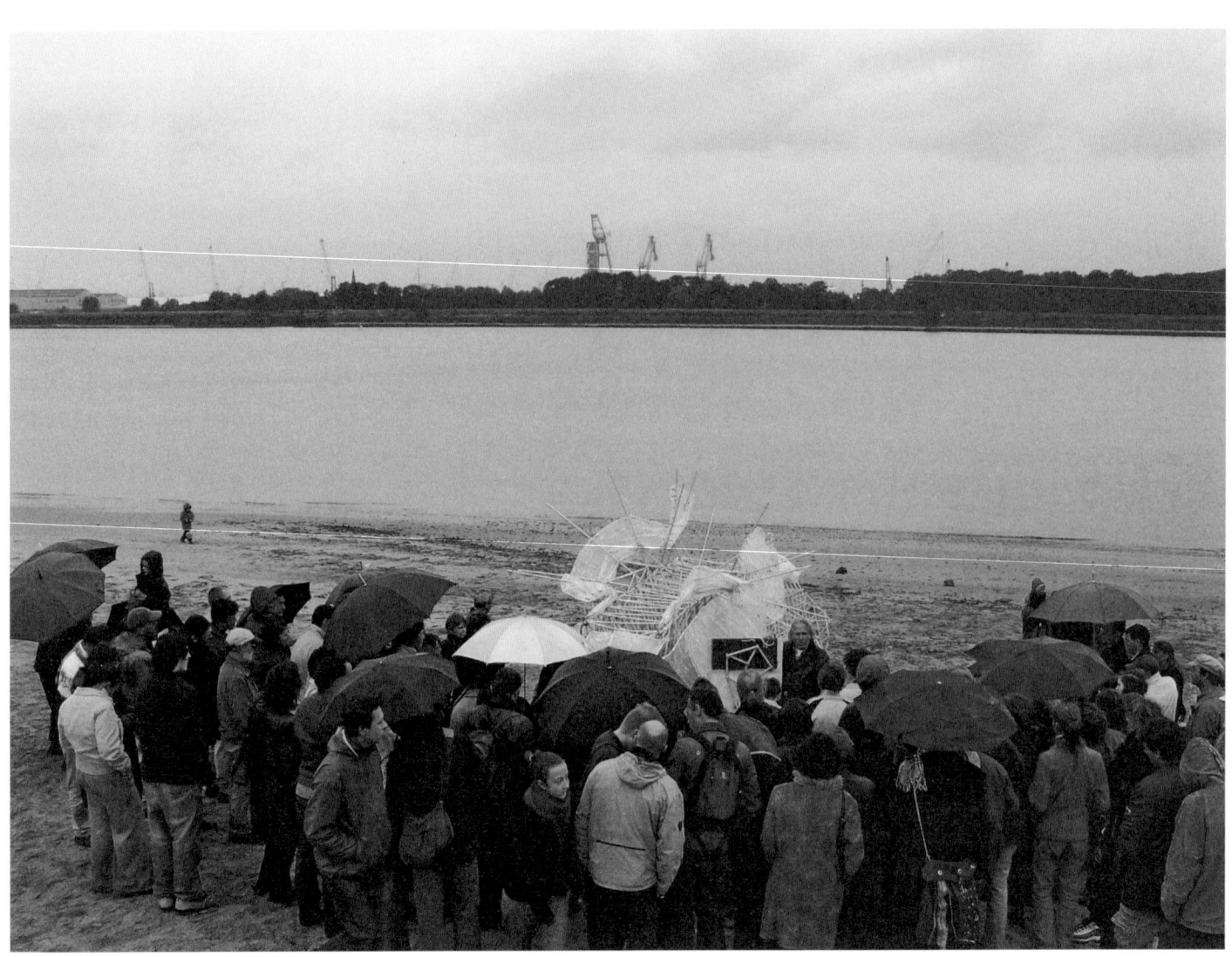

St. Anna beach, Antwerp, 2004

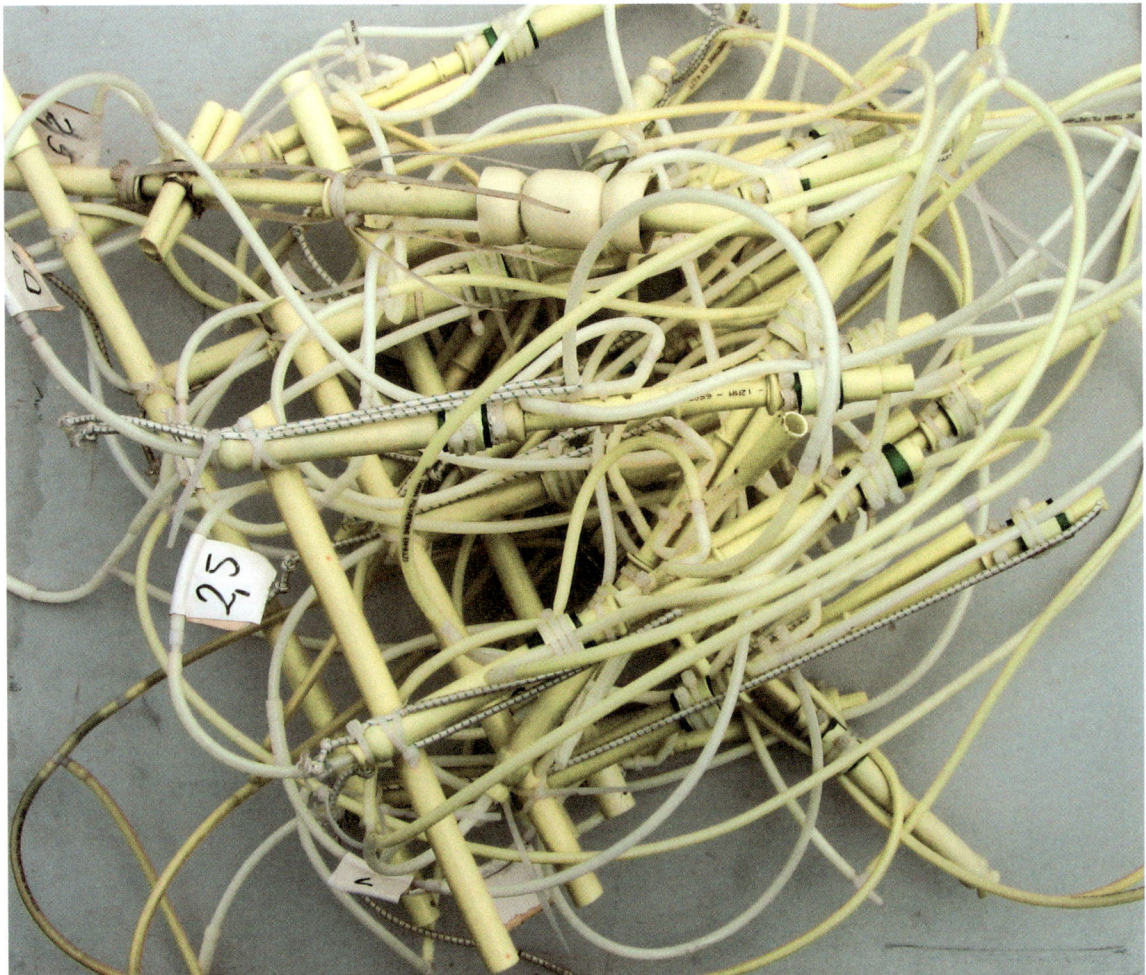

pedometer

Automatic reproduction

In the Tepideem, Animaris Geneticus could be made to reproduce by replacing rods. The lengths of the 375 rods determined the animal's properties, its rigidity, its capacity to walk and its susceptibility to blowing away in the wind. These were its 375 genes. A practical disadvantage was that I had to help Animaris Geneticus in a big way with the business of reproducing. I had to replace tubes, organize walking races to get the selection process under way and afterwards reanimate the corpses by implanting a new genetic code (more tubes to replace!). And all this dragging bodies around and general drudgery might need to go on for years before there was an evolutionary change of any significance. In short, I just haven't the time. So I devised a new creature, the plug animal. This has to be able to reproduce unaided. For this the following is required:

a. The rods must be able to change length themselves;

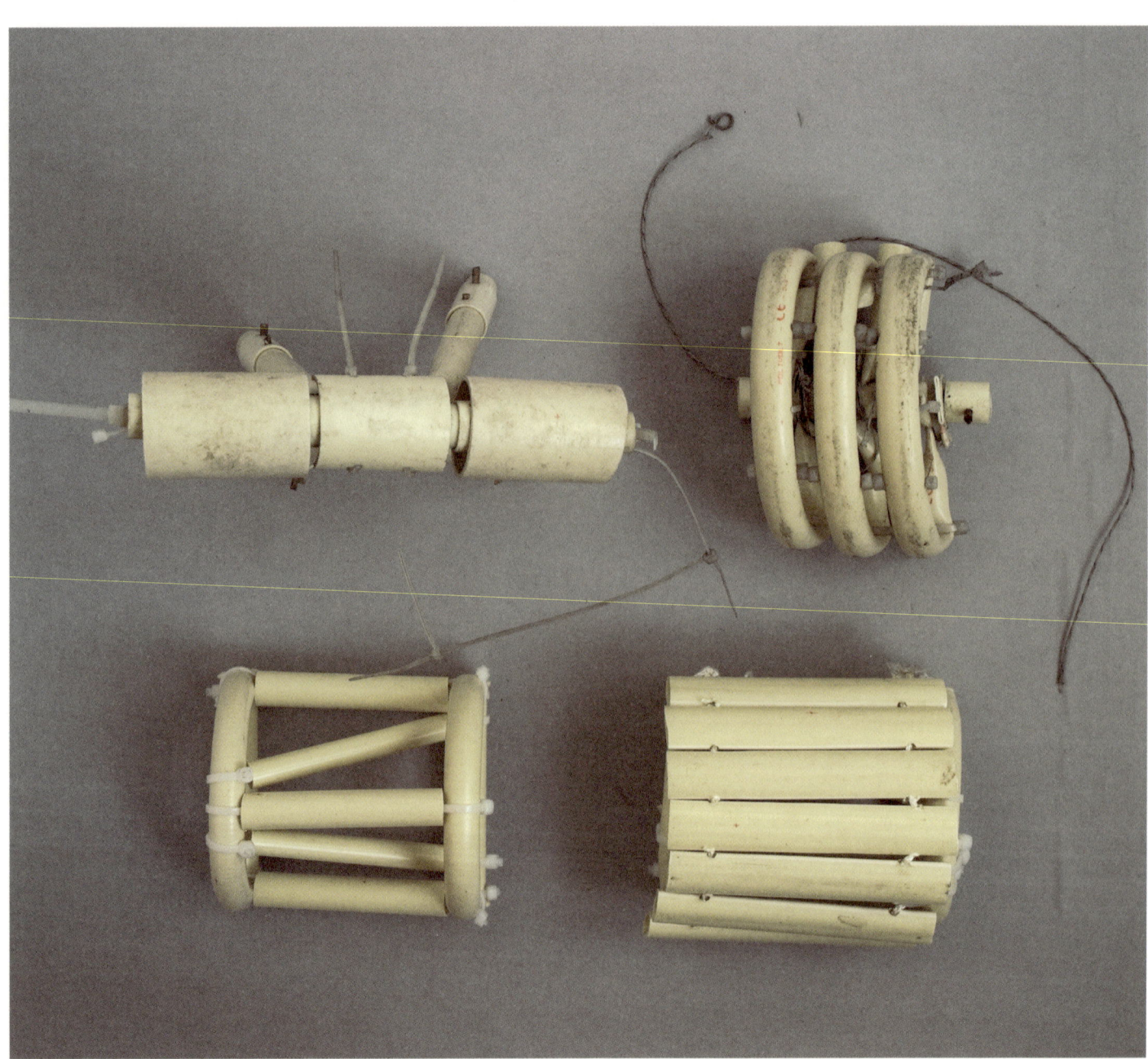

evolution of the foot

b. They need to select themselves on the basis of walking speed;
c. The winners have to transfer their genetic code to the losers themselves.

Variable length

As we saw earlier, the walking properties were determined by the lengths of the rods, in other words by the distance between the joints. The number representing this distance I have chosen to call the gene. Before now, the gene consisted of a small rod fixed between two joints. The new gene is slightly more complex, being a combination of three tubes of different length.

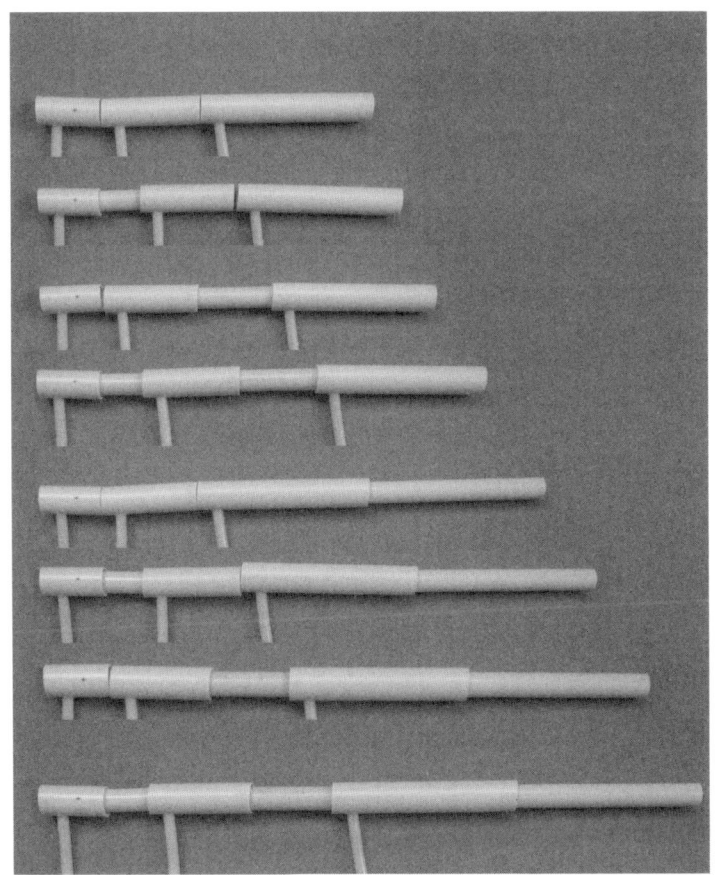

three-bit muscle

These tubes contain pistons. Tubes and pistons together constitute something resembling muscles. When air is blown into the muscles, they extend and the aggregate gets longer. If all the muscles are activated, the gene reaches its maximum length. When no muscles are activated the aggregate stays short. Activating

Pitch

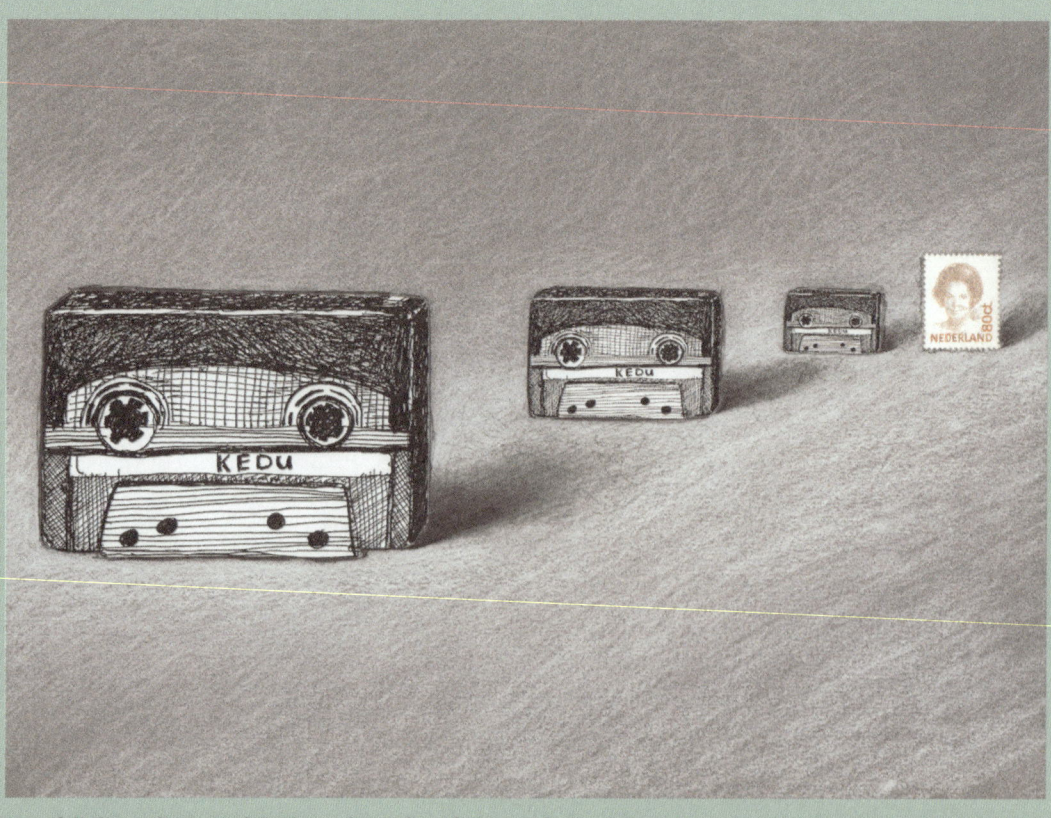

The party was getting out of hand. Your respectable types had already departed. Just riff-raff left. The dregs. Fridges were ripped open in search of chicken legs or other snacks, even a bag of crisps. Not a thing. This meant that the fatal effects of drink could continue unabated and the evening would inevitably end with the arrival of the police. It was a great evening. Music-wise too. The host drummed on the table and sang an African song. It sounded jungly, tree-barkish, from the bowels of the earth. It was as if the booze had opened the gate to his soul. A respectable gent in a suit during the day and still so in the early evening, he was later seen stripped to the waist. His drumming was gradually taken up by other revellers who seriously dented the table top by bashing it with the big ends of beer bottles. It sounded great. I had never felt the excitement of rhythm quite as forcefully as this. Nor felt such musical shock waves running through my bones. It had never struck me before that I was able to sing three-part harmony, by myself. It sounded so good that the host had, it later transpired, surreptitiously recorded the proceedings on a small tape machine. He shouldn't have. When we listened back to the recordings a few days later, it seemed more like a workshop for amateur carpenters. What a cacophony of thumping and banging. The vocals had a wavering, searching quality to them. 'But just listen to this,' said mine host. And using the pitch control on the tape recorder, he cranked up the speed. This had a strange effect. The rhythm not only became faster, but also more focused and tauter. The singing rose in pitch but also became more in tune. The faster the tape was run, the closer it came to the feeling of that evening. From this, one could conclude that mistakes in rhythm and singing are lessened when the whole is speeded up, so that they no longer register. Both rhythm and vocals seem more precise. So I would advise amateur groups to play their music slowly and a couple of tones lower when recording it and speed the tape up when playing it back. It then sounds a good deal more professional and upbeat. You might compare it with a drawing. Suppose you wanted to make a drawing the size of a postage stamp. Then it's advisable to make it large, say on a sheet of A4 paper, and reduce it to postage stamp size using a photocopy machine. It was comical to hear the police winding up the party at top speed... a degree of confusion set in during the African harvest song... a new instrument seemed to have joined in ... it was the doorbell ringing non-stop ... more confusion ... apologies came thick and fast, especially fast ... it was all the fault of the absent chicken legs. Then silence ... the speeded-up variety.

some muscles and not others gives you the lengths in-between. The small muscle has a deflection of 2 cm. The second muscle has a deflection of 4 cm and the third of 8 cm. When all three are activated the gene becomes 14 cm longer (2 + 4 + 8).

Some possible lengths are given below. A one means that the muscle in question is extended and a zero that it is retracted and at rest.

Shortest muscle	0	1	0	1	0	1	0	1
Medium muscle	0	0	1	1	0	0	1	1
Longest muscle	0	0	0	0	1	1	1	1
Length of gene in cm	0	2	4	6	8	10	12	14

The genes are attached by lengths of air hose to a gene administration centre. This is a series of memory units, like the bits in a computer. Bits can be either on or off. If turned on they activate the muscle they are attached to.

Self-regulating walking races
Each beach animal is equipped with a pedometer, an instrument for counting paces. Made of tubing, air hose and rubber rings, these pedometers keep track of the distance (in kilometres). Animals that get caught up in fencing along the dunes score poorly, as they have stopped walking altogether. Animals that have been blown over don't rate highly either. But animals that proceed smoothly carry a high score. I call these high-scoring animals *dominants*. It is the dominants that need to reproduce. Their gene code has to end up in the gene administration centre of another animal, one with a lower score. This is how beach animals are to reproduce. First of all, they must find themselves another of their species. Then the pair have to decide which of them is the dominant. The dominant has to insert its code into the other animal's gene administration.

Self-operating plugs
Let's start with the most important part, inserting the code. As we saw earlier, the transfer of genetic information between beach animals is the same as that between humans: with a plug. The plug is a series of small tubes in two possible lengths: short and long.

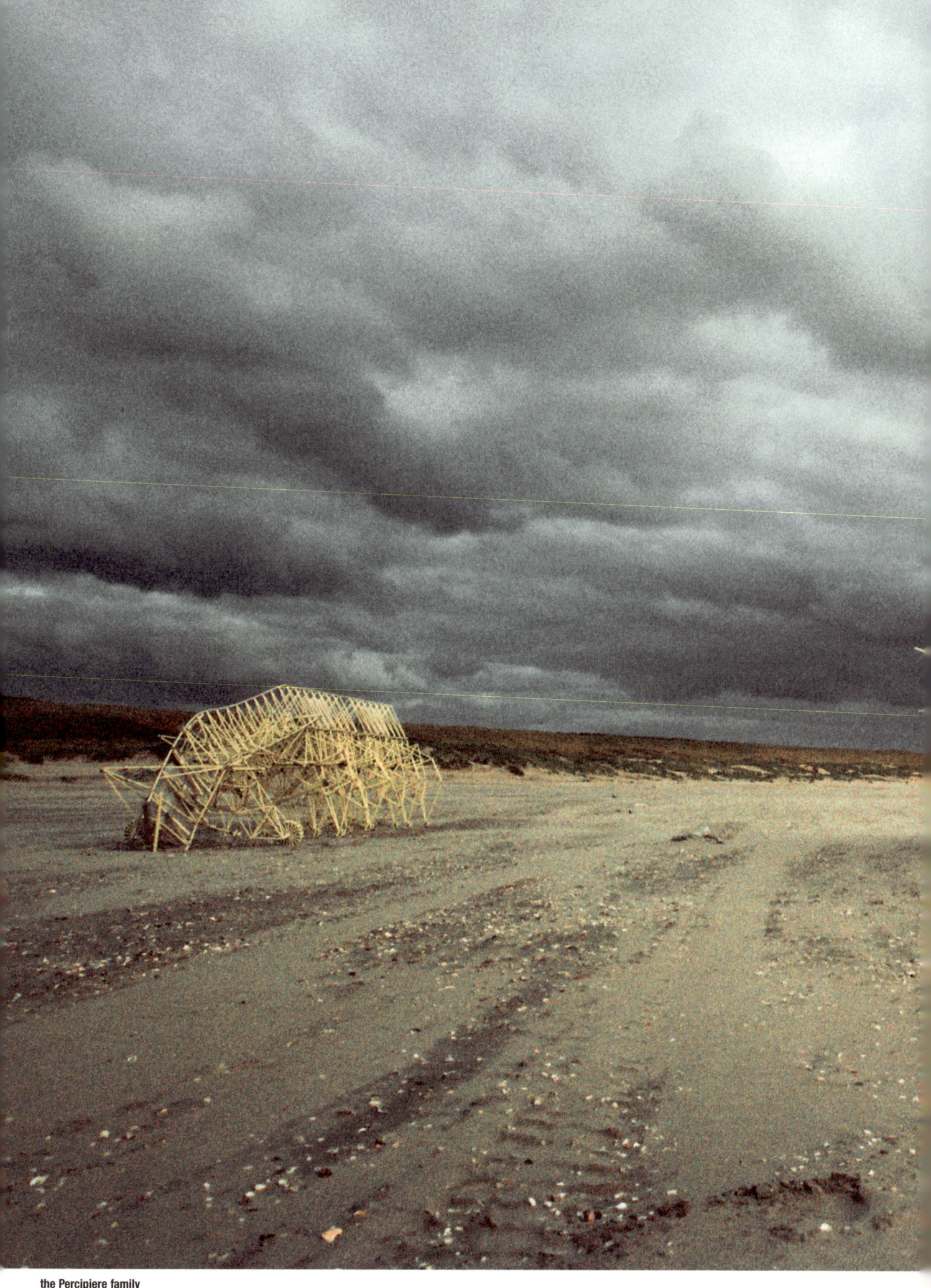
the Percipiere family

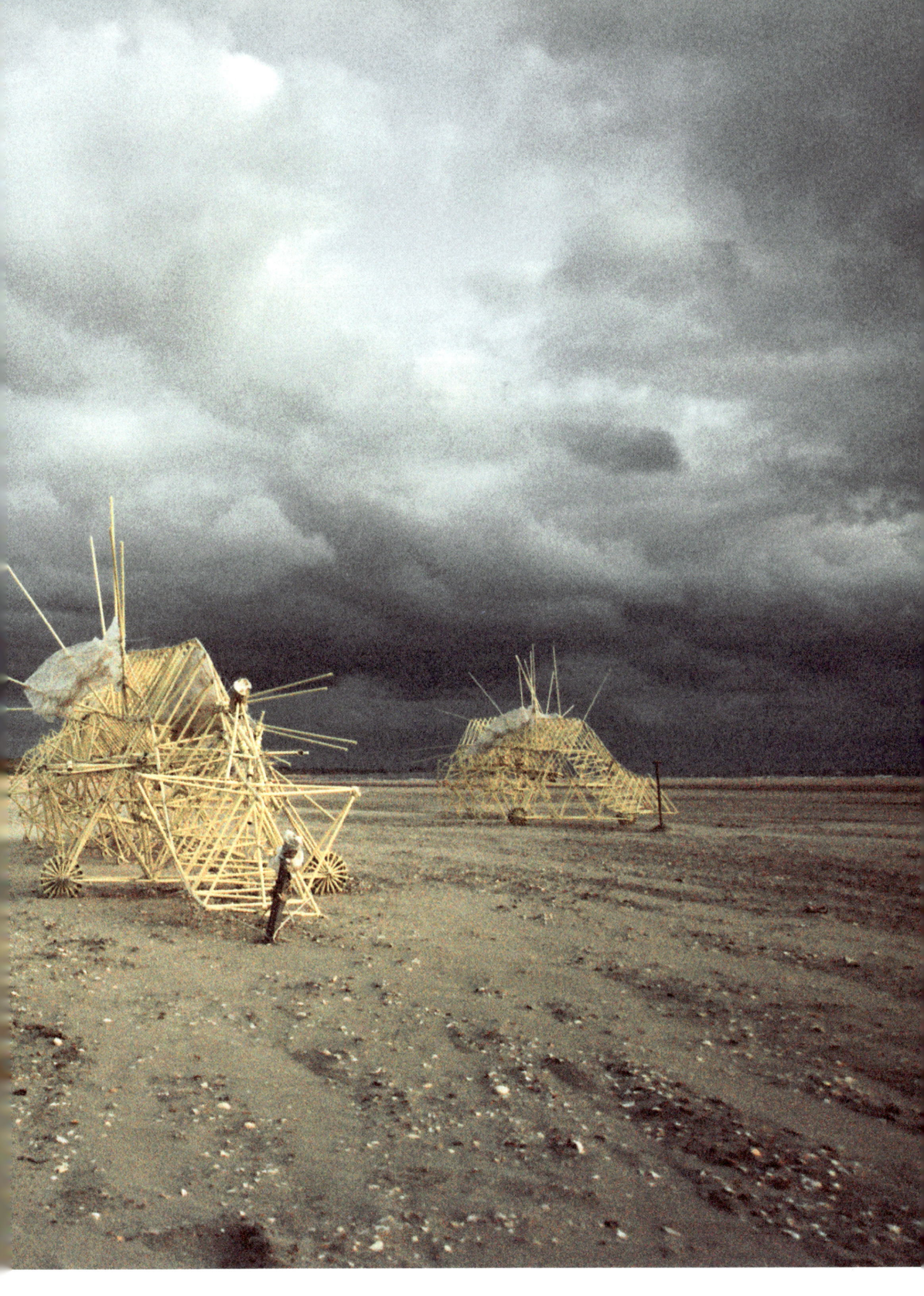

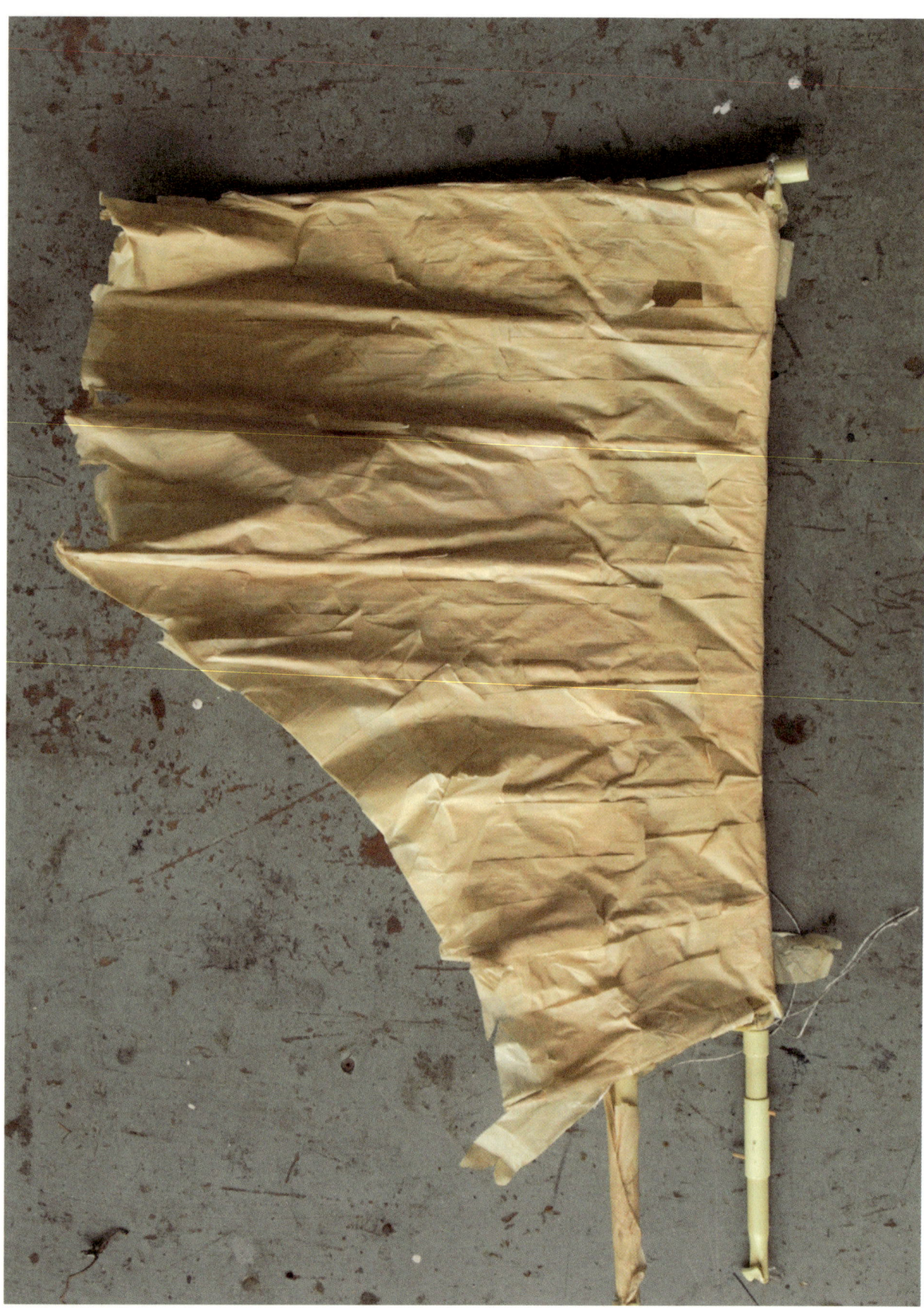

dorsal fin of Animaris Properans

When a plug is inserted into another animal, a long tube turns on a memory unit in the insertee. Air then passes through a length of hose to the relevant muscle in the gene, which gets longer as a result. A short tube in the plug switches off a memory unit that had been on. Memory units that are already turned off are left alone. Once a plug has been inserted into an animal, that animal can suddenly change its shape completely. Imagine you have inserted a full plug – that is, a plug of long tubes – into an animal. Then all its genes suddenly take on the maximum length. On insertion, the animal swells up and then probably collapses as the rods have become too long to carry the weight. Of course, mutations are a part of the reproduction process. There have to be mistakes so that evolution can keep developing. Well there are bound to be mistakes now but this doesn't worry me. A memory unit may seize up, for example. But if it all works, we have an automatically evolving beach animal. We'll put a pair on the beach and when we return years later chances are that the animals in the herd will have quite a different form. They may be bigger, or smaller, and probably less symmetrical. They should move around better. Until now, I've been claiming that reproduction is the same thing as copying genes. I take it back. The difficult thing about reproduction is that more examples arise in the course of time. Not so with the plug animals. The number in a herd remains the same. From every plug animal that had been declared a corpse there comes a new one. The total number remains the same. Maybe it's possible to make a wooden machine that would go into the forest to chop wood to make a machine that would go into the forest to chop wood. In truth, such a machine already exists. And who happens to have been the first person to build it? Father Walter de Capir, a missionary in Venezuela and an amateur in the field of science. He invented it in 1970. The machine was not made of wood but of paper. On it, atoms of carbon had been arranged in such a way that they formed letters. In fact it was an ordinary sheet of A4. At the top were Chinese characters and according to Walter de Capir this was a prayer. It was a very special prayer, one that would bring salvation. In the text that followed, the reader was advised to make ten copies of the sheet of paper and drop them in the letter boxes of those friends and acquaintances who could use a prayer. This was to be done within twenty-four hours. If they didn't, the most dreadful things would happen to them. The letter gave a number of examples: a stillborn child in the Philippines, a collapsed building in Mexico, a case of total dehydration in Madagascar. On the other hand, if the reader did comply, great fortune would

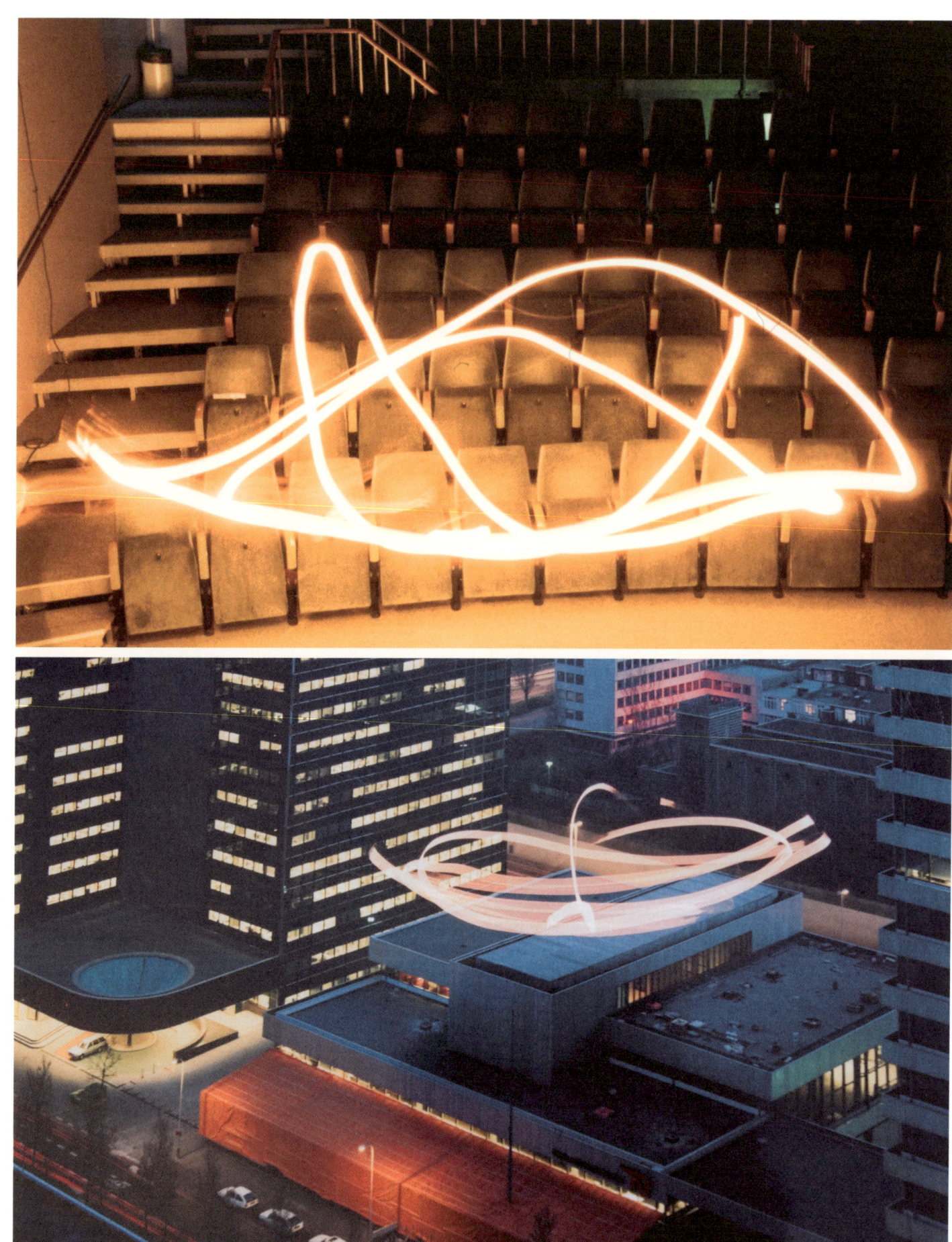

Lissajous (1990); swinging lamp in RO Theater, Rotterdam and swinging drill (with propeller instead of drill bit) above Utrechtse Baan, The Hague

come their way. This letter came to me in the post. I heard from acquaintances who had also received it. Evidently it was reproducing itself. There must be millions of those letters in circulation and this figure is increasing all the time. I know what you're thinking; you're angry with the missionary and it's your opinion that the sheet of paper isn't reproducing itself but that people are reproducing it. I'll tell you why this is your opinion. It's because you're a human being. Despite Galileo, despite Darwin, humans still insist on seeing themselves as the centre of creation. Look at it from the paper's point of view for once. To that sheet of paper, a human is no more than a pile of protein molecules. Its atom configuration was evidently interfering with the order of the letters on the paper in such a way that the A4 sheet was multiplying. Let's say the missionary only needed to throw one sheet of paper from the tower of his church in Venezuela to get this process of multiplication under way. Look at humanity as an antheap. Toss in a note with the letters in the right order and it copies itself. There are many examples in nature of one species inadvertently helping another. There are bushes with berries so tasty that the birds keep coming back for more. They expel the seeds miles away so that that species of berry-bearing plant is guaranteed a wide dispersal. So birds are merely pawns in the bush's dispersal policy. We humans have been attributed that role in disseminating the letters. Man is just a pile of protein molecules in the shape of a person. These molecules do their work as molecules.

Plastic tubes entered my life in 1990 on a fine September day. Since then, the beach animals have ruled my life. It became an addiction, a disease if you like. It's a virus that refuses to leave my body. I am a victim. The beach animals are forcing me to make them. I'm happy but utterly dependent. The animals and I live in symbiosis. We profit from one another. They give me a place in this world. I earn my living from them. In their eyes I'm just a pile of protein molecules, an unresisting victim. Because of me, they make a good job of reproducing. There have been photos galore in magazines. In May 2005 the Reuters press agency filmed them on the beach at IJmuiden. A week later, the results were shown in 170 countries. That news item was on BBC News but also on Al Jazeera. Millions of viewers across the world came to know of the beach animals. *Through* me, not because of me. The beach animals have been using yours truly. First they made me ill, ill for love of them. I lay awake for nights on end. I wasn't much of a com-

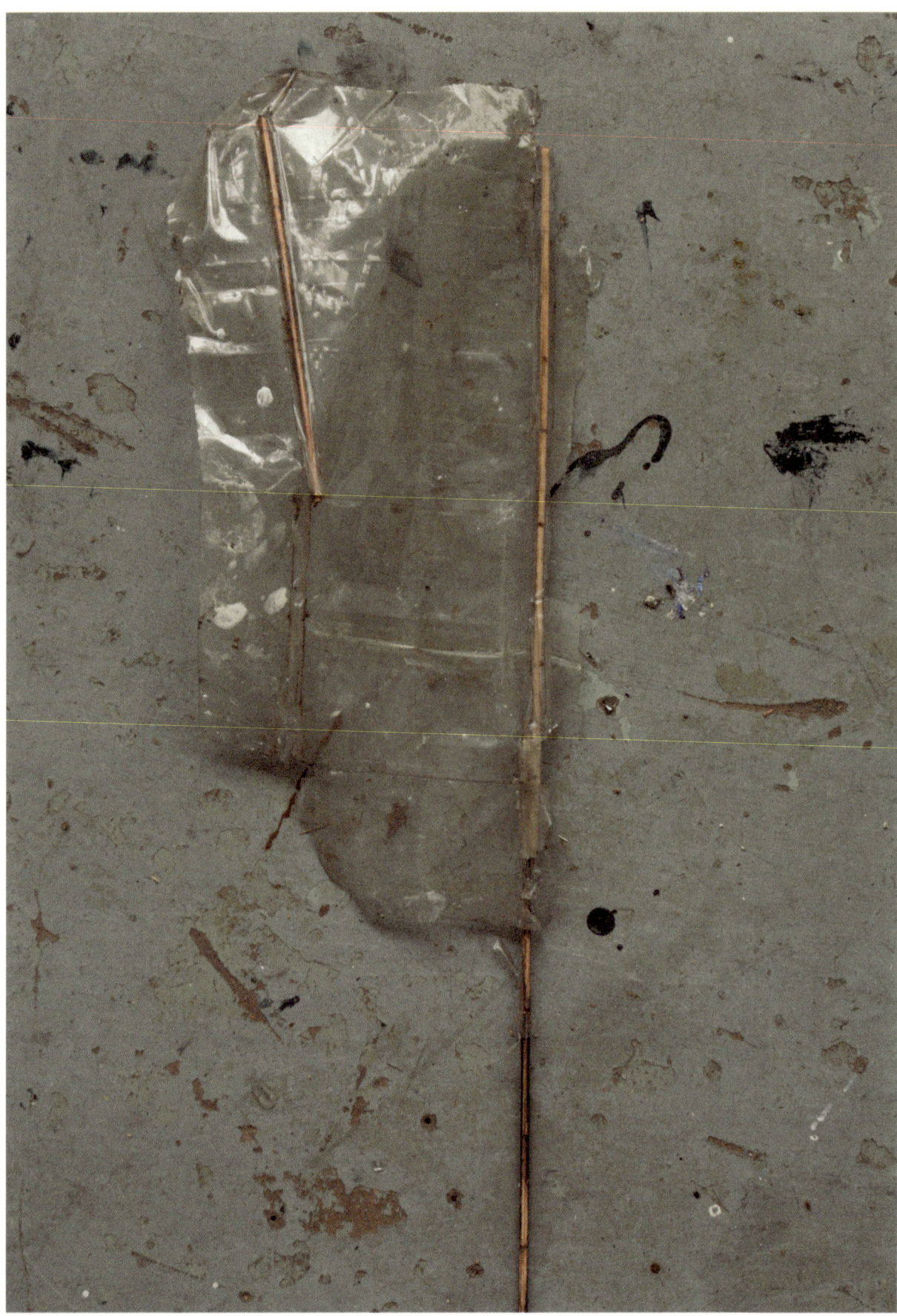

propeller fin of Animaris Rigide

panion for my wife and children, sitting there at breakfast staring into space. I was a pawn in the beach animals' campaign to conquer the world. I would have most liked to have made a beach animal that could climb the dunes and head into the city streets in search of rubbish skips. It would take out all the plastic tubing and from it make a new beach animal. That for me would be the finest thing. But what *I* want doesn't come into it. The real beach animals opted for a much more effective means of reproduction. They used the media. There was a commercial for BMW in which the beach animals were on camera for 95 per cent of the time. This ad was televised every evening for months. Now the beach animals are hoofing it in millions of minds.

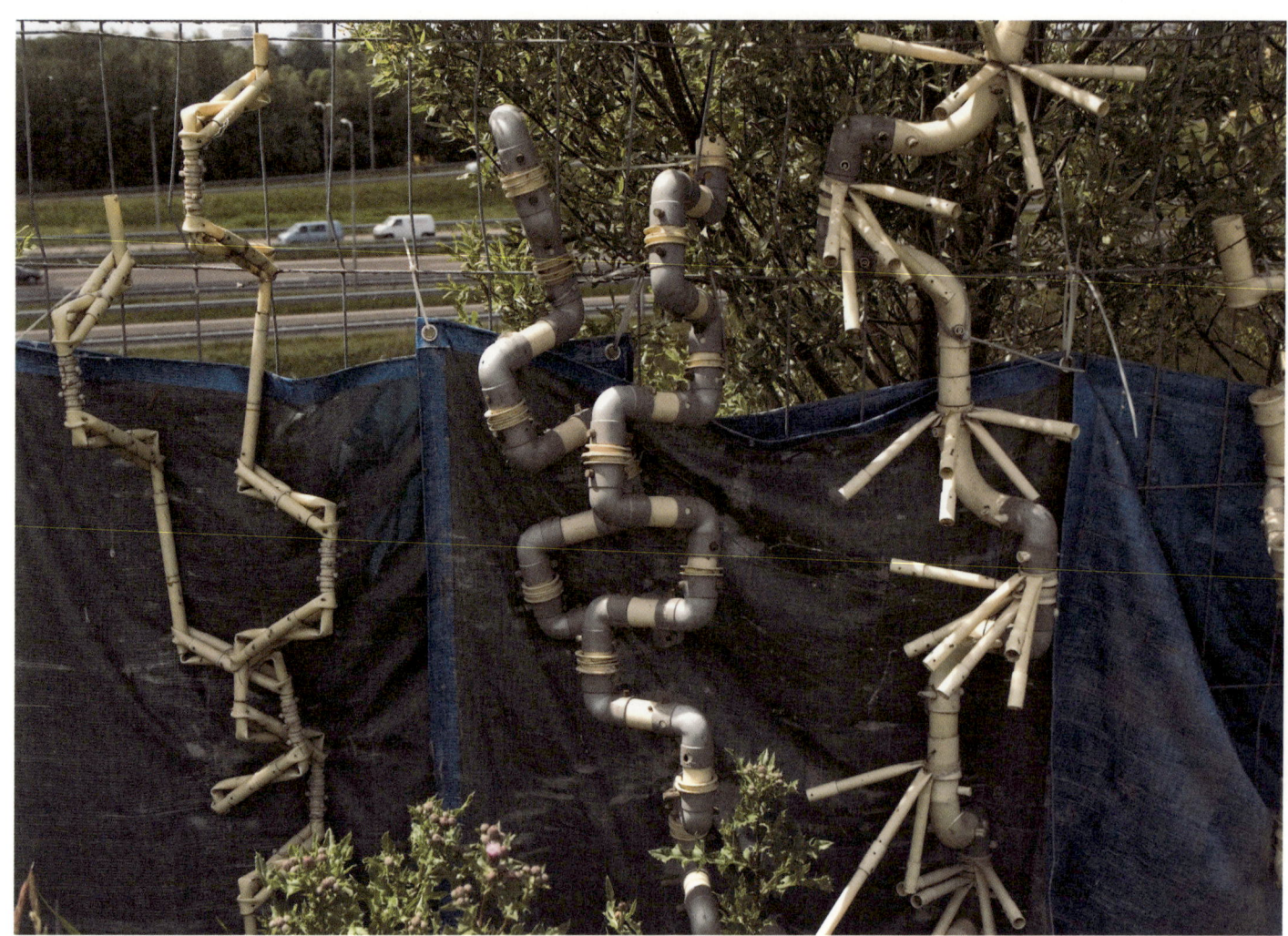
vertebral columns of Animaris Currens (left) and Animaris Geneticus

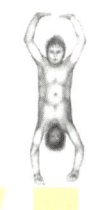

CEREBRUM
the brains period (2006-present)

Brains are basically vast accumulations of nerve cells. We needed nerve cells to activate the muscles of the beach animals. Once we had them, they turned out to be handy material for constructing a thinking machine. I can imagine the existing nature followed that same order of development; first muscles, then nerves, then brains.

Pedometer
After a time the animals began to exhibit paranormal gifts, or so it seemed. Everything was there, I just had to put it all together. Until then, the animals had only been able to walk with difficulty in the loose sand of the berm. And on the beach face they ran the risk of being engulfed by the waves. The best place for them was the stretch of hard sand between the water and the flood line. Where to go from there? The animal had to be able to feel the water from a few metres away without actually coming into contact with it. It had no eyes, no ears, no antennae, no remote control, nothing. Amazing but true. The new animal had at its disposal a kind of witching rod. To understand how this works, we have to return to the first stage of evolution. At that time the animal had two sensory devices, the *water feeler* and the *soft-sand feeler*.

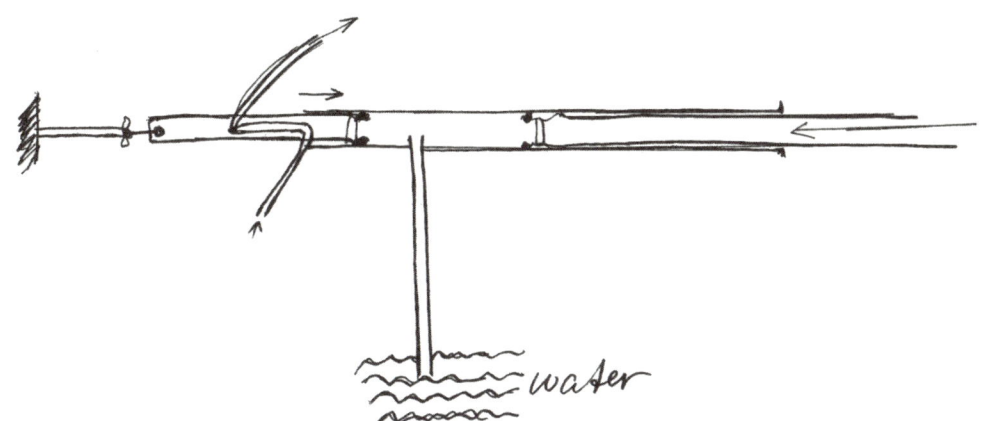

action of water feeler

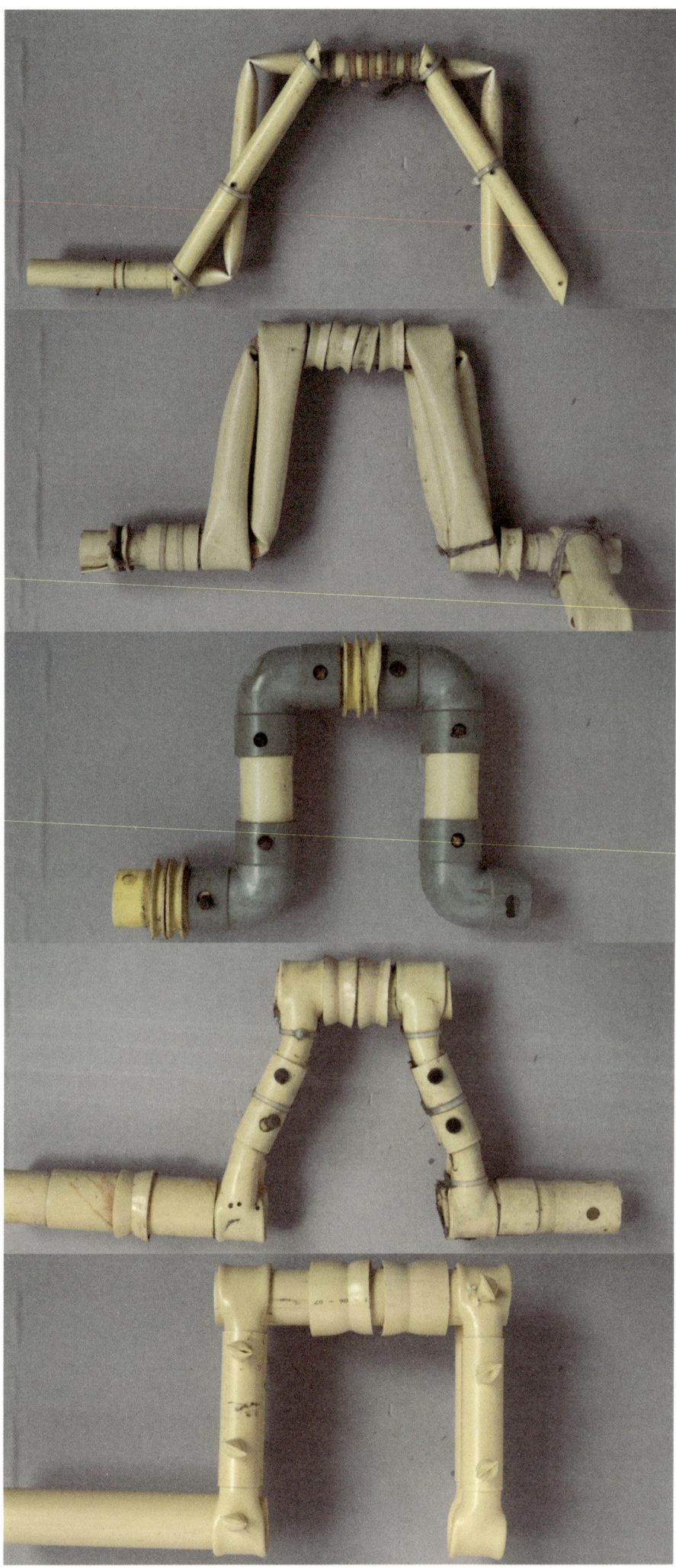

evolution of the crankshaft

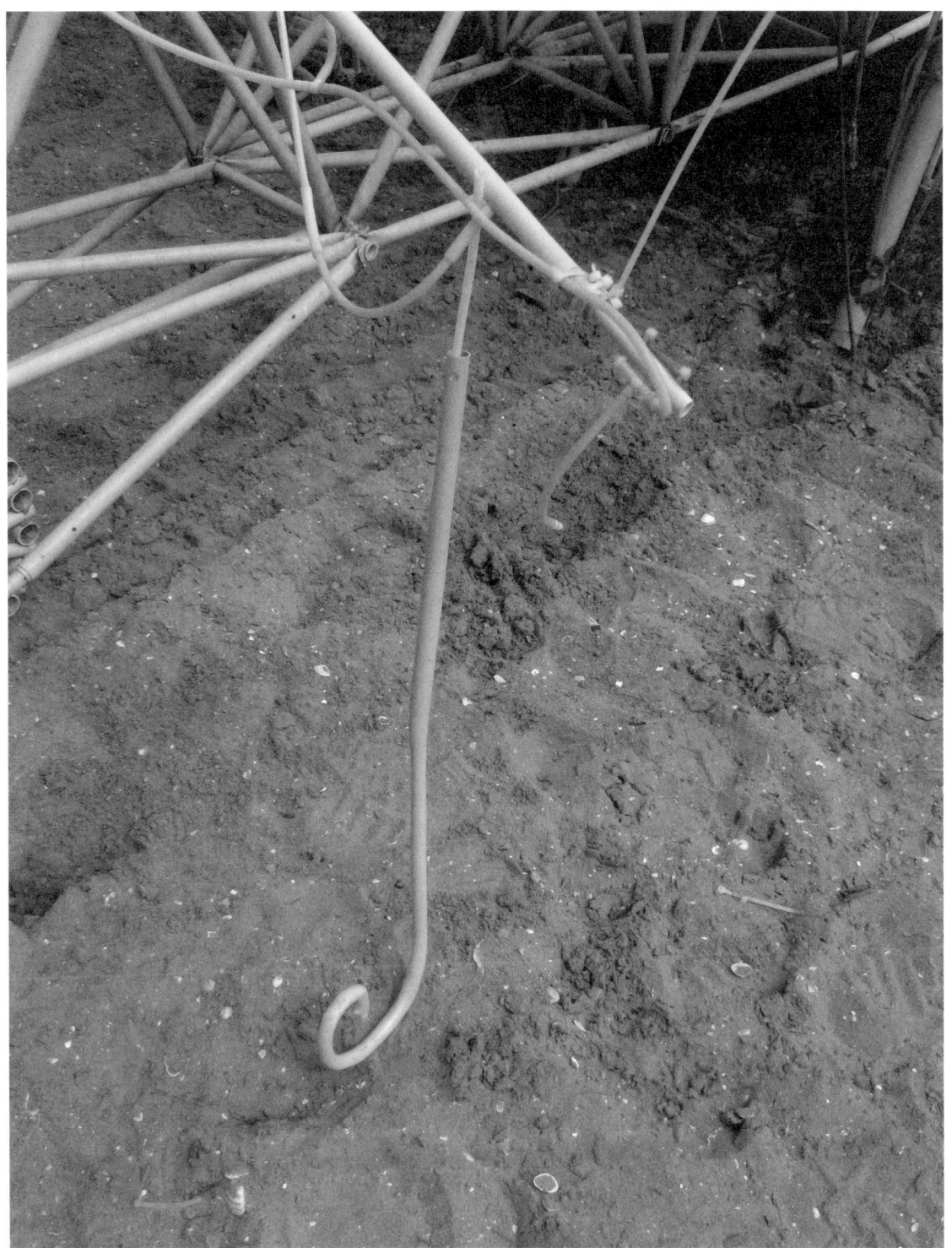

water feeler

The water feeler consists of a length of hose which trails along the ground. Under normal conditions it sucks in air. In that case all is well. But as soon as it takes in water, its resistance increases and the animal feels this. A nerve cell switches direction whereupon the beach animal turns round.

Time map

Old marine clay, bog, fen, sandy ground, I still know exactly where these soil types can be found and what colour they have. Marine clay is green and is found in Zuid-Holland province, fen is purplish-pink and is found in the Giethoorn area north of Zwolle, diluvial sand is yellow and is found south of Hilversum.

This knowledge was literally hammered into me, so literally in fact that I was justifiably astounded to later discover that marine clay, sand and even fen and bog were actually grey. Again, the dotted lines on the road map turned out not to be the staccato bursts of road that had embedded themselves in my imagination.

You are doubtless surprised that I could be so naive, but you can't begin to imagine what kind of punishment was meted out at the institution where I was taught the basics of geography. When your palms are a mass of weals, you'll believe anything you're told. The school atlas is imprinted in my memory as a rack, a guillotine. Political maps, relief maps, climate maps, rainfall maps, drought maps and of course unmarked maps clouded my view of the world as it really was.

The torture I underwent is probably what put into my head the absurd idea of a time map. A time map looks at first sight like an ordinary road map. The difference is that the length of a road in a time map relates not to its physical length but to the time required to traverse it by car. This can be useful when you need to know the quickest route. In this hectic age, a time map is a must. In it, the highway along the IJsselmeer barrier dam has the same length as Overtoom avenue in Amsterdam.

Here the Netherlands is distorted beyond recognition. Its cities bristling with traffic lights take up most of its surface area. Ring roads round those cities pinch together the urban streets into a crinkled pattern. Crossroads in particular register as an inextricable tangle of twisting roads. Indeed, the question arises as to whether such a map is mathematically possible; would all the twisting lines fit onto it?

This is certainly questionable if verges and berms are no longer seen as obstacles. Then there's nothing stopping you from driving through a field. It's a slightly slower process but you can do it. In an even field, all distances are in proportion to the time needed to cover those distances. So the ratios in the time map are the same as those in an ordinary map. As you know, a plane is a collection of infinitely many lines. This means that the plane of the field is itself a collection of straight and winding roads.

If there were to be a ditch in that field for your front wheels to get stuck in, that ditch would take up an infinite amount of space and spill out over the edge of the map. The map is blocked-in in flat colours. Each colour is for a different soil type.

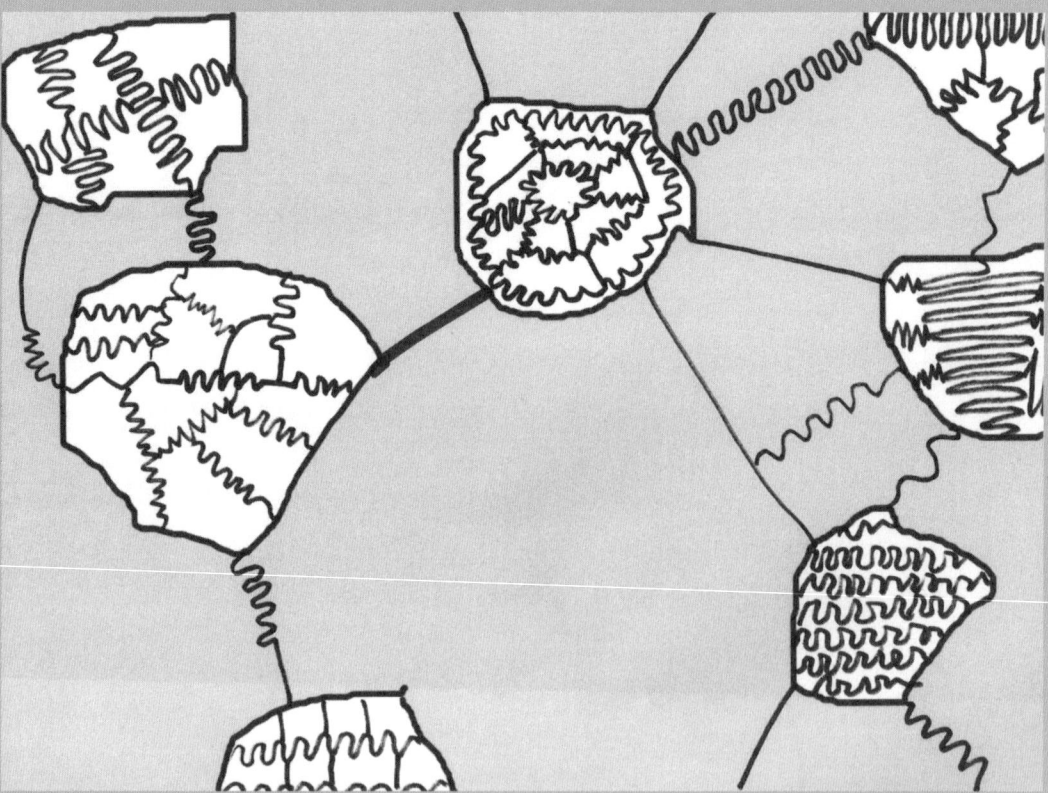

The soft-sand feeler senses the pressure building up in the muscles. That happens when walking becomes laborious. When one of these sensory devices is activated, the reflex is to turn round; the walking direction of the legs is reversed. The result of this reflex is that the animal moves to and fro between the beach face and the soft sand. This works well. Even so, the animal's life is in danger on occasion. As it approaches the water, there is a chance of it being hit by a wave. The witching rod is there to minimize that risk. It is nothing other than a pedometer, with the nerve cells from the Vaporum linked together in such a way that they count the paces.

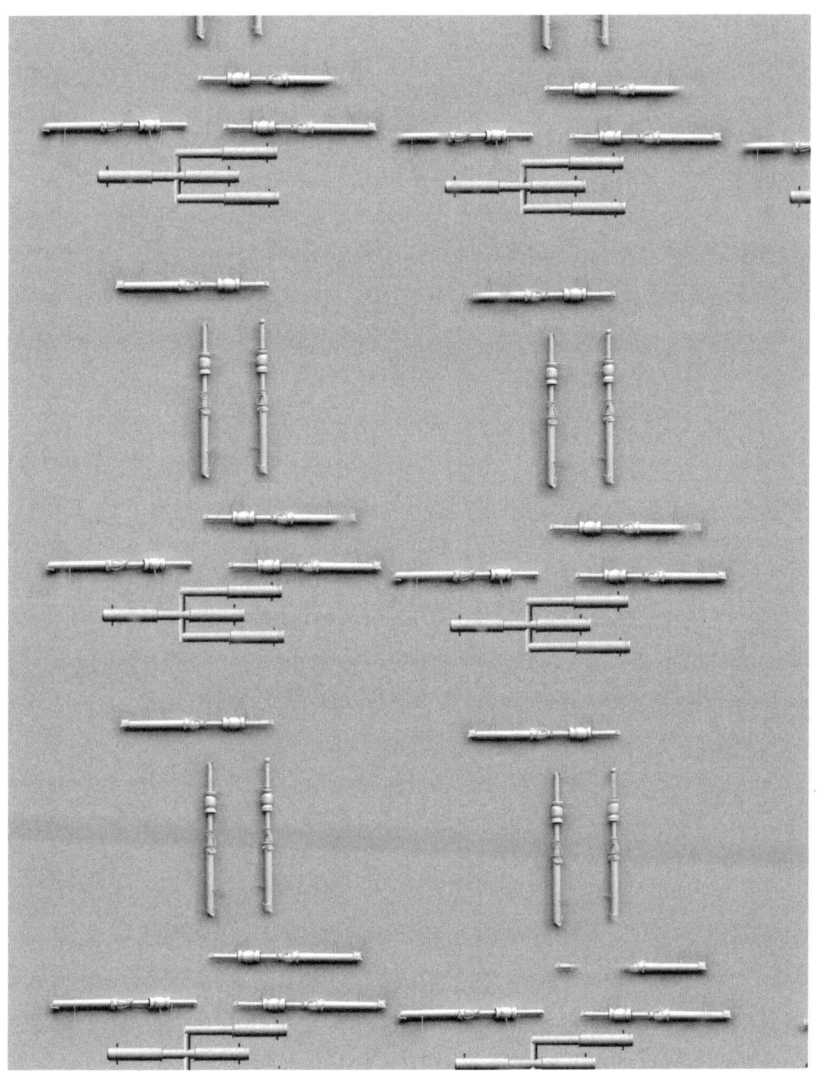

diagram of pedometer

This pedometer is a binary mechanism of zeros and ones. When it registers a particular number this may produce a reflex. There are now two reflexes on

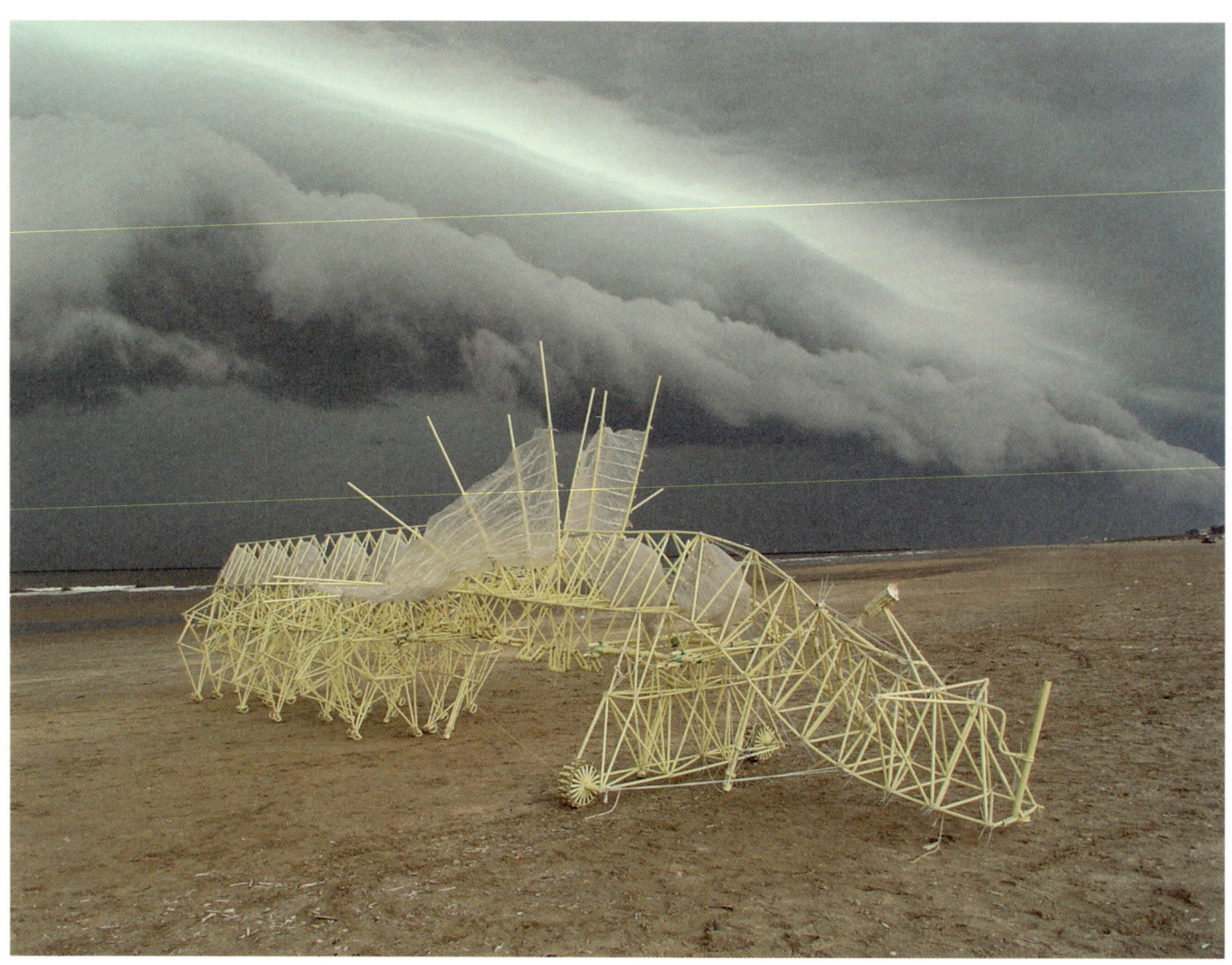

Animaris Rectus

board. When the counter switches from 4 to 3, the animal turns round. Again, when the water feeler is activated, the animal reverses direction and the counter returns to zero. The package of reflexes is as follows:

1. touch water ⟶ turn round + counter back to zero
2. counter from 4 to 3 ⟶ turn round
3. soft sand ⟶ turn round

You must have got the picture by now. After having come into contact with the sea once, the counter is reset at the water's edge. The next time the animal approaches the water, it stops at just three paces from the waves. This greatly reduces the chances of drowning. It doesn't have to venture a second time into the sludgy beach face. It's always risky at the sea's edge. The animal's feet can get sucked into the wet sand. Then it's really in trouble. Look at the counter as a brain. In that brain is an abstraction of the world. The counter tells the animal where it is at that moment. The animal's image of the world consists of issues crucial to it – the limits of the hard sand – primitive but an image all the same. If you look at the pedometer, there is no way of telling that it contains an abstract image of the beach. It is a coded world, as in the book *The Code of Things* (see page 133). The animal's paranormal gifts can be traced back to a number of reflexes. Normally, the reflexes work through a nerve, for example when the water feeler is activated. It feels water, the switch is thrown. The reflex is that the animal reverses its walking direction. At the so-called paranormal reversal three steps away from the sea, the animal's response is the result of a package of reflexes (1-3) involving an entire series of nerve cells in the pedometer. This series I call the brains. The beach animal's brains are getting bigger all the time. As you all know, the edge of the sea shifts with the tides. These can be simulated in the animal's brain with a simple timepiece. So it's not just the hard sand it imagines but also the moon and the sun, the heavenly bodies that cause ebb and flow. So the brains contain a coded cosmos. The beach animal's timing mechanism works like an hourglass. The space beyond the entrance to a nerve cell is pumped full of air using a valve. This causes the cell to close down; no more air leaves by the exit. This would remain the case if there weren't a leak in the air hose at the entrance to the cell. The air gradually exits through the leak at a rate depending on the leak size. Once the nerve cell is empty, it is reactivated and air pours in again. And so the clock keeps ticking. The frequency with which it ticks

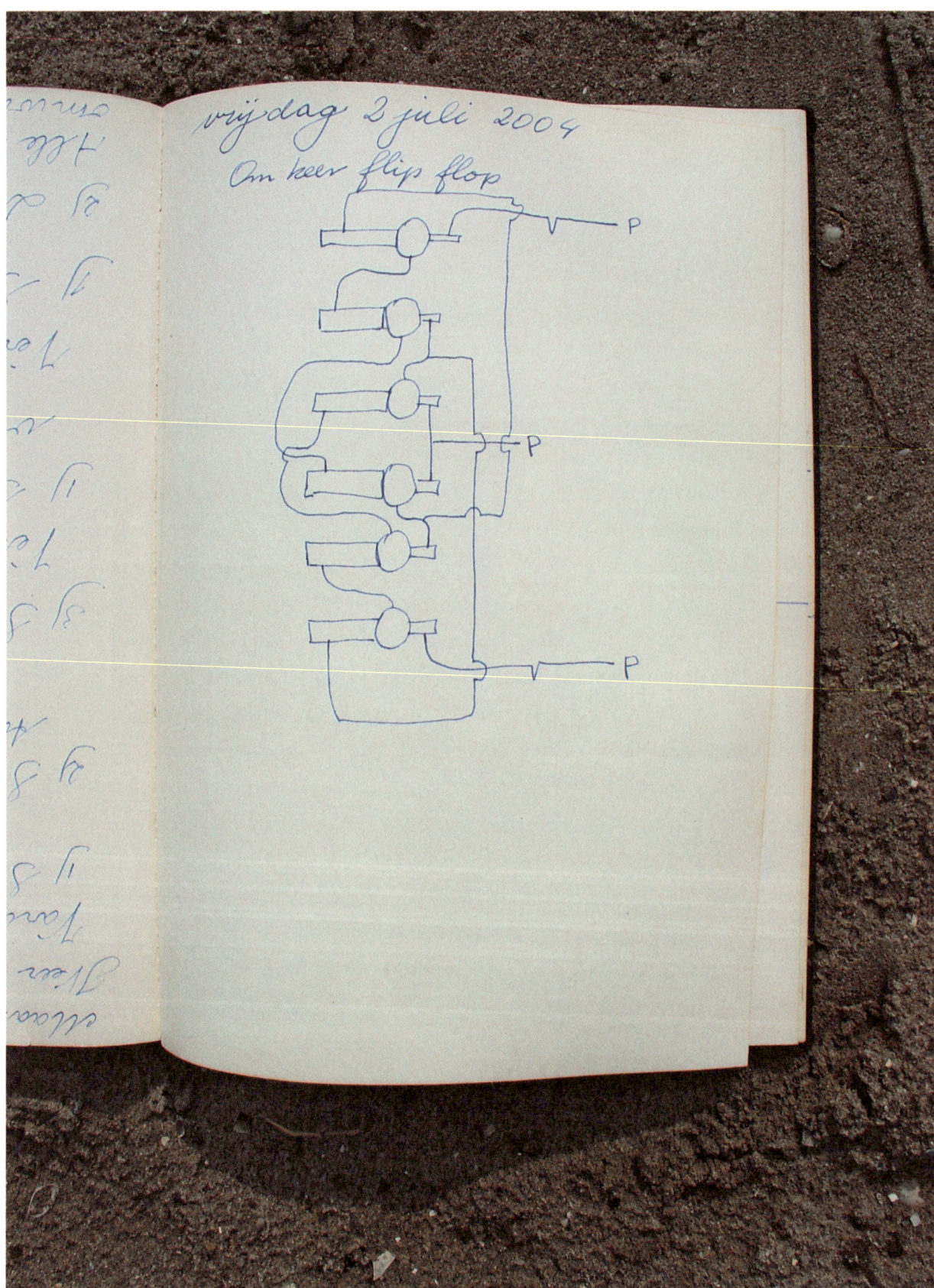

page from the log

depends on the size of the leak and the volume of air the nerve is able to receive. If it leaks only a little, the clock will tick slowly.

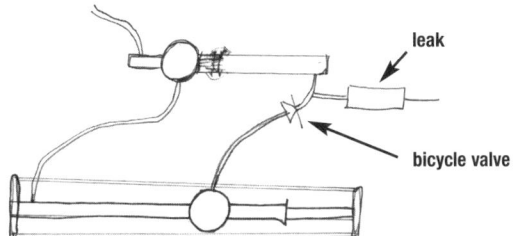

timing mechanism

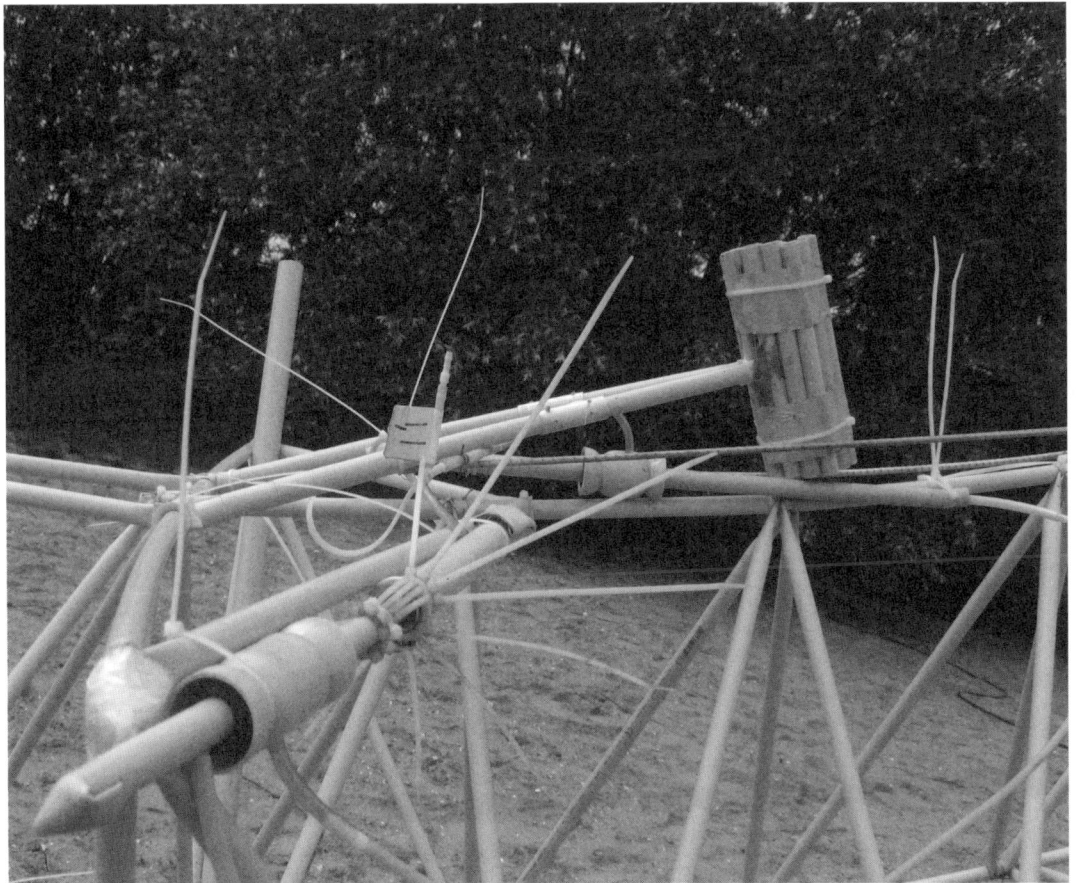

The hammer on the nose of Animaris Rectus is connected to the timing mechanism which causes it to strike the peg attached to the front of the animal at regular intervals.

The 'leak' consists of a connecting piece with a bamboo satay skewer pushed into it. The air escapes through the fibres of the bamboo. Let's call this a resistor. It can be compared with electrical resistance. The size of the leak can be influenced by having several resistors in a serial configuration. In this way, the

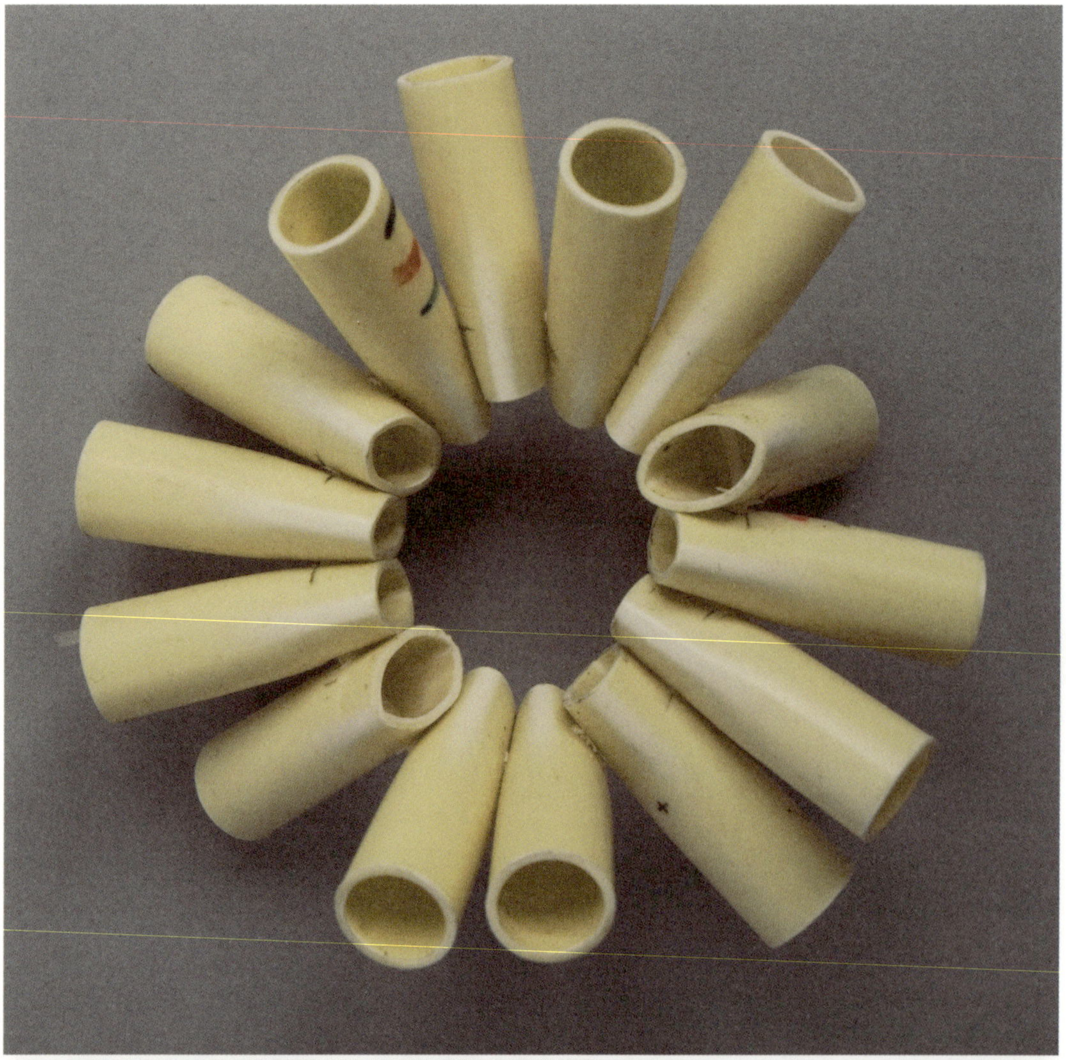

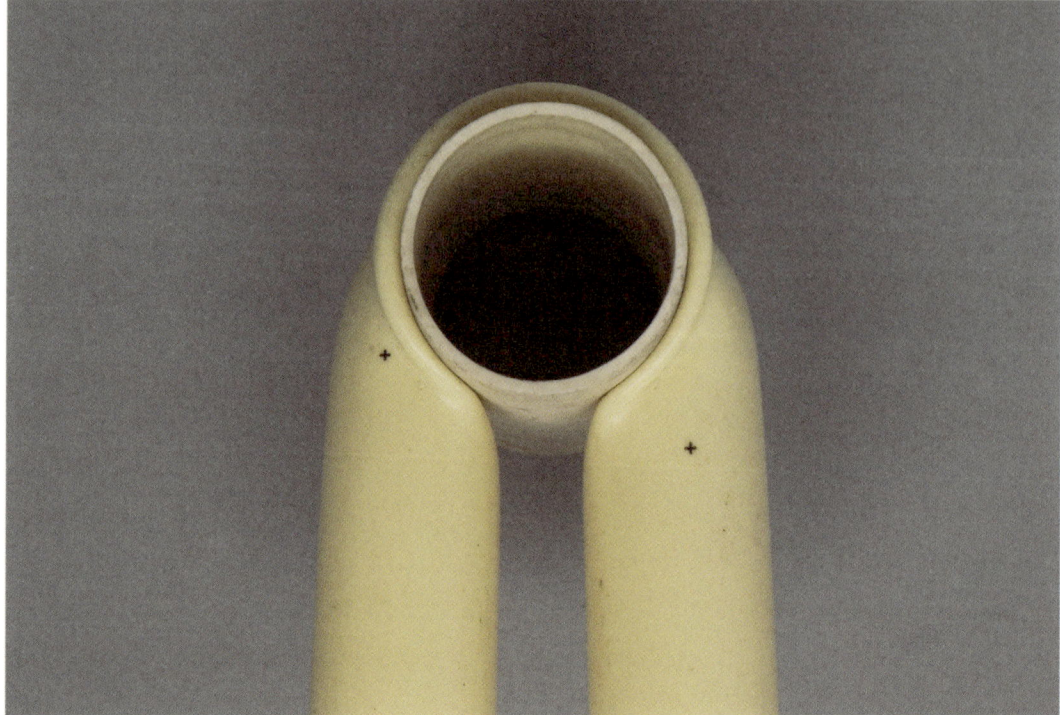

inner circle of wheel (Animaris Circodentis),
upper side of crank (Animaris Modularius)

clock's ticking frequency can vary from one second to an hour. Its ticks can be counted using a counter comparable to the pedometer. In the future the beach animal's reflex packages will swell. Entire clusters of nerve cells will populate the animal's back. It will be an explosion of nerve cells, the way this has happened with the human spine during the past 100,000 years.

Soap, DNA and SM
In the 1970s a number of ladies in my circle of acquaintances were up in arms against the systematic rape in duck populations. Not that they did anything to stop it. They just felt that something of this kind shouldn't be tolerated. But to take to their boats and coerce the drakes into an acceptable form of foreplay with oars or sticks was going too far, in their opinion. Instead, they brought up the subject of the suffering female ducks at birthday parties, which was better than nothing. I should point out that lovemaking between ducks is anything but subtle. Every spring it's one hell of a racket in the water near my house. Sometimes a female is had at by two males at once. The harmony in nature is often hard to equate with the injustice done. During the mating season for deer, the vast majority of males are the losers. They have to battle it out with other males in an ungentlemanly brawl. They clobber their opponent, get clobbered themselves, suffer humiliation and eventually take to their heels. The losers watch balefully from a distance as the hinds blindly follow the victors into the woods. I must confess I have half a mind to chase those victors with a stick, give them what for, and that way provide equal opportunities for us males. A hind apiece, now that strikes me as a fair deal. The way we humans do it. That's the best way. One to one. Complete rubbish of course; if there's injustice anywhere, it's in human relationships. Handsome folks are clearly at an advantage. Most of us don't fall for your ugly types. I realize that beauty is subjective, but there are many cases around of absolute ugliness or absolute beauty. You have ugly couples and handsome couples. Partners enjoy equal footing on the rungs of beauty. You hardly ever see beautiful and ugly combined. When it does occur, you find that wealth, a sense of humour, the capacity to listen, an air of congeniality or virtuosity on some musical instrument tends to compensate for the ugliness. If you have no such capacities and you're ugly to boot, you might as well give up. Injustice first reared its ugly head a long way back in evolution with the division into two sexes. Before then, there was no arguable difference between male and

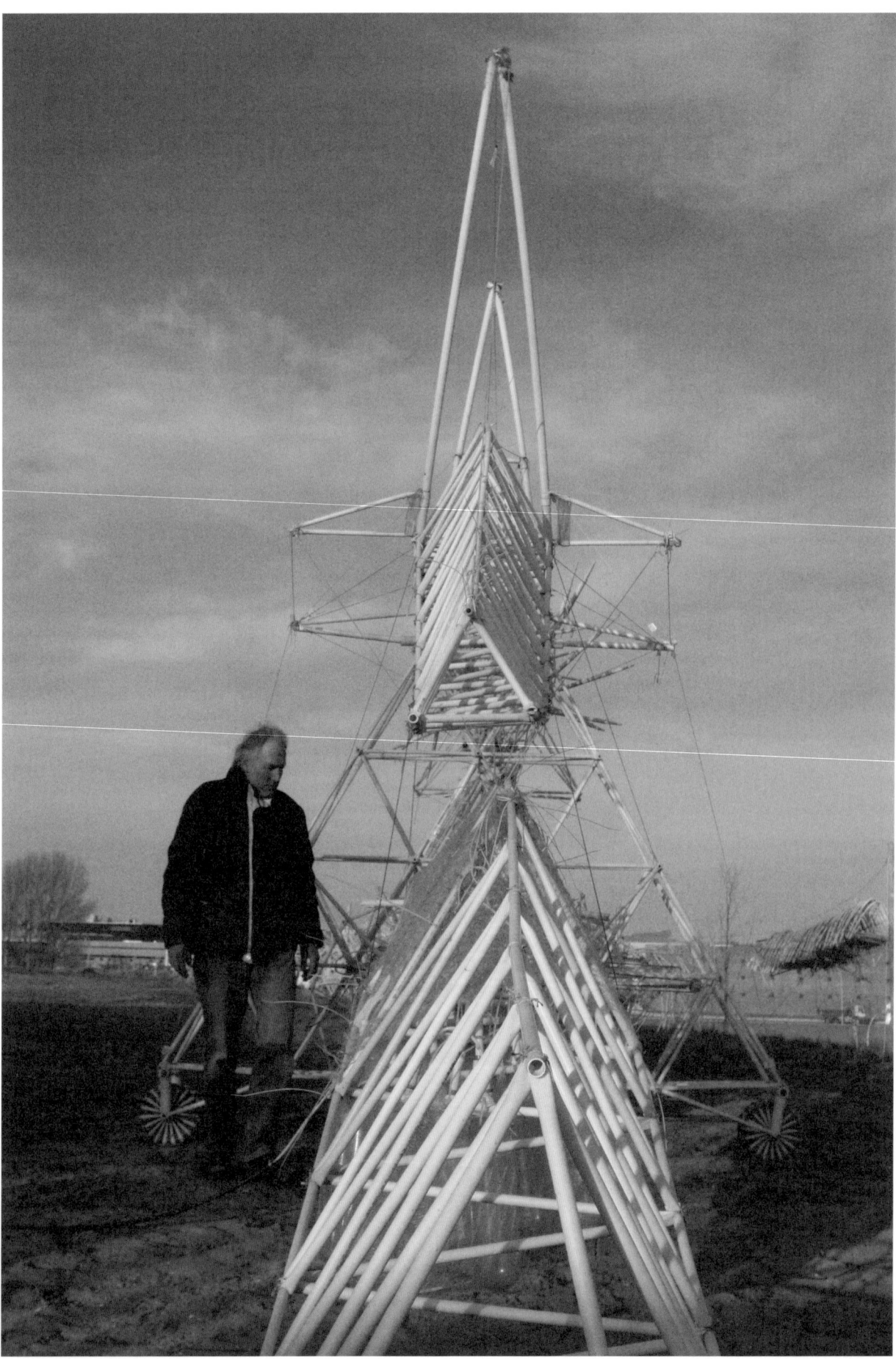

forked boom on Animaris Excelsus

female. None between sperm and egg. There are life-forms today where this still applies, hermaphroditic species of mould that are able to fertilize one another. That division into sexes must have occurred way back along the evolutionary trail. The one branch specialized in transporting the DNA and the other in developing it into young, or breeding. As breeding was a time-consuming business, the breeders concentrated on a single thing, the egg. The transporters by contrast were into quantity. Women look for quality in the DNA, and have to be selective with incoming sperm. Men are only interested in quantity, and splash their sperm about liberally. And yet men and women have exactly the same number of descendants. This contradiction is the cause of a complex game of deceit, violence, power, cooperation, give and take. And of course rape. That rape was an accepted means of procreation among our distant ancestors can be deduced from the fact that men are bigger and stronger than women. On the one hand men had to fight each other, which was a means of selection in itself, and on the other they had to be capable of subjugating their better half. Fertilization could only be achieved in a lopsided balance of power. Today, rape is a rare form of fertilization among humans. Maybe the number of rapes decreased when we began walking upright. The female reproductive organs became more concealed. Rape became more difficult. Men had to find other ways of getting their DNA delivered to the right address. They had to convince women to let themselves be fertilized of their own free will. Women developed reflexes that made them fall for men. Or rather, their reactions to sensory stimuli were decisive in choosing the man who was to lay the DNA on them. Selection, which originally had nothing to do with the woman, had passed on to her. She herself did the choosing now. The word 'herself' shouldn't be taken too seriously here. Self didn't exist yet. The self was a bunch of brains with connections. Perception by the senses evoked, via that so-called self, a response from the limbs. So the woman herself didn't choose. Her brains did it for her. This was an important stage in evolution. Until then, it had been the surroundings, the circumstances, the mechanics (strength, muscle power) that had done the selecting and moulded animals into shape in accordance with the principles of evolution. For example, the surroundings gave fish their shape. This was such that they glided through the water almost without trying. Dolphins took on more or less the same shape independently. A question of aquadynamics. Porcupines grew spines so that their enemies got pricked. A question of mechanics. Each point has a tiny surface area and applies great

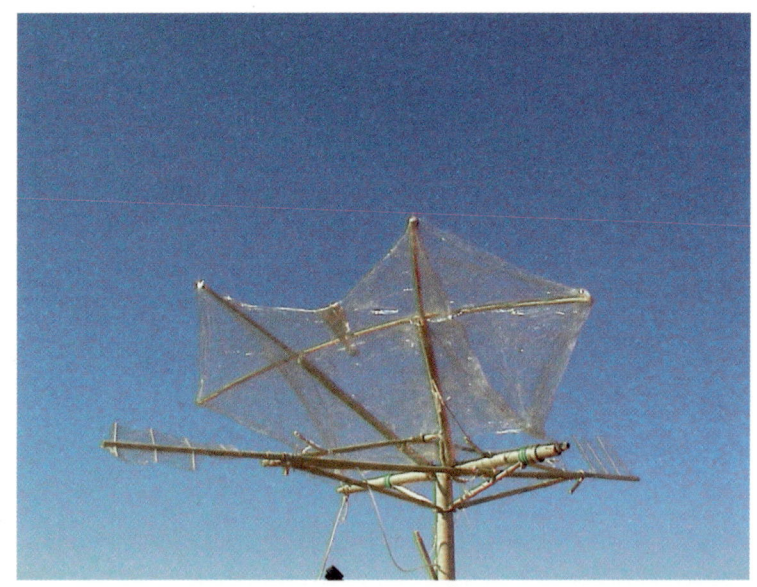

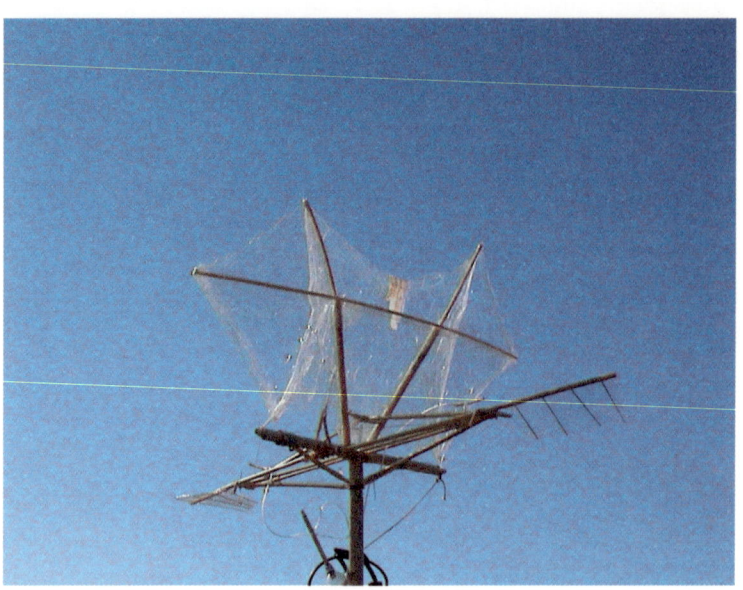

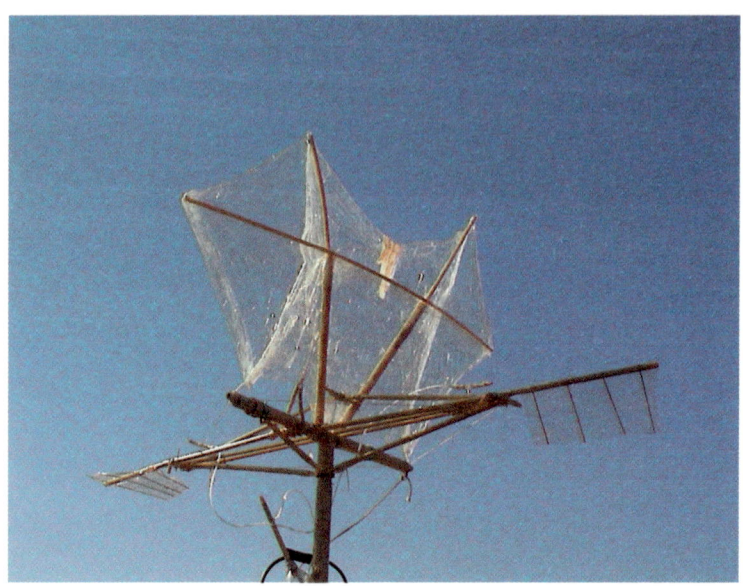

sails atop Animaris Currens Vaporis

pressure, too great for the skin of an adversary. It was the surroundings that had given animals their form. From the moment the brains started doing the selecting, the species itself did the creating. Peacocks created their seductive tails. Pigeons created their exorbitant crops. In mechanical terms, these attributes were superfluous. The peacock's tail is troublesome when its owner runs. The pigeon's crop is troublesome when its owner flies. These two random examples happen to have nothing to do with function. In contrast, the selection system of women developed functional reflexes. Women became susceptible to the hunting achievements of the male or his capacity to protect the offspring. Selection became more refined and at the same time more complex. So much has happened in the meantime. The game of the sexes has become exceedingly complicated, thanks still to the old opposition between quality and quantity. Men and women share offspring though they are programmed in opposing ways. This discrepancy is the motive force behind the intrigues with which life is so liberally sprinkled. Such DNA intrigues comes across most clearly in soap operas on TV. You see males acting as if they were attentive family men while they seek their pleasures elsewhere. They splash their sperm about. The wife is not pleased at this, since there is a chance of her husband going off with another woman. Then she'd be left alone, which would not benefit the quality of her offspring. There are also women who dip into some extramarital DNA and then deceive their husband into thinking that his DNA is responsible for the ensuing offspring. She's combining the DNA of the one with the well-meaning attentiveness of the other. Favourable in an evolutionary perspective perhaps but rather risky; if hubby finds out, he's off. Which is why attentiveness is still a quality that has an effect on women. Men use their attentiveness as a weapon, women their capacity to seduce. A soap opera is a chain of events almost exclusively related to sexual selection. I don't mean that everything can be reduced to sex, but certainly to reproduction. If a soap opera is fun to watch it's because the events are familiar, as though we've been through them ourselves. They evoke reflexes, revive memories of our forebears. Does the thrill induced by a scantily clad or nude female body derive from a memory of times when we spent the whole day without clothes? Can this be compared with an odour that brings back memories of our childhood, a déjà vu sensation? When we undress, we are returning to the good old days of Adam and Eve, of wandering round in the buff. It's something I certainly don't do. Besides the fact that clothing keeps out the cold, it also protects against incoming sperm.

Impossible!

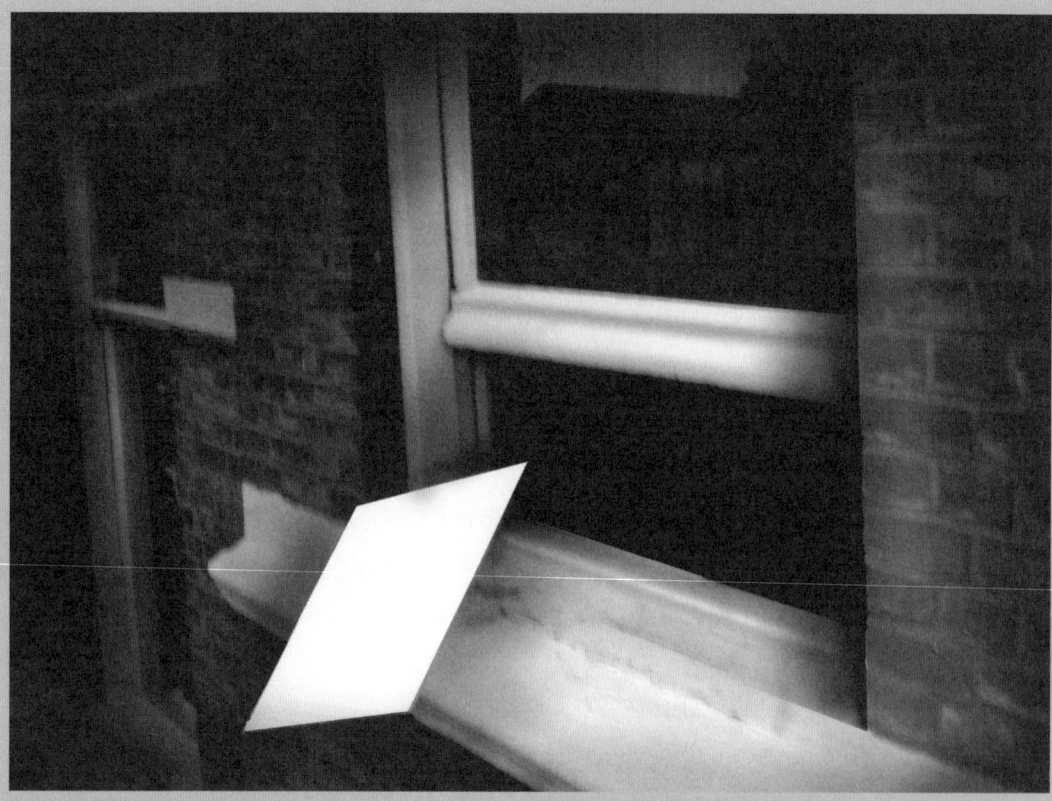

For some time now I have been living in a house twenty metres away from a post box. What a blessing. I don't have far to walk to post a letter. More to the point, I don't have to walk at all; I just throw the post out of the window. Remarkably, nine out of ten letters reach their destination. Miracles do happen, at least in ninety per cent of cases. All the same, this Saturday will be a nerve-racking time for me. Once a fortnight, I have something important to post. This is a large white envelope containing a drawing and an article for *de Volkskrant*. 'Off you go. See you on Saturday, I hope,' I thought before tossing the envelope out the window. Now you must be familiar with Escher's ingenious lithographs. You see a stair endlessly leading upwards, bolt-upright pillars that seem to stand on the ceiling as well as the floor. In the picture two points of different depth are joined in a seemingly logical fashion. But the image jars, it looks all wrong; you could never construct Escher's buildings in reality. I've tried drawing something similar, only replacing the spatial dimension with that of time. It is time, not space, that is in conflict with itself. You can see this drawing here. It shows the instant the envelope is being thrown out of the window. What's in the envelope? A short text and a drawing. What drawing? The drawing shown here. Impossible, I say! That envelope may contain absolutely anything except the drawing you see here. After all, you can't take a photograph of an envelope containing the photograph you're taking. So the envelope in the drawing could never contain the drawing itself, assuming it contains anything at all. There's a conflict in time. The great thing about such a conflict is that you can also do it in language. There's an example in this article. Some lines earlier in it you can read '"Off you go. See you on Saturday, I hope," I thought before tossing the envelope out the window'. This passage was already in the envelope the moment it went through my mind. There's no way you can throw an envelope out of the window and commit your thoughts to the paper in it, speeding out of the window as it is. A conflict in time then. Other examples: I'm peeling the potatoes, kneading the dough, opening a bottle of beer. Impossible, unless I'm able to open a bottle of beer with one hand and write that I'm opening a bottle of beer with the other. In every sentence where actions are described in the present tense and when the personal pronoun 'I' is the subject, time is in conflict with itself. There is one exception: I am writing.

Taking off your clothes can signify the onset of fertilization. That sadomasochistic thrill of watching films in which people get tied up must come from somewhere. Stuff like this must have been common practice in the age of rape. Those were the circumstances in which our ancestors produced the next generation. Along with the DNA they passed on the reflex of excitement that occurred in just such an asymmetrical balance of power. That excitement has percolated over the centuries into the heads of today's SM enthusiasts. When all's said and done, we all originated from those few thousand people that lived in prehistoric times. Their perilous adventures have fragmented and the splinters showered over us. History is in our heads, as the memories of events that happened before our time. Such events are acted out in films. You might say that it is not just memories that have been reproduced through us but events as well. They inhabit our brains as if living things. Fragments of surroundings with their reflexes live in our brains. If we want to know how a situation will unfold in reality, we first let the reflexes in our brains do their stuff. A few thousand years ago we entered into a new phase, although the old mechanisms are still active. The onset of language and writing has placed us at an even further remove from the mechanical world. We now have reflexes whose usefulness can no longer be ascribed to procreation. The peacock's tail might have been clumsy and useless in the mechanical sense but he could use it to seduce the ladies and so it did serve a purpose in the reproduction stakes. We are transcending the earth with its unfair mechanics.

In and out
For me the Dutch verb *uiten*, which means to express or utter but also to emit, has a claustrophobic ring to it. It says that there is something inside and this has to be thrown out (*uit*). *Uiten* is not the same as *uitscheiden*, to void. We discussed voidance in the chapter on the Tepideem. By voidance is usually meant material output, faeces, urine, sweat, offspring too. But there is also a non-material output. The body is perpetually giving off signals. For the voiding of signals we use the term expression. Screaming is a form of expression. Talking is the ultimate form. Dancing is a form of expression too. Sometimes material voidance combines with the voiding of signals, such as odours relating to fear or sex. These are tiny particles that are voided, they are chemical signals. Call it a material voidance with a non-material message. Territories are in many cases

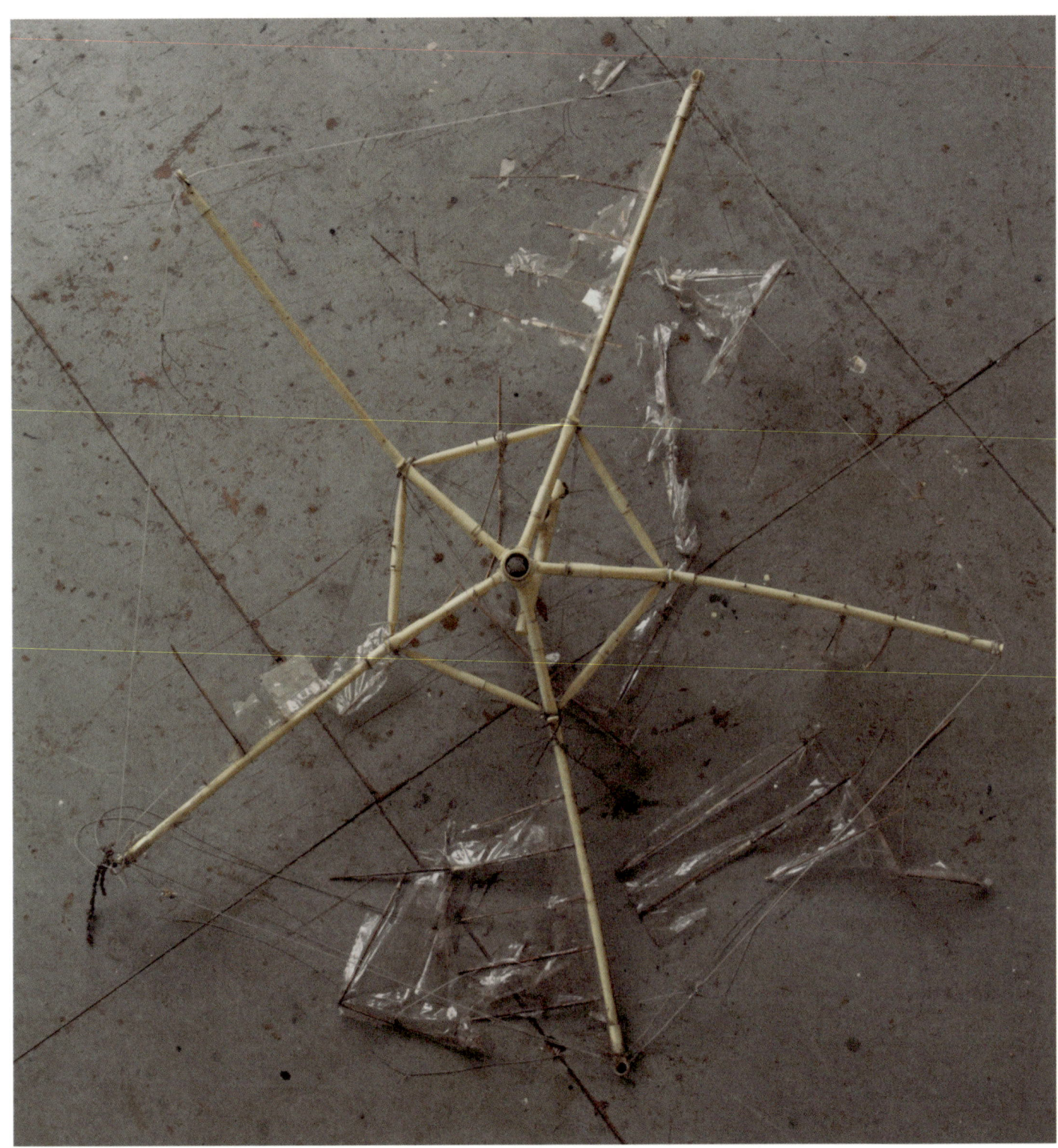

fossil of propeller from Animaris Rigide

marked with urine. So messages can be wrapped in matter, but by and large they are wrapped in physical phenomena such as sound and light. Talking and screaming obviously fall under auditory expression. Dancing is rather more difficult to place under physical phenomena. And yet dancing to my mind can be categorized under 'light'. The dancing or courting animal takes up the light of its surroundings and processes it so that it impacts on the visual perception of another animal in a particular way. In displays of courtship, one animal uses light to manipulate the visual perception of another. It sends a signal. The verb *innen* has to do with money and means to cash. Odd, you'd think that it would have the opposite meaning of *uiten*, in other words to receive signals. Receiving signals is of course done through the senses. Let's call these in-organs. So there are out-organs and in-organs. Out-organs include vocal cords, limbs and perspiratory glands. In-organs consist of the senses. The point I'm making is this. The body is a transformer on different fronts. It originally had the form of a shell with a way in and a way out. In went the food at the front and out came the waste matter at the back. Between those two stages, the food got digested. In principle the body still works the same way. Only the shell now has bits sticking out; a head, arms, legs. Further along the evolutionary path the body seems to have taken on the shell-form once more, this time at a non-material level, once again with an entrance and an exit. Here, the collected information gets turned into forms of expression. This, too, is a kind of digestion. The in-signals need processing into out-signals. Now this can be done very simply using reflexes. A reflex is the act of processing an in-signal into an out-signal. Doctor hits knee with hammer, lower leg shoots up. The digesting of signals has become a little more complicated over the past 100,000 years. Now you have virtual bits sticking out of the shell. You could regard thinking as the complex processing of an in-signal into an out-signal. Thinking is a chain of countless reflexes.

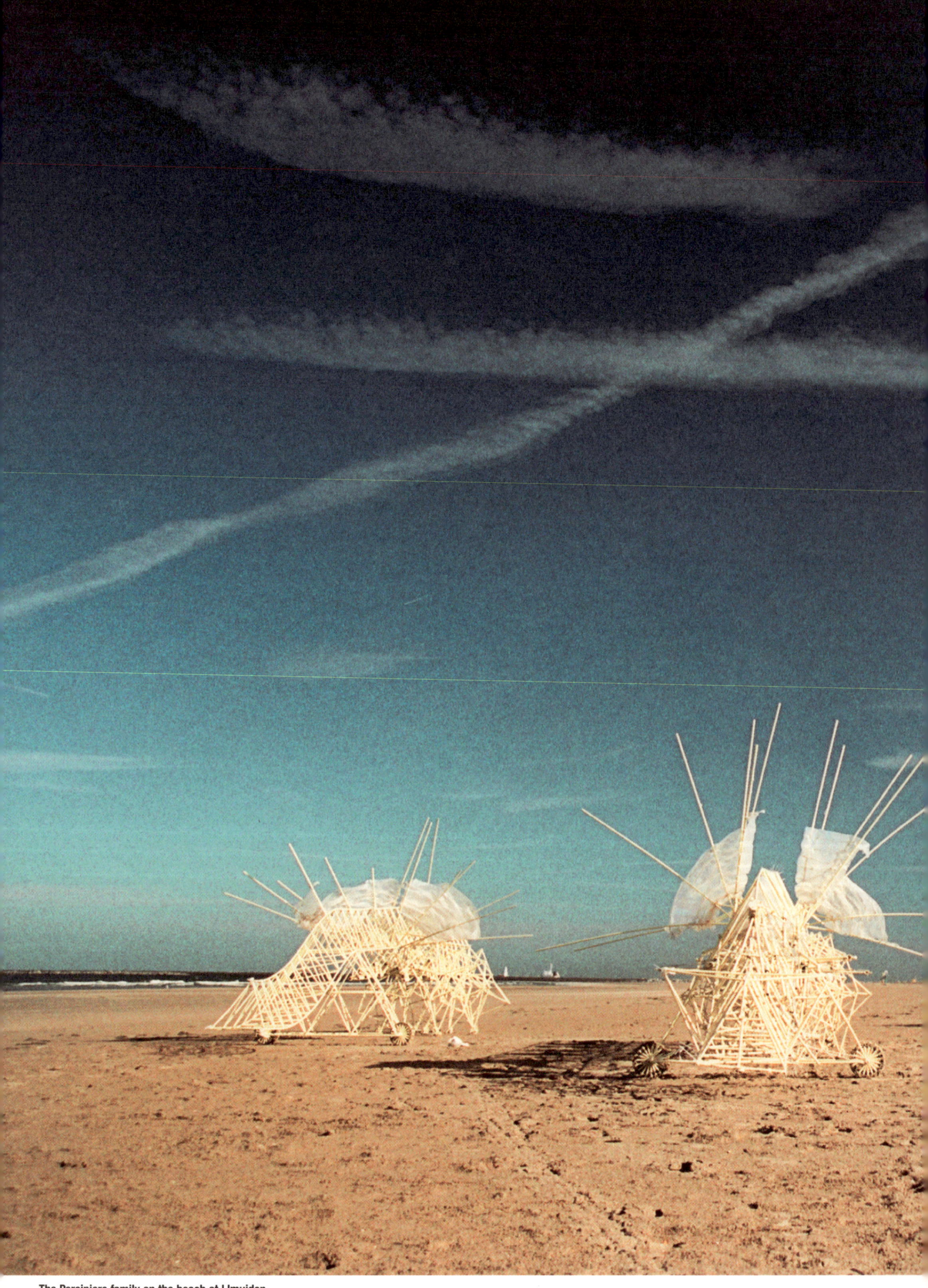

The Percipiere family on the beach at IJmuiden

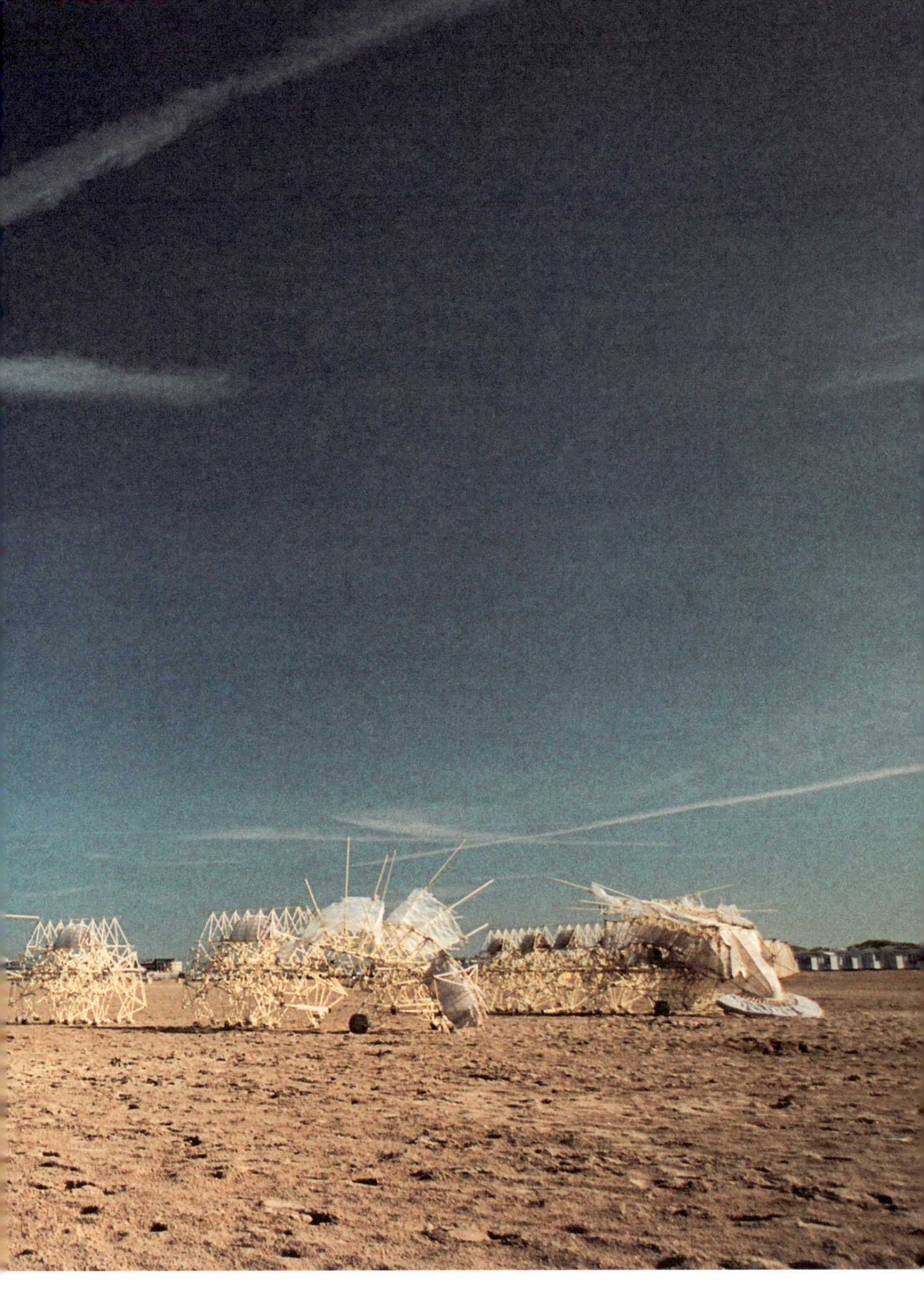

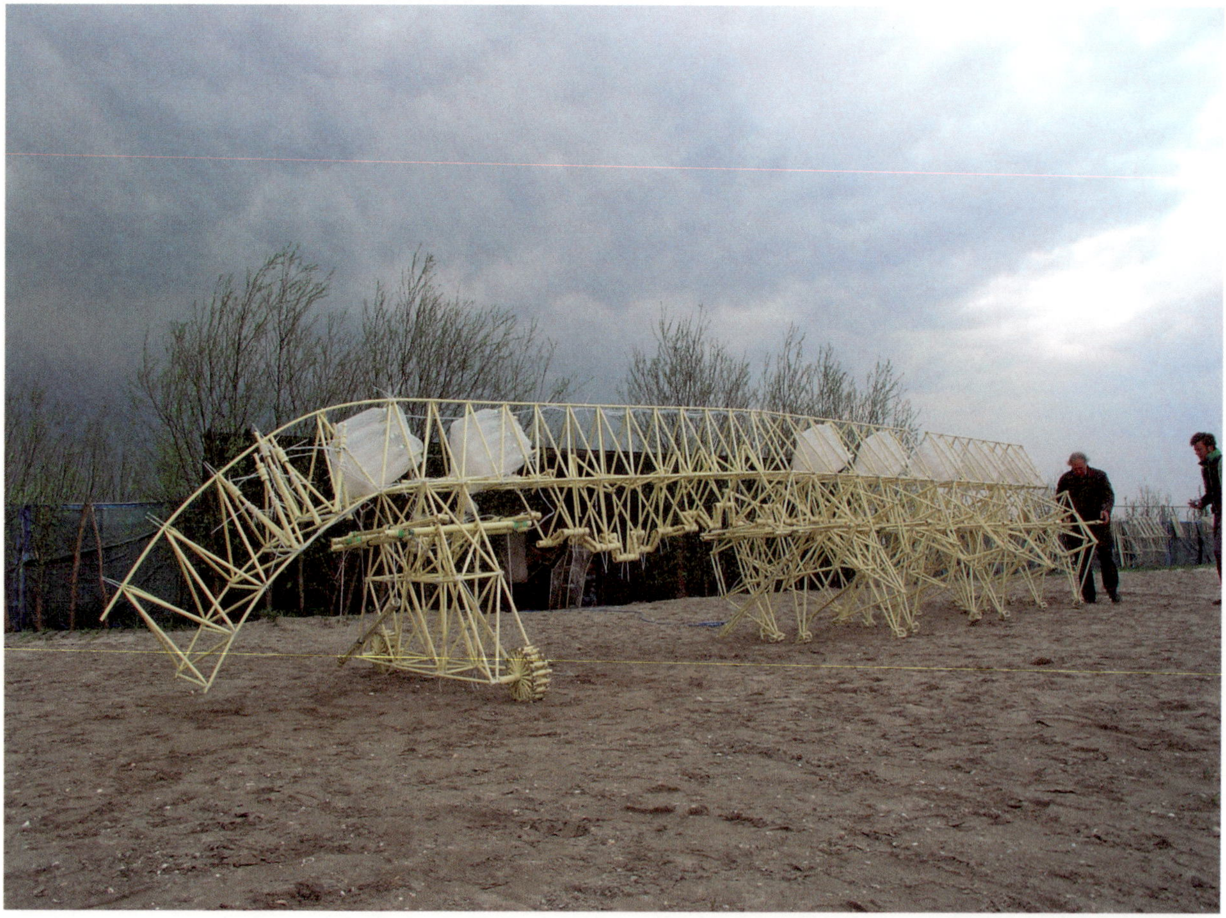

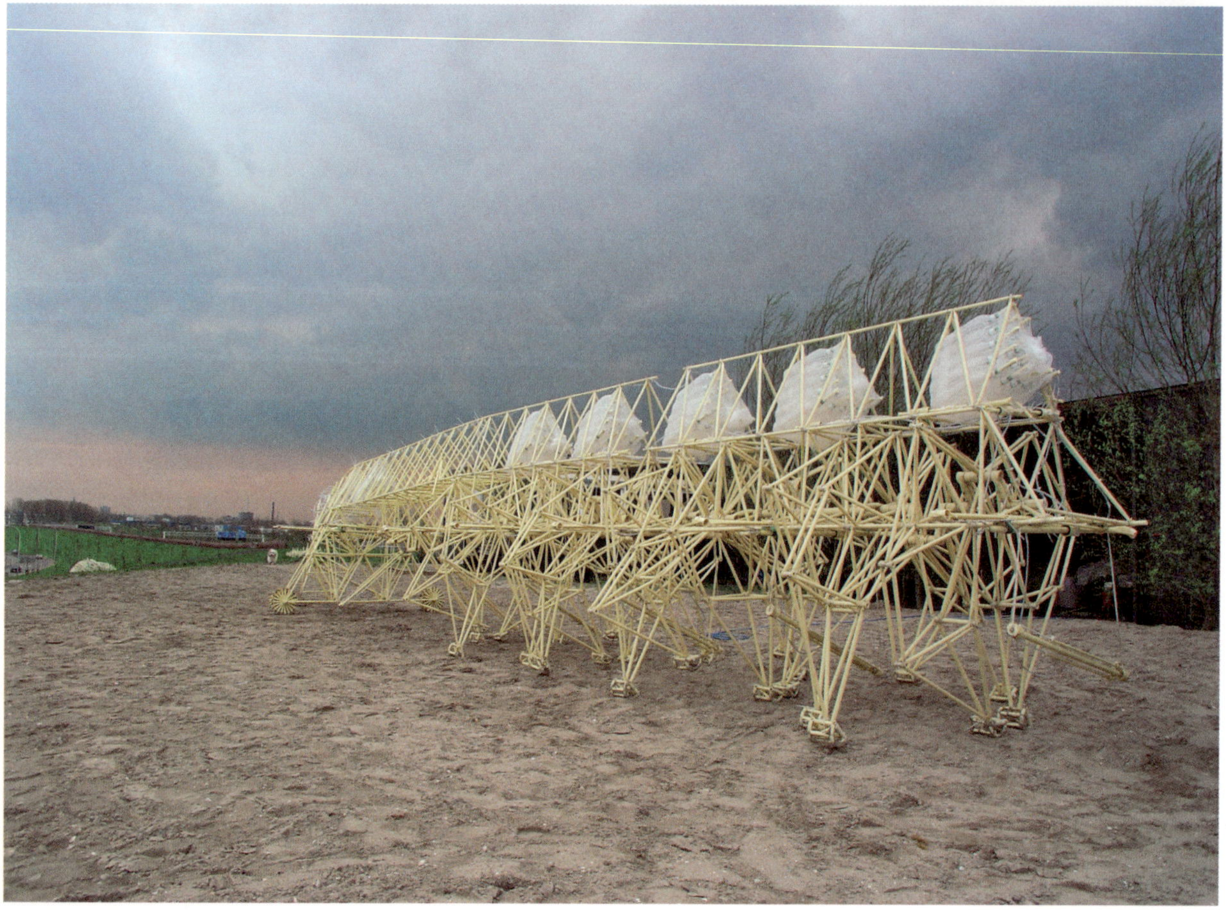

Animaris Percipiere Secundus

Between one shell and another

A sensory perception can also evoke responses within the body. For example, something that happens may cause an adrenaline rise. I call that a reflex too, for the sake of convenience. Nor is it only perceptions by the senses that can elicit reactions within the body. It can just as easily be a thought. You could regard a thought as an event in the mind. For example, a thought can give you goose bumps or wind you up sexually. Every one of us possesses thousands of such reflexes. Every human interaction is the accumulation of toing and froing of reflexes among those involved. It is an intricate board game of attack and counterattack. Each move observes the rules of the reflexes, much like the rules in a game of chess. The difference being that everyone has their own rules as every individual has their own reflex patterns. That makes every one of us unique in the world. A large share of one's reflexes is inherited: like father, like son. We are blessed with reflexes that are millions of years old. There has been a major shift since then. You can get by better with some reflexes than with others. Selection and reproduction have caused our reflexes to develop. The events have simply kept on repeating, though minor changes cause them to develop along with the reflexes. You could see this as a co-evolution of the reflex and the event that goes with it.

Signals to the world

In the animal kingdom, a signal in most cases is intended for another animal. An obvious example is the mating call: 'Let's reproduce!' But there are also forms of expression which are not intended for another animal. Building a nest is making a change in the world as it is, but it isn't a signal. Yet I would still call it a form of expression. It's a change in the world and the bird's beak is the cause. Just as dancing is manipulating another's visual perception, to build a nest is to manipulate the world. The immediate environment is manipulated to the extent that it is experienced differently thereafter. The nest is nice and warm and keeps its occupants dry when it rains. The expression of building the nest brings a change in the data arriving via the senses: nice and warm, nice and dry. Building a nest is more a signal to its builder, a reflected signal. It's a signal to the environment and this returns a signal of warmth. Devouring another animal, besides being the greatest possible pain signal to the prey, is also a signal to the devourer: its stomach has stopped rumbling. So the body once again proves to be a trans-

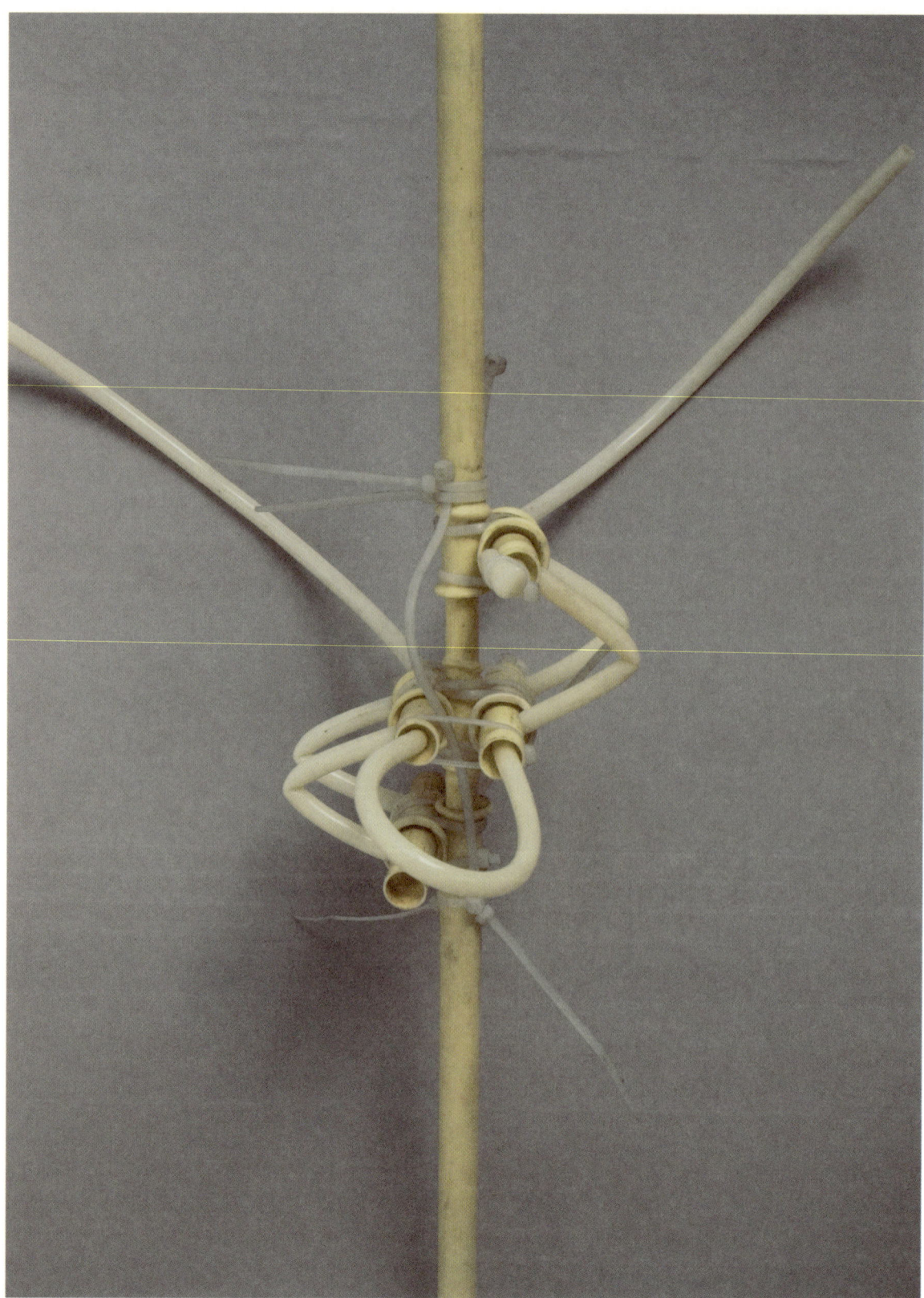

tight-bending nerve cell

former. It expresses itself and gets back signals as a result, which in turn give rise to a new act of expression.

Organs of expression (**output**): mouth, perspiratory glands, limbs
Senses (**input**): ears, eyes, nerve cells, nose, tongue

The physical environment works the same way as the body. It processes the out-signals and makes new signals from them. The environment is also a transformer. The body is made in the image and likeness of its surroundings. From dust we come and to dust we shall return. It's the same with the houses of Siena in Italy, whose reddish-brown colour comes from the ferruginous earth around those parts. Of course this was for practical reasons, as that earth was simply the nearest to hand. Our bodies are similarly assembled from stuff around us: hydrogen, carbon, nitrogen and oxygen. But besides our bodies being made of materials from the physical environment, we are like it in a non-material sense too. We process signals in accordance with the same principle. The world processes signals in the same way we do. How can an environment process a signal? you ask. I'll try to explain. Push against a mirror with your hand and it pushes back. Or rather, the hand of your mirror image pushes back. The moment your hand approaches the mirror, another hand appears in it that starts pushing against yours. You can feel that hand pushing. The power it uses to push back is equal to the power you yourself exert in pushing. Which is why the mirror stays where it is. If the mirror were to be a bare wall, the same thing would happen. From this we can conclude that in all matter there is an invisible hand that pushes back the moment you push against it. Action and reaction is the most primitive reflex in mechanics. I used to practise judo when I was young. Two opponents stand face to face and pull or push at each other. In judo the thing is not to check or counter your opponent's pushing and pulling but to reinforce it. You have to go with it. This is no easy task for us mortals. To every action a reaction is deeply ingrained in our nature. You always push back. Objects act just like animals and give a reaction. Another term for this is cause and effect. A sheet of paper in time turns brown from its exposure to the light. You can also say that the paper perceives the light and reacts by changing colour, the way lettuce leaves turn green as a result of light (the innermost leaves are yellow). The lettuce leaf perceives the light and reacts by changing colour. The relation between

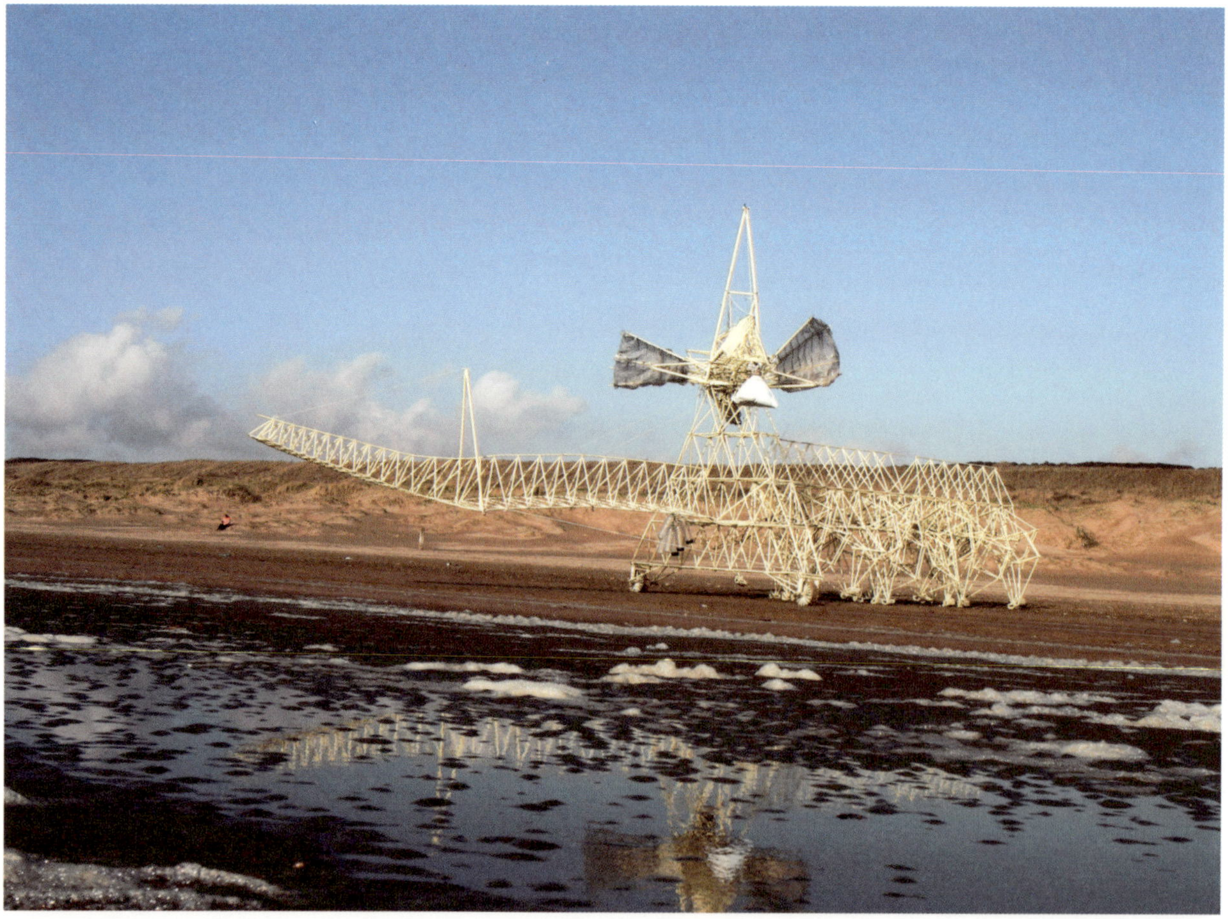

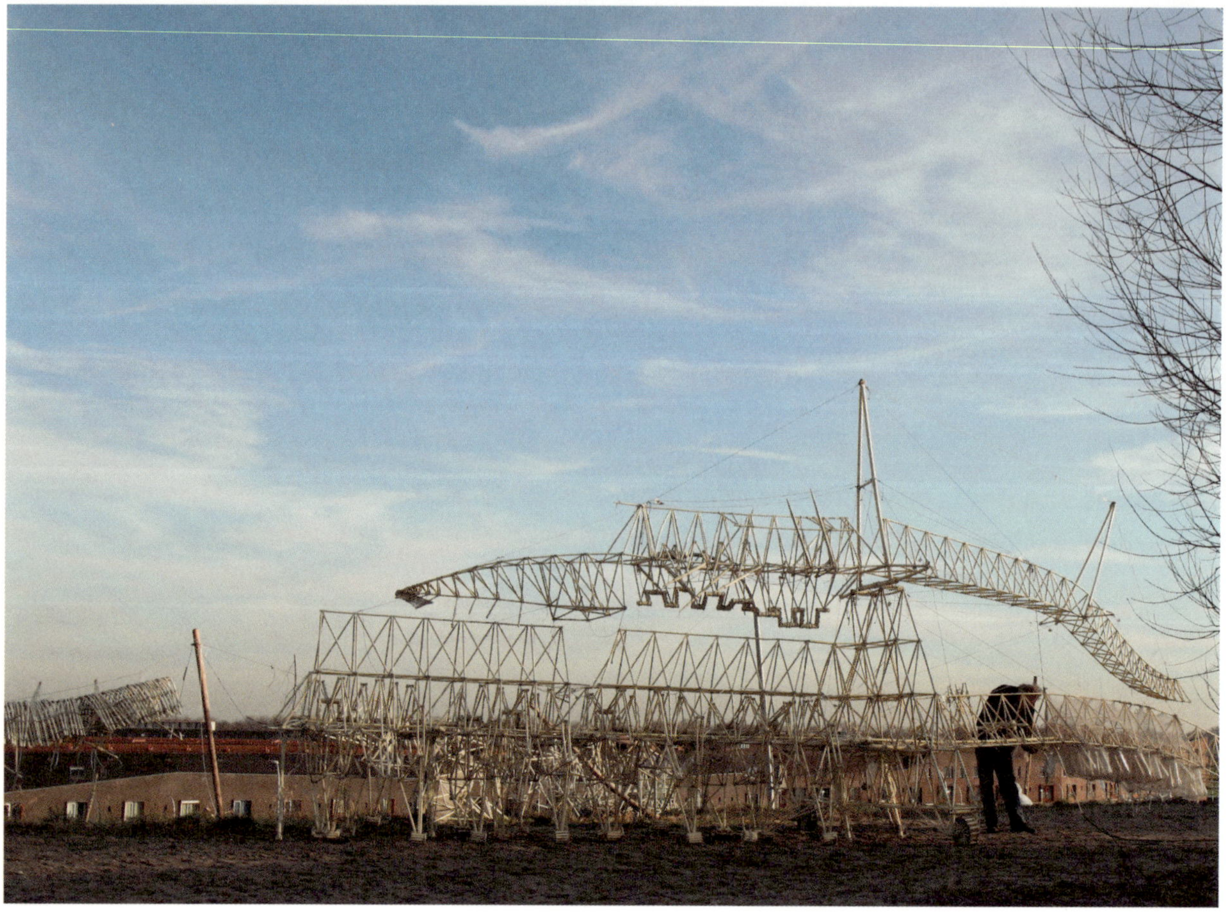

Animaris Excelsus

stimulus and reaction may be more complicated at times. A chameleon changes colour when danger threatens. This makes the chameleon a complicated object. During the course of evolution, objects have become more complicated. Between action and reaction lies a whole string of intervening stages. Our eyes must have begun as a single cell. Like a sheet of paper, this cell gave a particular reaction to light. This reaction generated a reaction from a nearby cell. This reacted in turn and ultimately the entire organism reacted. That's how I see it happening. Since every population admits to a degree of diversity, different examples will react differently to light. The reflexes with the greatest chances of survival will nestle in the genes of subsequent generations. Both the perception system (the eye) and the response system (the brain) became more complex. The eye is sensitive not just to light but also to a combination of light particles striking the eye to form an image. The eye is sensitive to dance. An image perceived by the eye can also signify danger. Possible reactions to such images include running away fast (gazelle), freezing as if dead (stick insect), rolling into a ball (hedgehog), creeping into a hole (guinea pig) and acting aggressively (dog). These are all responses that have advantages relating to the build of the animal in question. Statistically, that is. The perceived image can evoke different reflexes, such as fear or excitement. The most telling example of all is the reaction of the stiff penis. A nude or scantily clad woman looms into view and hey presto, an erection. A dry sponge does something similar. It responds to fluid. It gets bigger when it absorbs fluid. Of course I know that there's more to an erection than that, but still I'd like to class the two phenomena together. Reflex sounds similar to reflection. It is the reflection of signals. In the case of the body we call it a reflex, and in the case of the environment we call it an event. An event is the transition between cause and effect.

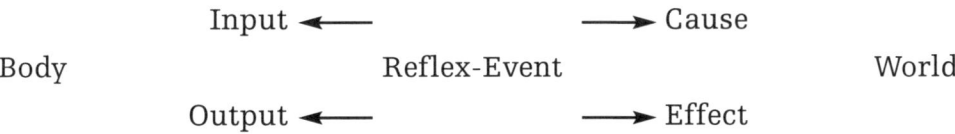

Let's take an example of an event: a man and a woman meet at evening classes. The are attracted to each other, live together for a year, have a child and a few months later the man meets someone else. Nothing new, these things happen. You could regard an event such as this as being assembled from a number of sub-events. Each sub-event in turn is assembled from sub-sub-events. The birth

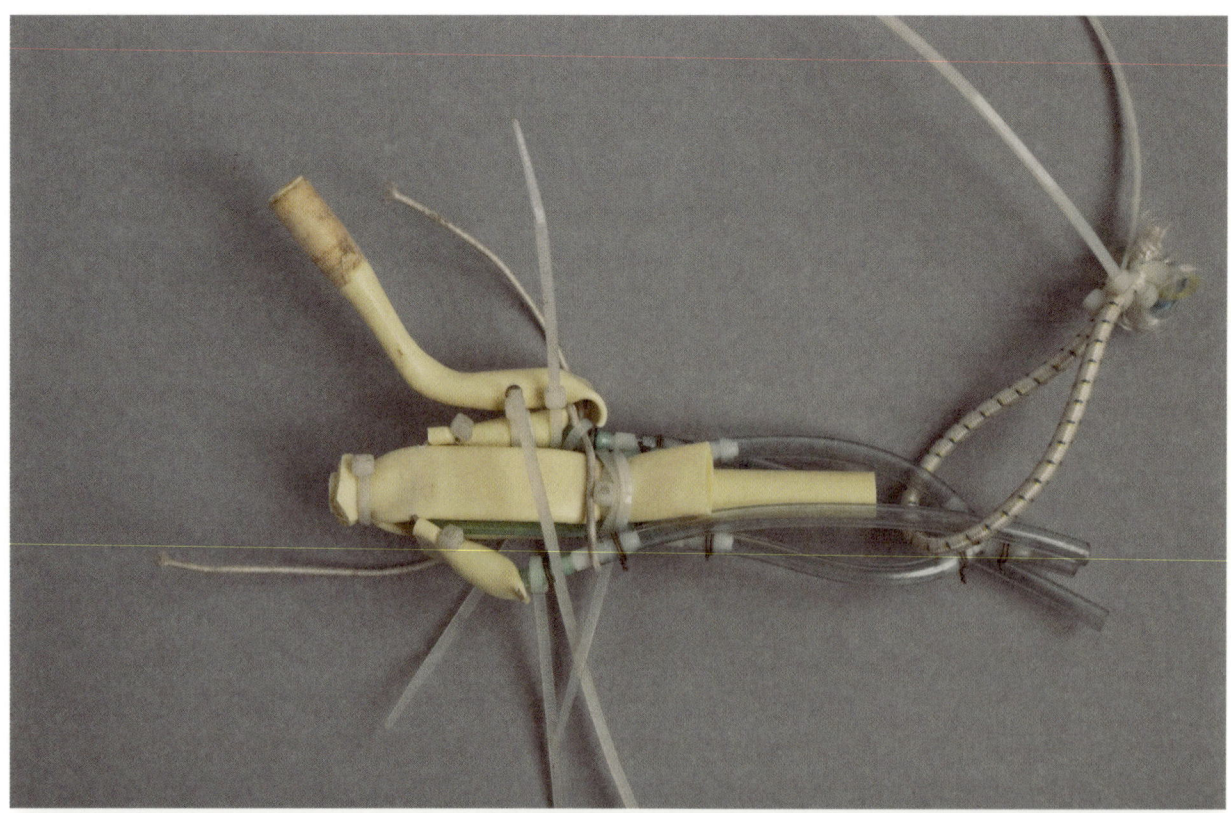
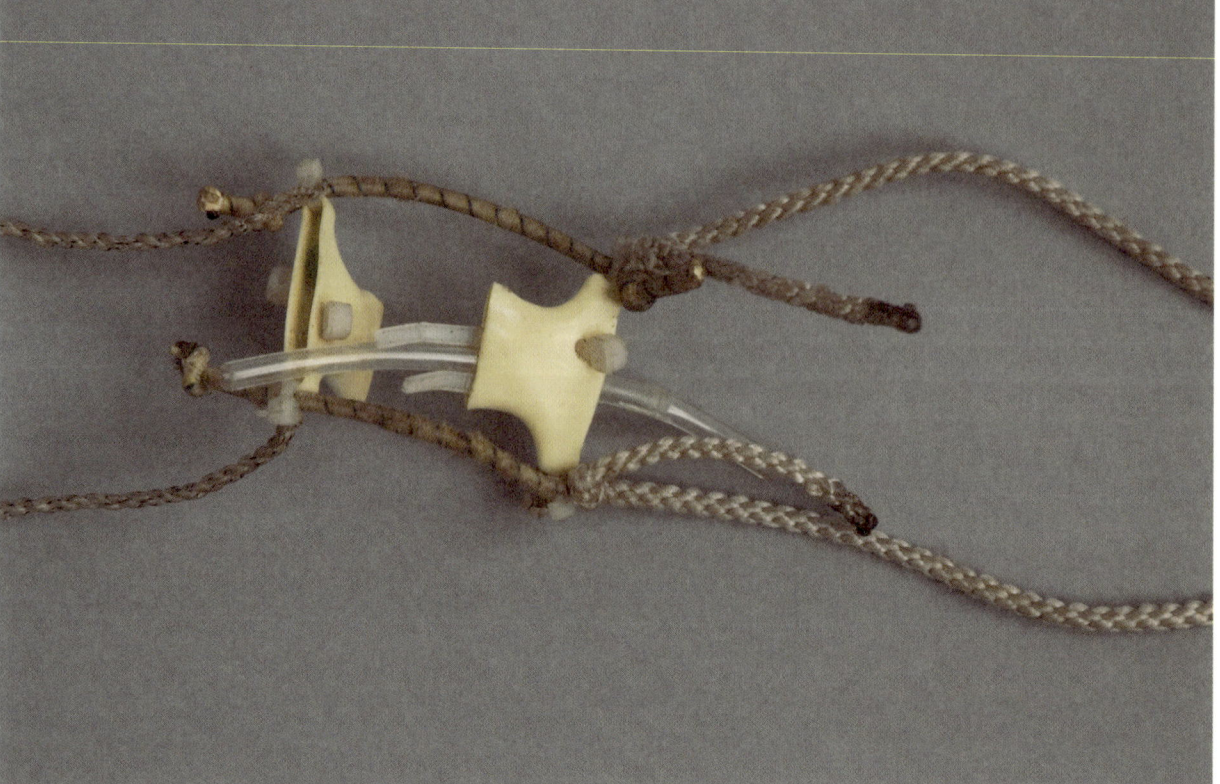

Animaris Percipiere Primus, bending mechanisms

of the child is one such sub-event. The discovery that it's a boy is a sub-sub-event. The smallest building block in a history such as the above is an event that is just big enough to be perceived. Such a building block triggers a reflex, a response in or of the body. That reflex itself influences the event in turn, so that another event takes place that again triggers a reflex, and so on. Events and reflexes ping-pong back and forth. Physical environments are like animals in a way. Their way of processing signals is analogous to ours. Bodies and environments exist by virtue of each other and react to one another. Unlike our bodies, environments are interlinked with no clear boundaries between them. They lie like a patchwork quilt over the earth's surface. Each environment consists of sub-environments. A sub-environment in turn consists of sub-sub-environments. This is comparable to our bodies, which consist of cells. Each part of the body can be regarded as a sub-body, each cell as a sub-sub-body. Let us now treat physical environments along exactly the same lines as bodies. In these terms, reflexes are the same as events and causes the same as sensory perceptions. Effects are the same as forms of expression. Think of the image we have in our minds of our physical environment as the offspring of that environment. It's a copy, a coded copy. The environment lives in our heads, where it shows more or less the same reflexes as the real environment. It reproduces itself in our heads as different scenarios. Scenarios are also like animals in a way. They are films, dreams. These dreams live in an environment we call the brain. Each and every one is played through, projected. Some die, others reproduce themselves as new dreams. And we think it's just us reproducing in the environment, when the reverse is equally true. Physical environments reproduce in our heads. They are codes that reproduce. They reproduce just like genes. Some codes die out and some survive. The dreams that survive are converted into actions. The bottom line is that the codes are translated into reality. They are converted into deeds. The physical environment is influenced by the acts of man. You might say that the environment reproduces itself as a new environment through the codes in our brains. It's not only genes using our bodies to multiply, environments do that too though their 'codes of things'. Cars use our bodies to multiply. Some, such as the five-door variety, multiply more readily than others. Cars, computers, houses, forks, computer viruses, lettuces, sheep, letters from missionaries, beach animals – they are all using us, believe me.

A tinful of screws from heaven

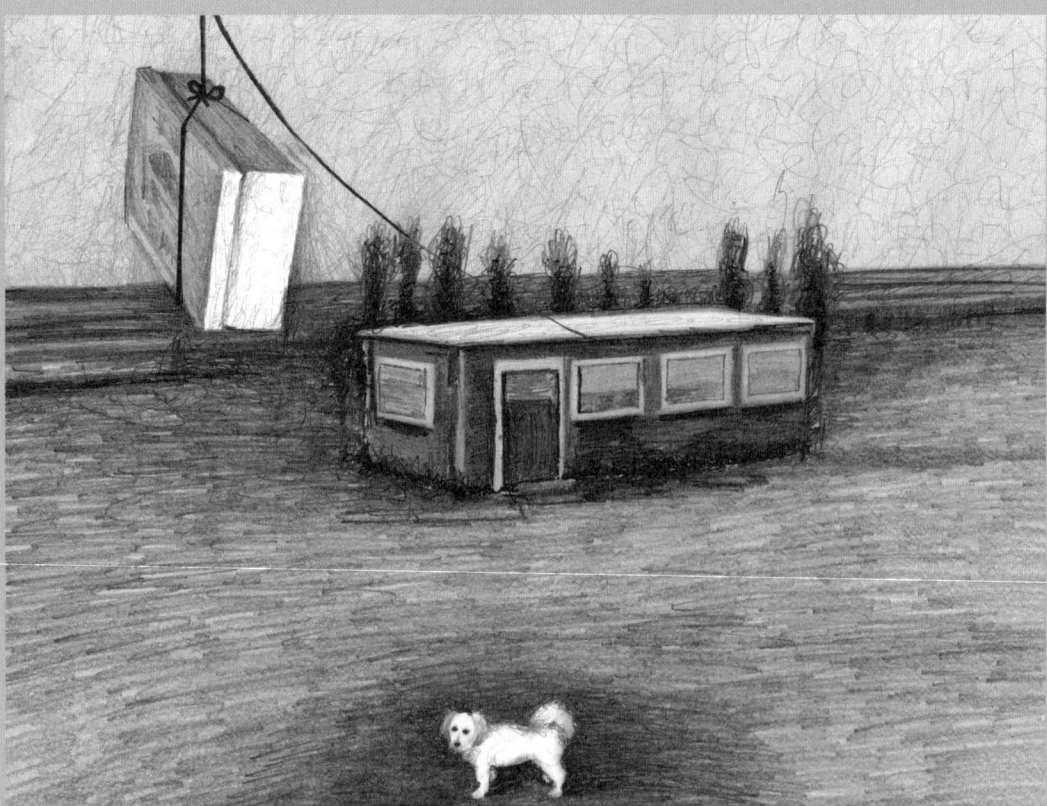

My place of work is nothing to write home about. It's a portocabin next to the motorway. Cold and damp in winter, sweltering in summer. Thistles, stinging nettles, daisies, poppies, it's all growing there. And there's enough life along the motorway: dragonflies, spiders, butterflies, moles, rabbits, hares and strange creatures I have yet to find in the encyclopedia. One of these creatures is my dog. A little mongrel with the friendliest eyes. As I go about my business in the portocabin, he goes about his outside. His business being poking around in the bushes. He keeps an eye on everything. Don't think you can approach unnoticed. He can smell you before he sees you.

Well, I've noticed that my dog believes in God. For years now, a cigar tin full of dodgy-looking screws has been lying in the corner of the cabin on one of the shelves. The tin has tape round it. How it got there I don't know. I never use it, that tin. It's just been lying there. Now, finally, it looks as though it may prove useful, as if all that time it had been waiting for this moment.

My dog can be quite industrious at times. He can bark for a full quarter of an hour at cranes chugging in the distance or at a falcon hovering above us. You can shout at him to stop, but does he? It's only my master, you can hear him thinking, a man of flesh and blood with no supernatural gifts. But then suddenly a clattering tin drops down next to him, as if cast down from heaven by a reproachful God. You can see him wondering in astonishment where on earth that tin came from. He shuts up immediately. I feel he shouldn't notice that I was the one who threw it. It wouldn't impress him any more. I would like to sustain his belief in God, as kings and potentates used to do. Let's keep the people in line with a god. I've suspended the tin on a string leading into the cabin from a pole standing outside. I only have to give the string a pull and it's quiet outside. A little anonymous rattling does the trick.

Supernatural powers inspire awe. This is their great strength. As there is no earthly link between cause and effect, there is no earthly defence against them. Such powers put us in an impossible position. Silence, obedience in other words, the way my dog reacts, is the only solution. Kneeling, begging for forgiveness, doing penance, spending hours on a church pew – you do it because it's at least worth a try.

God's punishments kept the nation in its place for centuries. There was no defence against rattlings from above. It's different now. We humans claim to understand the Creation. For dogs it's still a fairy tale.

Invention of the 'I'

Once I was a regular visitor to Rotterdam Zoo. They had this beaver, a marine beaver or snow beaver or something, in a small pool. His plaything was a ball with a rope attached to it. It was just him swimming around with that ball. Whenever I was there, the beaver played with it the whole time. Though it looked more to me like he was giving the ball a good screw. He was at it all day. It was pitiful to watch. He had no better way to illustrate his desire for a female than this pointless ball game. An acquaintance of mine has a budgerigar. He regularly lets it out of its cage. And what do male budgies do when they're sitting on your head? They imagine your head is a female budgie and start going at it like the clappers. It's not just we humans who are able to simulate a partner, animals can too.

The capacity to pretend is not to be scoffed at. Being able to fantasize is one of mankind's major talents. You might almost think that the evolutionary development of our imagination parallels that of our hands. The motor skills of the human hand far exceed that of an animal's foot. Humans are much more adept at cranking their fantasies into action. You can safely assume that the mind's eye of humans sees the contours of a partner that isn't there much more clearly than the mind's eye of animals. Not just that, the image goes beyond the partner to include the world around us. The perceptions of our senses enter as packages of information and from that limited quantity of data we construct our image of the world. We think we live in the real world but that's simply not true. We live in our image of the world. We dream our lives. The dream is fed by the senses on one hand and by an internal source of thoughts and ideas on the other. In that dream we can imagine a partner, even one that doesn't exist. There is someone else we can imagine. There's a very special someone circulating in that dream, the ego or 'I' figure. We see it circulating there. Self-awareness could be said to come from observing the experiences of the I-figure in our imagination.

We earthly creatures are cursed with a certain measure of egotism. There are even evolutionists, among them Richard Dawkins, who contend that all actions can be reduced to advancing the multiplication of our genes. 'Own genes first' seems to be the slogan of the software in every creature. How does this work in practice? Prior to every selfish decision there is a calculation. First, the possible scenarios are acted out in the dream world, much like a simulation in a computer. To see what happens. No aspect is left unconsidered. In the end you choose

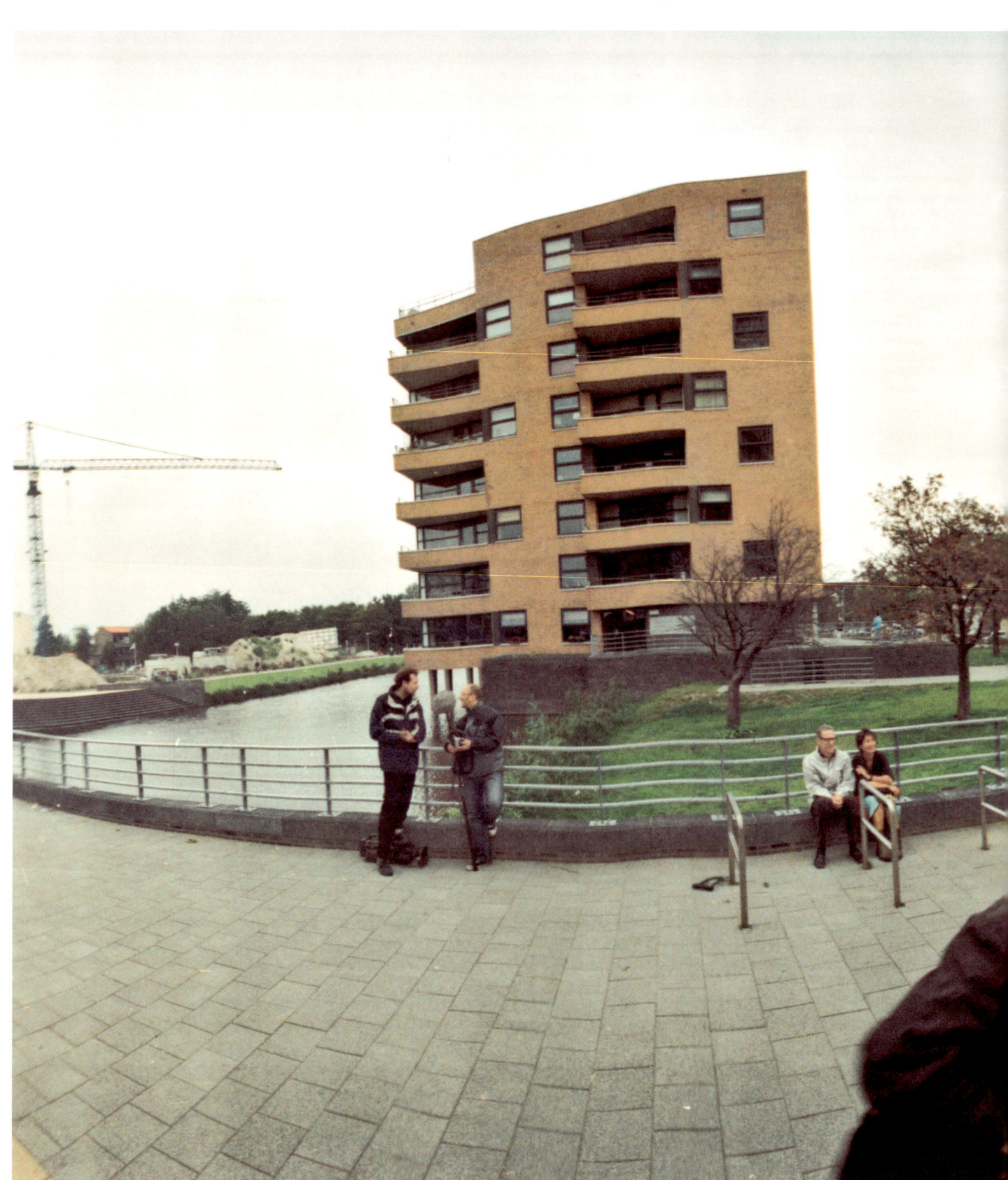
Animaris Rhinoceros Transport in Amsterdam

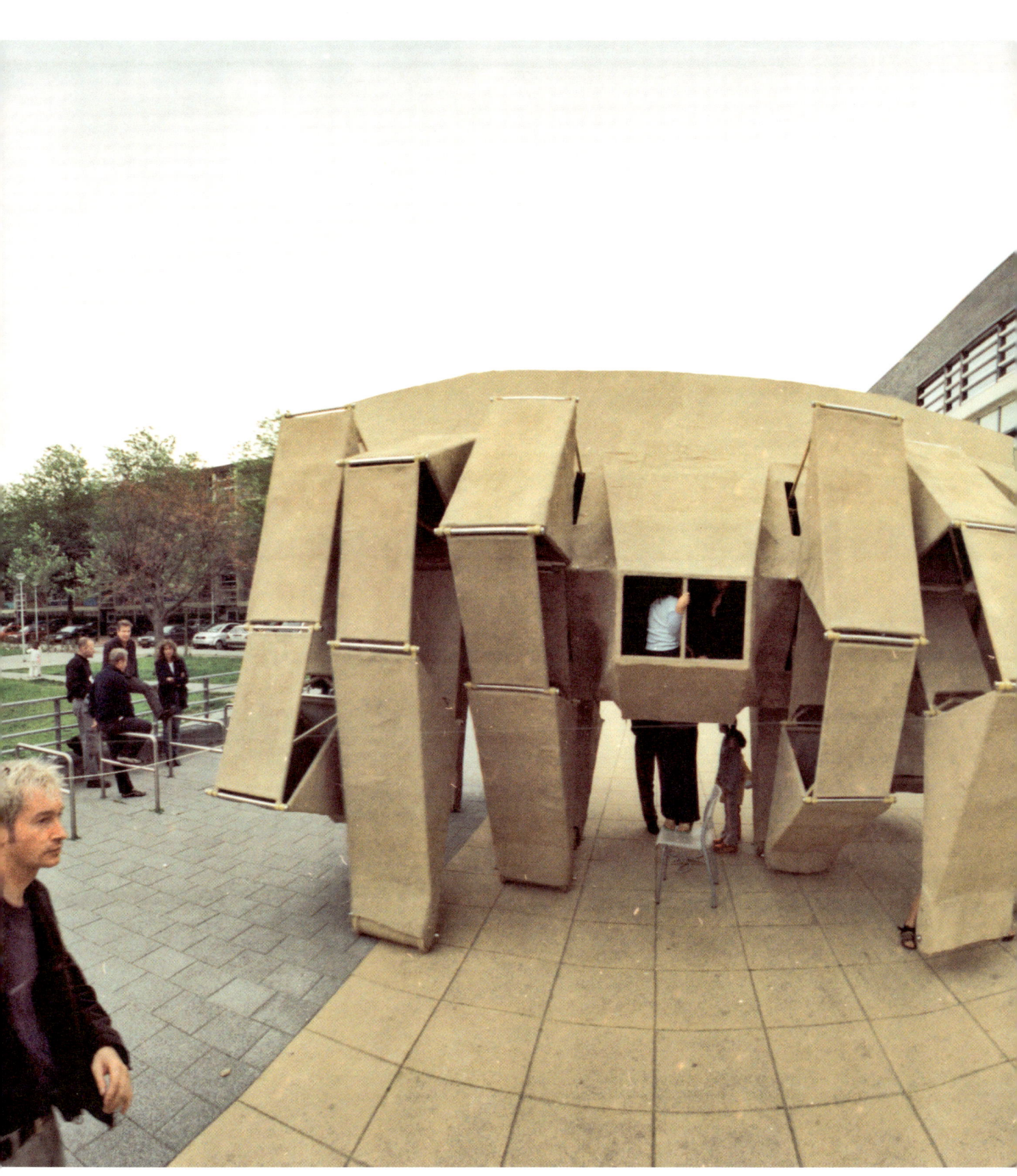

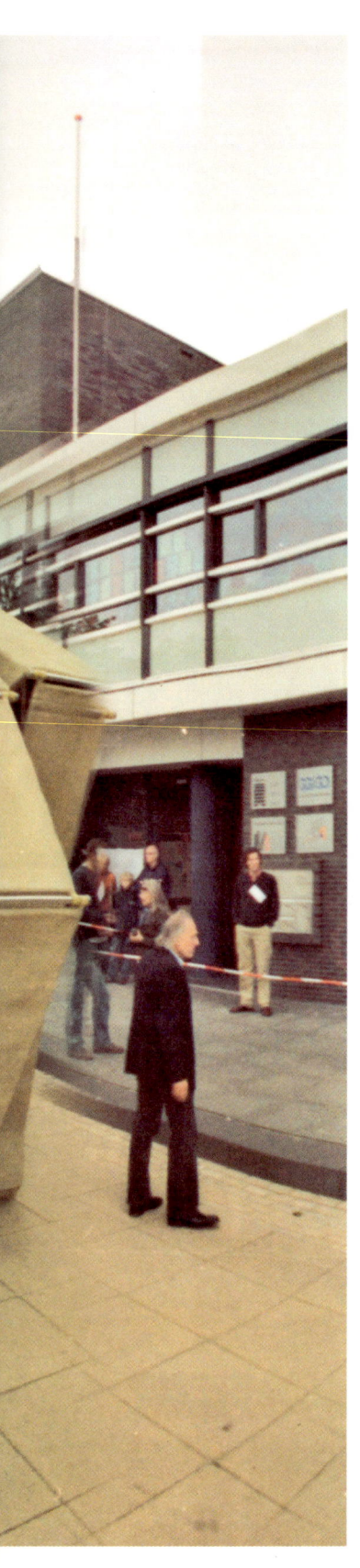

the scenario that most benefits the I-figure. That scenario is then transformed into deeds. The I-fantasy is a brilliant discovery made by evolution. Self-awareness is a tool for giving shape to the selfishness of our genes. The clearer the contours of the 'I' in the fantasy, the greater the chances of survival. Belief in the existence of the 'I' has assumed disproportionate forms. We are convinced we exist, and we exist because of it.

The Great Pretender
Crocodile Dundee – you know him, the wisecracking hero from the Australian bush who ends up in New York – now he's an all-round survivor. He gets by not only in the bush Tarzan-style but also in the snob-infested circles of the New York elite. He's strong, a good fighter and, above all else, cunning. We men identify with him because of his ability to survive wherever he is. The Crocodile Dundee feeling is deeply ingrained in our genes. Heroes of the silver screen appeal to us, since they draw on our instinct for survival. Wherever women go looking for survival genes men try to give the impression that they possess them. What is remarkable about Crocodile Dundee is that he survives in the bush *and* in the city. I can't see the humourless Tarzan doing that. After roaming the streets of New York for a fortnight he'd get run over by a car or succumb to pneumonia. On the whole, primitive peoples are unable to make it in the city. Aboriginals live in poverty in Sidney, Indians and Eskimos in Montreal. You need particular skills to get by in an urban society. There, 'survival of the fittest' takes on a special meaning. That word 'fittest' is the source of a deep-seated misunderstanding, particularly in countries where English isn't the main language. We Dutch immediately think of fit as in strong; 'only the strongest survive'. But Darwin's conception of fittest was as in fitting, the best adapted. Darwinist survival is not the same as physical strength. A stick insect isn't strong, but when you're a tiny creature living in a tree it's best if you resemble a twig. A lion by contrast is big and strong and so it gets by fairly well in the dry grasslands of Africa. But it wouldn't last an hour in the streets of Amsterdam. The urban landscape is above all a social landscape. A lion has no sense of humour and no vocabulary to speak of; all it can do is roar. Survival in the city requires the ability to communicate. Enemies, predators, occupy that social landscape too. These come in the shape of depression and apathy. It's easy to fall prey to depression in today's society. This depression-predator polishes off 1500 of our species each year in

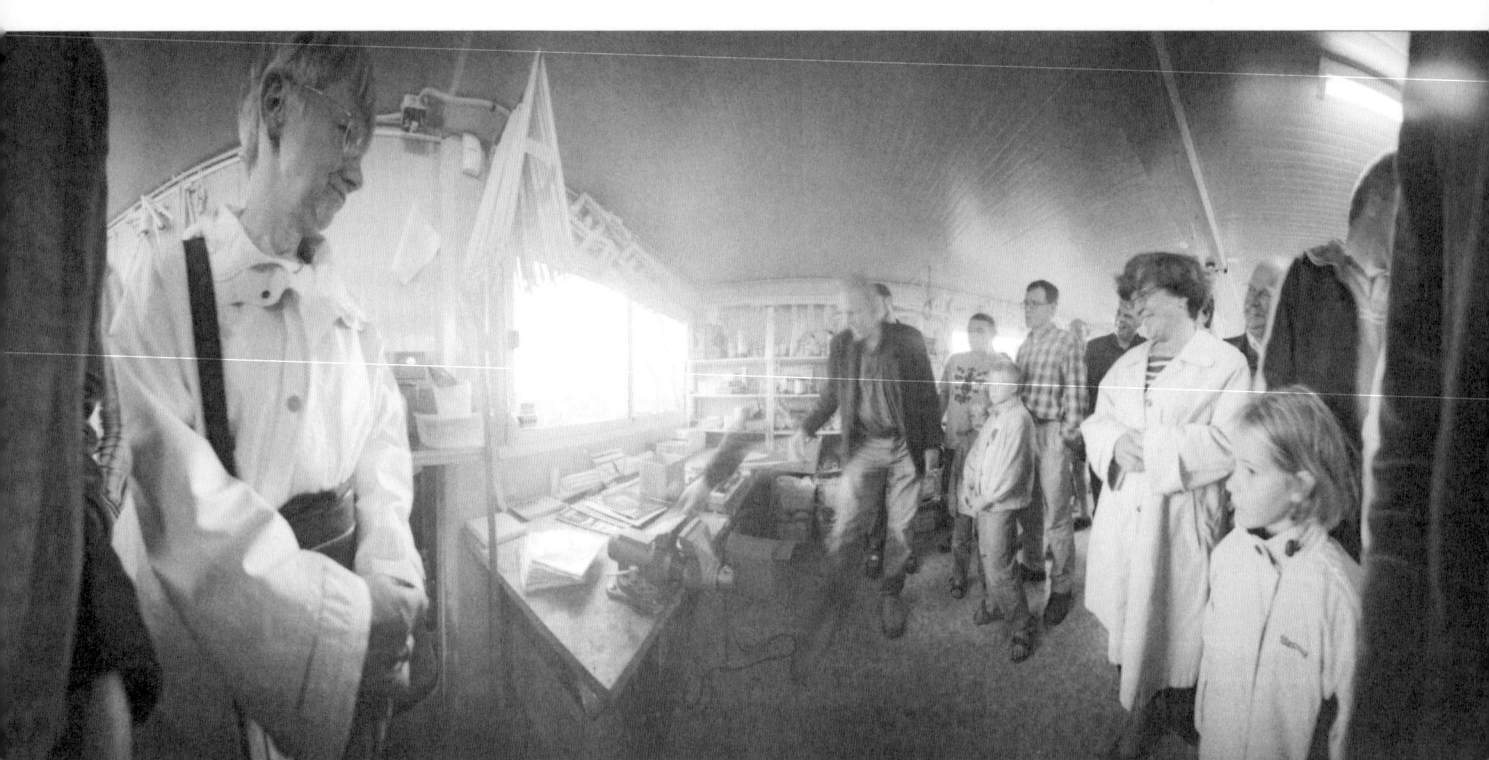
visitors in the cabin

the Netherlands. Cars are predators too: 800 victims a year. We respectable types, not too deep, well brought-up and used to the social landscape, are pretty good at surviving in the city. We wouldn't survive in the bush. Every landscape has its own rules. Survival is a lovely word. It doesn't just mean that you don't die but that you look beyond your own life and reproduce. You refuse to die, as you want to give yourself the time to produce offspring. The longer you live, the greater the chance. Not that it ends when your offspring arrives. That's just the start. Then you have to see that it's trained – brought up, that is – so as to be able to survive as an adult in the social landscape and produce offspring itself. At the end of the day, it's down to reproduction. It's not just about having kids. It's about having kids that will be able to have kids themselves. This cycle is a big deal for man and beast alike. It costs an incredible amount of energy. During the past several thousand years we have been working on a new mode of reproducing. The rock paintings in the caves of Lascaux were the first visible expression of this. Such art simulated a world, cattle, deer. By making marks on the rock, our ancestors were not only reworking that material but also reworking the visual perception of their fellows. That way, images from the brain of one person could be transferred to that of another. This act of transplantation saw the image copied at the same time. To copy is to reproduce. It is the reproduction of thoughts, ideas, music, art, science, technology, amusement, type, written words, spoken words, emails, gestures, body language and sex, lots of sex, a jungle of sex, virtual sex – in short, data, masses of data. We have arrived in the data jungle. This means of reproduction has been growing at a furious pace in the past thirty years. Data breeds faster than rabbits. You are now reading the word rabbits, but you're not the only one. A further thousand people are reading the word rabbits somewhere. The word has reproduced itself faster than real rabbits could ever do. Richard Dawkins is the inventor of the word memes. Memes are something like genes. They are units of cultural information. Instead of such things as building construction plans, these units might, for example, contain the data of a tune. Take the melody of the song 'The Great Pretender'. You know the one: 'Oh yes, I'm the great... pre-te-hen-der'. A chart smash from the fifties. That melody is in a lot of heads. You are better equipped in the social landscape if you concentrate on disseminating and multiplying memes rather than on attaining immortality by multiplying genes. The last-named option is very energy-consuming and proceeds slowly. It's not just genes that have been using our

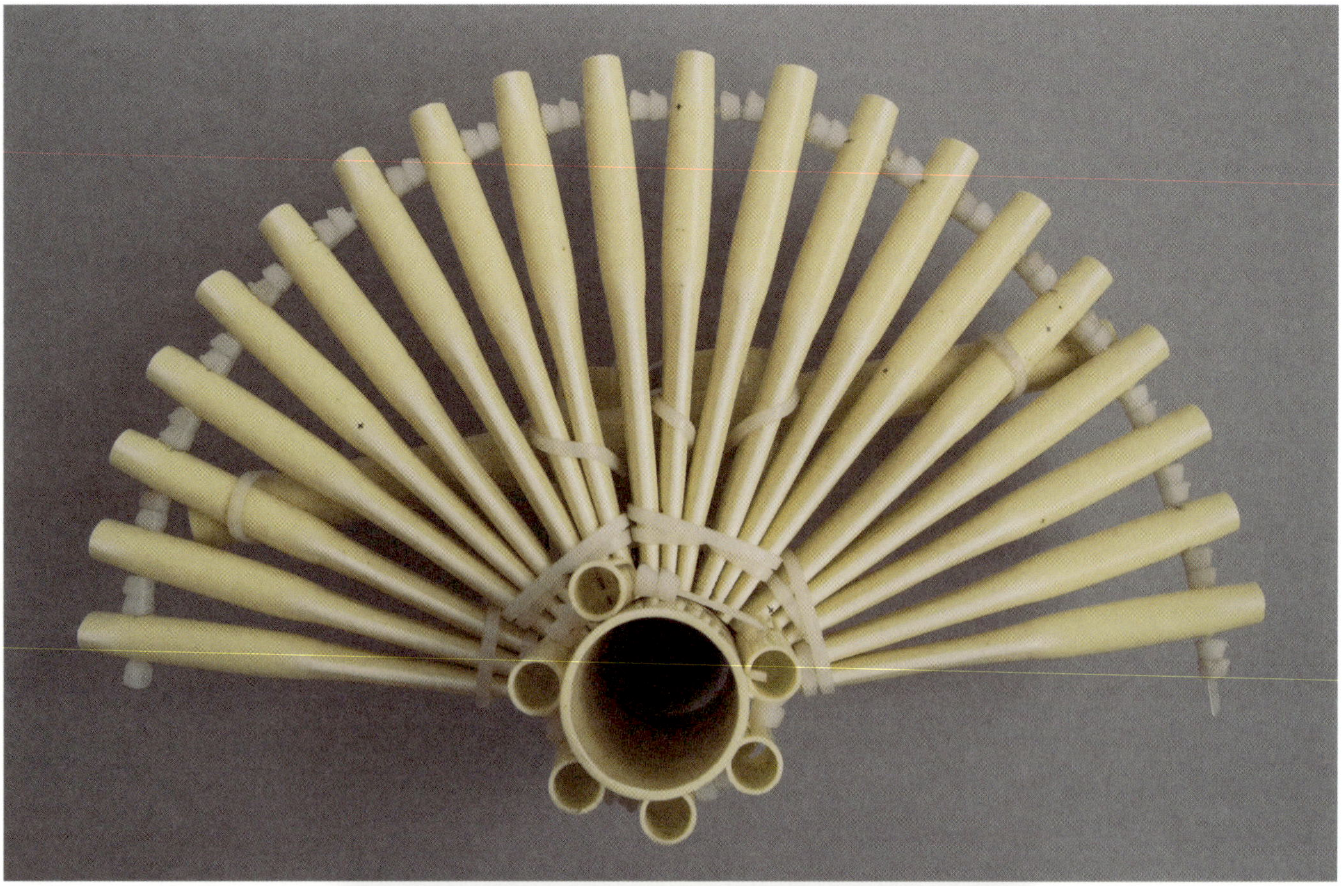

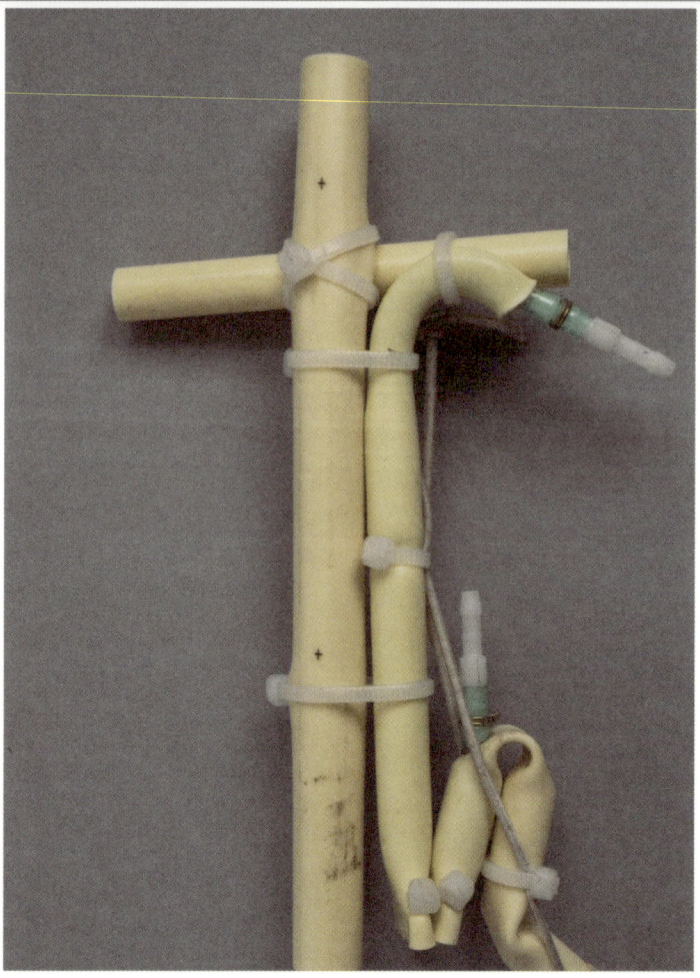

**fan from Animaris Circodentis,
end of feeler**

bodies over the past ten thousand years. Memes have been using us too. We are familiar with reproduction above and below the belt, both non-material and material reproduction. To my mind, the centre of activity is now at chest height and rising fast. The locus of reproduction is on its way to the top. The ratio of brain weight to genital weight (in both sexes) is rocketing. The emancipation of women can be seen as confirmation of this change. Women today want to make their own way in life and not just be tied down to looking after the kids. If this trend continues, we'll be largely ignoring physical reproduction a hundred years from now. It costs too much time and trouble. A pregnancy is exhausting and a birth is painful. All that old-fashioned stuff. Obviously children will still be born but far fewer than now. The number of offspring per person will dwindle to something like 0.5. We are now somewhere around one per person. We'll still go through the old-fashioned survival program but in a virtual world. In fact it's already happening. I mentioned Crocodile Dundee. Films take the place of real adventures in the jungle of earlier times, when our genes were still in the bodies of primitive man. A major perk of films is that you're not really exposed to danger and that you're in charge. The need for violence is deeply ingrained in us. Men in particularly are drawn towards fighting, shooting and clobbering. This must have had its uses in former times. On the other hand, resorting to violence can be a risky business. Another tendency is to distance yourself from violence, in other words, from the fear brought on by violence. You can enjoy adventures at home on the couch with a beer and a bag of crisps while the hero struts his stuff on TV. What you're doing in fact is rerunning age-old programs. The question is how people will rerun the programs on physical reproduction. Sex, as we know, is simulated to a large degree. The body below the belt has been the victim of deceit for centuries. The loins are easily hoodwinked. But what about pregnancies, births and above all upbringing? Every woman would like to be pregnant some time and give birth to a child. It's deep down in their program. All the same, pregnancy does have its disadvantages. Which is why I can see us having birth simulators. These are much like tampons that a woman can insert into her vagina. After an hour or two, the simulator swells and gives off a substance that causes labour pains. The swelling hurts a bit but not that much; this is a simulation, after all. At the moment of birth, out pops a package that inflates immediately like an air bag into a baby, nice and wet and warm. This is then laid on the woman's tummy. And that's that. The birth doesn't take long, five minutes at the

Shafts of air

Tubes enclose the air; the air sends the animals on their way; muscles and nerves draw in the air and blow it out. Air is the principal ingredient of the beach animals. They are in fact assembled from solid shafts of air protected by a layer of plastic.

I have made many animals from shafts of air without the protective plastic coating. It makes a world of difference weightwise but they do tend to blow away easily.

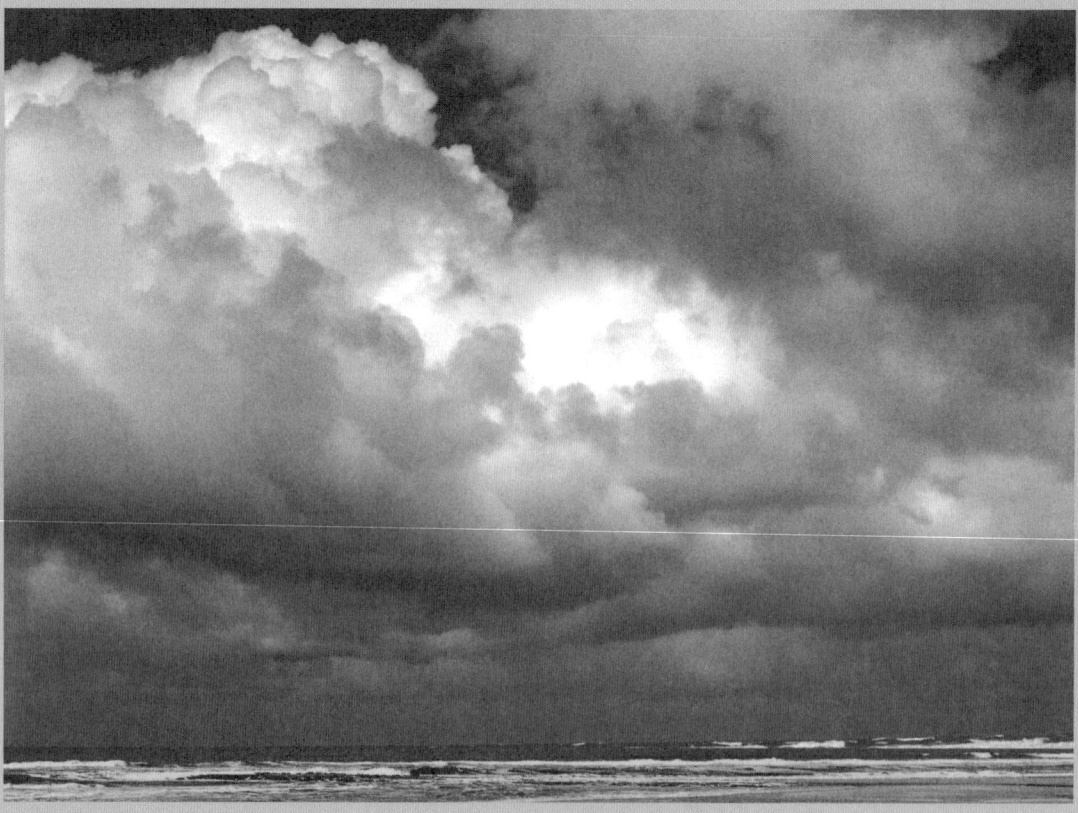
herd of twenty-two Animari Aerolae one second before they blew away

most. We might label it the ideal birth. It always ends well, no breech delivery, no caesarian and every baby looks great. These birth tampons will be on sale at every chemist's shop and subject to the commodities act. To be thrown in the dustbin after use. The body will relinquish its role. Today, handsome people have a better time of it than ugly ones. That's all going to change. Soon it won't be what you look like that matters, but how you're programmed. What your software looks like. It will be a sterile world. We won't feel at home there. We'll probably feel like those aboriginals in Sydney. The people of the future will be happy people. They will excel at rerunning their old programs. And so they should, since they are the product of millions of years of evolution, whether they like it or not. It will feel like being alive. The technology will let them experience everything as though it were real: virtual reality will get very close to reality in the years ahead. New programs will be added to the old ones. There will be new music, new art, new ideas. They will be the best of times.

Characteristics of the species

measurements: height x length x breadth

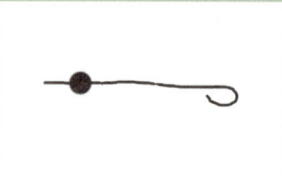

Lineamentum (line creature)
First life-form. Lived on the computer screen across which it moved as a straight rod. Able to reproduce (see Pregluton).
0 x 10 x 0.3 mm

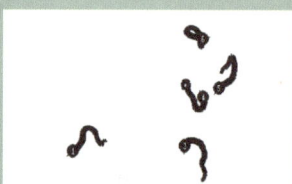

Vermiculus Artramentum (black worm)
Thick worm-like line creature that wriggled across the screen.
0 x 10 x 2 mm

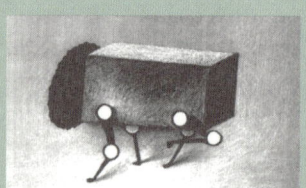

Quadrupes (four-legs)
Had two-part legs that could extend, bend and move backwards and forwards. Developed (through reproduction) a way of walking that resembled galloping.
0 x 20 x 30 mm

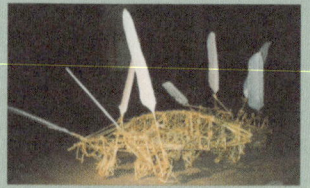

Animaris Vulgaris (regular beach animal)
The first beach animal. Had 16 legs which it could move only when lying on its back. Held together with adhesive tape.
0.6 x 2 x 2.5 m

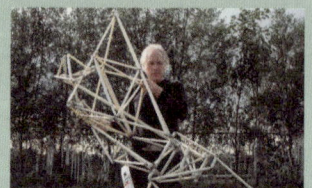

Animaris Derba (three-part-leg beach animal)
Only the leg is known. Its action corresponds to that of A. Vulgaris. The body must have been huge, judging from the leg's dimensions.
2 x 9 x 4 m

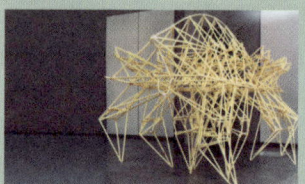

Animaris Currens Vulgaris (regular walking beach animal)
The first beach animal with a two-part leg. Walked with a jerky motion, pushing its heels into the sand.
1.4 x 2.1 x 2 m

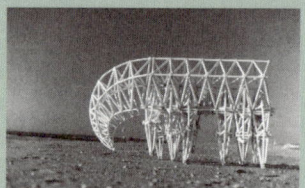

Animaris Speculator (scouting beach animal)
Attached to its mother by some kind of umbilical cord, the young Speculator was sent on ahead as a scout.
2.8 x 4.5 x 8 m (with a calf on each side)

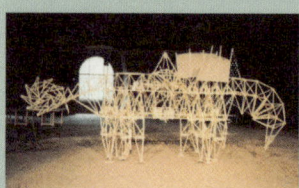

Animaris Arena Malleus (beach animal hammering in the sand)
Had a trunk that it could unfurl. A hammer on the end made attempts to drive a peg into the ground.
2.8 x 6 (with extended trunk and tail) x 2 m

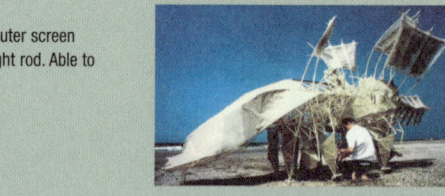

Animaris Sabulosa Adolescens (young sand-coated beach animal)
Driven by four fins at the rear. The legs were operated by a kind of gearbox. Was able to dig.
2.8 x 4.5 x 2 m

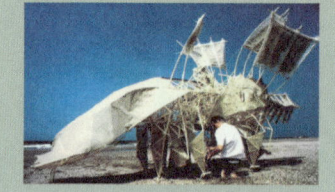

Animaris Sabulosa Cutis (sand-coated beach animal)
A spoiler at the front pushed its snout against the ground, fixing it there. Had a skin of see-through adhesive tape to which sand remained stuck as an unintentional camouflage.
2.8 x 5.3 x 2 m

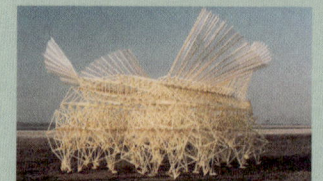

Animaris Currens Ventosa (beach animal running on wind)
Legs made in wooden moulds using a heat gun. Had 48 two-part legs. Two fins on its back rippled in the wind.
3.2 x 5.5 x 4 m; weight 160 kg

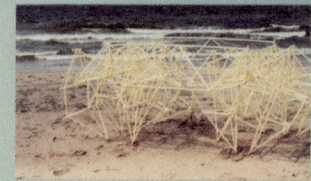

Animaris Filum (wired beach animal)
In A. Filum two diagonal wires replaced the diagonal tubes. Extremely lightweight but a poor walker. Moved forwards, not sideways like the other beach animals.
1.2 x 4 x 1.1 m

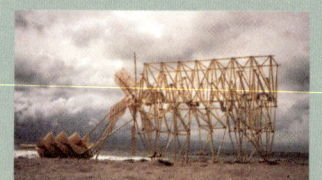

Animaris Rigide Ancora (sturdy anchored beach animal)
Had a propeller with a rolling anchor attached. This anchor caused Ancora to always face downwind.
1.6 x 3.5 x 2 m

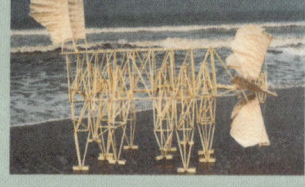

Animaris Rigide Properans (sturdy hasty beach animal)
Fast-walking propeller animal. Sturdy too, as the joints had been made under less-hot conditions than those of previous generations.
1.6 x 2.5 x 2 m

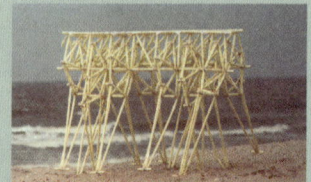

Animaris Propagare (reproducing beach animal)
Small mutant of A. Geneticus. Unable to walk, it could only stand. Living proof that tube lengths determine the animal's physical properties. The lengths are the beach animal's genes.
0.7 x 1 x 1 m

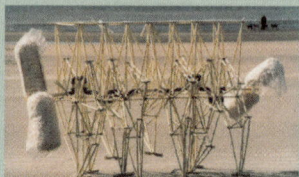

Animaris Geneticus (gene beach animal)
Had 357 genes, i.e. tubes. These tubes (and therefore the animal's genes) could be replaced. There has been a herd of seven Animari Genetici.
1.5-1.8 x 2.5 x 2 m

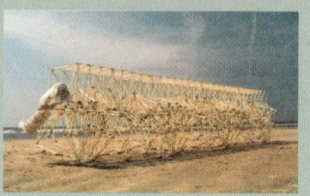

Animaris Geneticus Ondula (wave beast)
Large mutant of A. Geneticus. The cranks of the shaft were rotated to give a small difference in phase between legs (45 degrees) producing an undulating pattern.
1.7 x 8.5 x 2 m

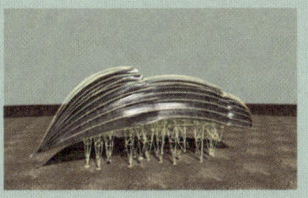

Animaris Gryllothalpa (virtual computer animal)
Has a genetic code of ones and zeros.
2.5 x 9.5 x 2 m

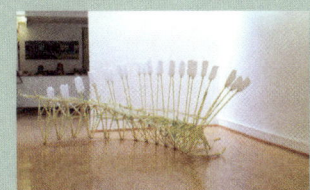

Animaris Vermiculus (worm animal)
Had 28 muscles, 14 nerves and some primitive brains for driving the muscles. 28 plastic drinking-water bottles acted as wind reservoirs.
1.6 x 5 x 0.8 m

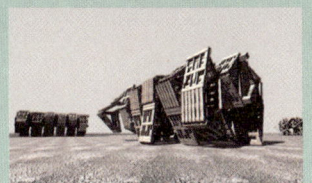

Animaris Rhinoceros Vulgaris (regular rhinoceros animal)
Had legs made of the sandwich elements of pallets. Could attain a great height due to the switch to wood.
2.8 (shoulder height) x 7.2 x 4 m

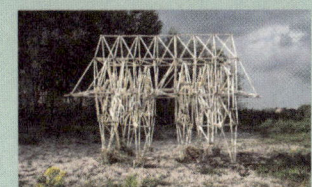

Animaris Currens Vaporis (walking steam animal)
The first animal with a crankshaft driven by pneumatic muscles. Makes a noise like a steam engine.
2 x 1.6 x 1.6 m

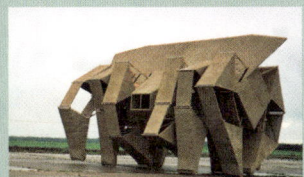

Animaris Rhinoceros Transport (transporting rhinoceros animal)
For travelling across the tundra. Can only move with the wind. A 'quiff' on its back was to catch the wind.
4.7 x 6 x 5 m (weight 3.2 tons)

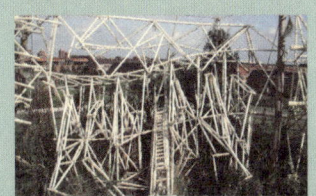

Animaris Circodentis
Animal driven by a cog ribbon and cogwheels.
2 x 2.5 x 2 m

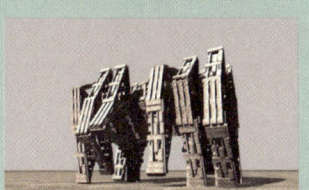

Animaris Rhinoceros Tabulae (plank rhinoceros animal)
An excellent wind-catcher due to its great height, with less friction at the joints due to its great size.
4.5 (shoulder height) x 6 x 4 m

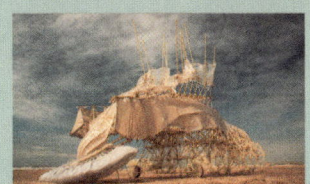

Animaris Percipiere Primus
First member of the Percipiere family. Wheels support the heavy wing structure.
3 x 10 x 2 m

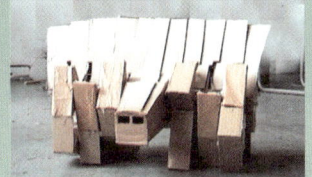

Spissa Carta (cardboard animal)
A miniature Rhinoceros Transport. Walked at almost 5 kph despite its smallish size.
0.3 x 0.4 x 0.4 m

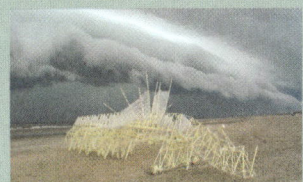

Animaris Percipiere Rectus
The first reasonably obedient beach animal. Lived for two years.
2 x 10 x 2 m

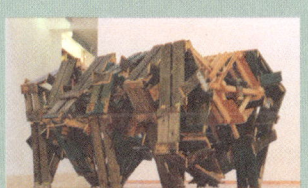

Animaris Lignatus (wood animal)
The first wooden creature to appear outside the computer. Has no hooves and therefore sinks easily into the sand.
1.6 x 2.4 x 2 m

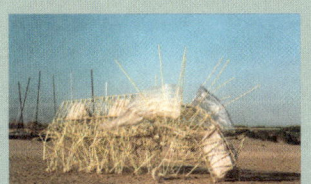

Animaris Percipiere Secundus
Rectus' naughty brother. Lived for one year.
2 x 10 x 2 m

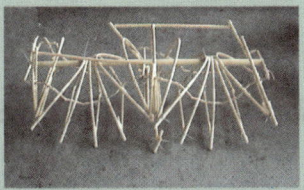

Animaris Rugosus Ondula (ripple wave beast)
The first caterpillar, it had no muscles. Was able to undulate with a little help.
0.6 x 1.2 x 0.4 m

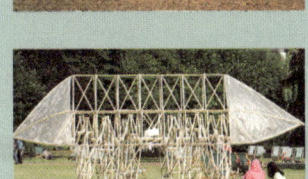

Animaris Ordis
Bastard form specially created for a BMW commercial to give a more varied image. An Animaris Secundus walking unit with sails on top.

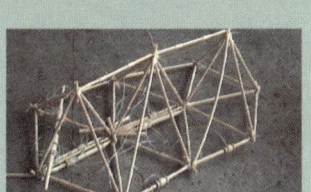

Animaris Vaporis (steam animal)
Makes a noise like a steam engine. The first animal to have muscles, four to be exact. One muscle when extended activates the next.
0.4 x 1.2 x 0.7 m

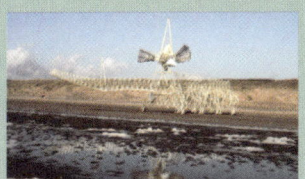

Animaris Excelsus
The tall one. Brains, wind-tank stomach, wing structure, hammer, everything was driven by the wheel mechanism.
4.6 x 10 x 2 m

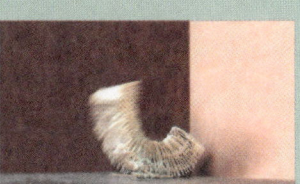

Animaris Rugosus Peristhaltis (peristaltic ripple-beast)
Made of bent tubes so that its back seems to ripple.
0.5 x 1.3 x 0.4 m

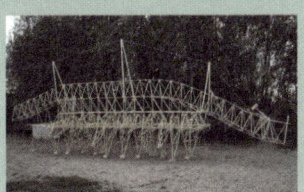

Animaris Modularius
Its weight was divided among the walking units by a crane-like structure.
2.6 x 10 x 2 m

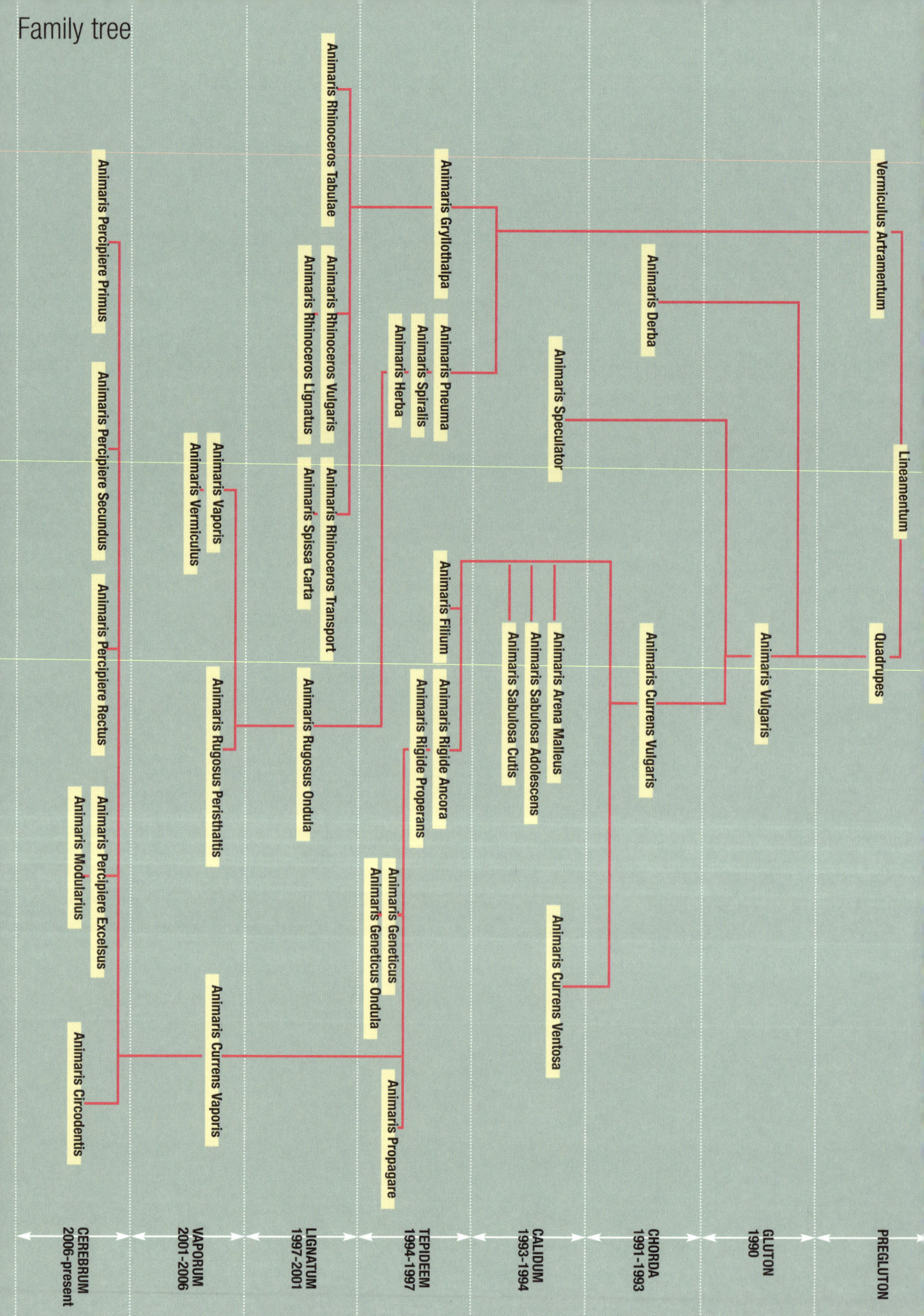
Family tree

BIOGRAPHY OF THEO JANSEN

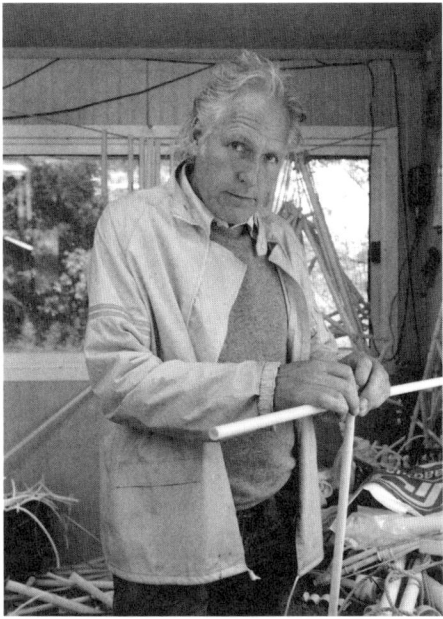

Born 1948 in Scheveningen, Netherlands

Studies physics at Delft University of Technology

1974
Exhibition at Galerie Lutz in Delft

1974-86
Works as student assistant at the Medical Faculty, University of Rotterdam

1975
Stops studying physics and becomes an artist

1975-1980
Paintings and drawings

1980
Flies UFO across Delft

1981
Flies UFO across Paris
Exhibits UFO in Lijnbaan Centrum, Rotterdam
Develops the painting-machine in Delft

1984-86
Writes columns for *Delta*, Delft University magazine

1986-present
Write fortnightly columns for *de Volkskrant* national daily newspaper

1986
Rotterdam Computerprint, 40 m long x 3 m high, Rotterdam Central Station

1987
'Signalementen', exhibition at Het Prinsenhof Museum, Delft

1988-2002
Teaches at the photography department, Royal Academy of Art (KABK), The Hague

1989
Shares exhibition at Het Prinsenhof with the Inventieven group. Makes large rocket, which was launched at the opening
Writes columns for Teleac ('TV university') course on mechanics
Drill machine project, The Hague
Light sculptures with an ultra-light plane, Delft

1990-present
Develops Animari (beach animals)
Engaged in designing a 'new nature'

1990
Gives a performance at Stroom (HCBK) in The Hague

1992
Exhibits at BMW pavilion, Korte Voorhout, The Hague
Takes part in 'Artificial Life' conference at TU Delft
Takes part in 'Wunderkammer' at Arti et Amicitiae, Amsterdam
Zogenaamd Ik, book published by Bzztoh, The Hague

1994
Writes columns for *Metropolis-M*
Exhibits at Studium Generale, Eindhoven
Exhibits at Galerie Akinci, Amsterdam
Awarded the Sandberg Prize by the City of Amsterdam
Shows video at 'From Perception to Action', ALife conference, Lausanne
Takes part in 'EUROPA 94' exhibition in Munich
'Mission Impossible' exhibition at Westergasfabriek, Amsterdam

1995
Exhibition at Stroom (HCBK), The Hague
Takes part in 'Babbage's Dream', CBK Groningen (computer art)
Magiorama '95, Martinihal, Groningen
Animaris Currens Ventosa purchased by Delft Municipality

1995-97
Animaris Currens Ventosa exhibited at the Faculty of Architecture, TU Delft

1996
Receives Max Reneman Prize
Klimmen in lucht, book published by SUN, Nijmegen

1997
One-man show at RO Theater, Rotterdam
Animaris Sabulosa exhibited at Artoteek, The Hague

1998
'Wilhelmina's Zomer', exhibition in Pakhuis on Wilhelminapier, Rotterdam
Animaris Sabulosa exhibited at KunstRAI
Exhibition at Galerie Maurits van de Laar

1998-2001
Takes part in 'Cafe de Wetenschap', VPRO Radio programme

1999
Awarded prize by Stichting Delftse Momenten
Exhibits Panorama 2000 at Centraal Museum, Utrecht
The Making of Panorama 2000, NPS film by Leon Giesen

2000
Artificial evolution of a herd of seven animals on the beach at IJmuiden
Takes part in 'Walking on Air', Zürich
Article in *The New Scientist* by Deborah MacKenzie

2003
Grand Exhibition, Kunsthal, Rotterdam
Ypenburg project. Construction of a laboratory on top of a sound barrier hill alongside the Rotterdam-The Hague motorway (A13). Includes a 50 metre wide sandpit and a bone-yard for deceased beach animals. With a portocabin as a workshop
Witteveen+Bosprijs Award voor Kunst en Techniek

2005-2006
Lectures in the US, several conferences

2006
Article in *Wired* by Lakshmi Sandha
Animals let loose in Trafalgar Square, London
Exhibition at Institute of Contemporary Art, The Mall, London
ETAT, exhibition in Taipeh, Taiwan
BMW commercial televised in South Africa

2007
Lectures at TED Conference, Monterey (CA), USA
Lectures at New York Academy of Sciences
'The Believers', exhibition at Museum of Contemporary Art, North Adams (Mass.), USA
BMW commercial televised in Germany
BMW commercial televised in Spain

2008
Souls and Machines Exhibition Reina Sofia Museum Madrid

2009
Grand Exhibition Hibya, Tokyo, Japan
Brussels: Prix Theo Jansen Award UN Environment Program
Grand Exhibition Garden5, Seoul, Korea
Exhibition Walcheturm, Zurich, Switzerland

For more information see:
www.strandbeest.com

This book was made possible by the generous support of the Netherlands Foundation for Visual Arts, Design and Architecture, Amsterdam, and Ernst & Young

Assistants

Cintha Bender, Anne Blair Gold Arton de Boer, Mascha Halberstad, Wim Ingenhoven, Ad Lagendijk, Hans Imthorn, Kees Jansen, Jeppe Kruiderink, Jouke Mellema, Jill van Orsouw, Allard Plaggenborg, Jaap Poot, Geert Kistemaker, Mark ter Mars, Margriet Vink, Architectenbureau Molenaaar en van Winden, Job Kneppers, Hubert Hoefsloot, Sander Hofstee, Christian Koks, Theo Zwanenveld, Coby Fritz, Toon Bongers, Quita Carlier, Divera Jansen, Zach Jansen, Iris Provoost, Wim Hoogerbrug

With thanks to Projectbureau Ypenburg and Stedebouwkundig Bureau Van Proosdij/Koster, The Hague. Special thanks to Loek van der Klis.

Photo credits

Judith Bender: pp. 10,33
Ingrid van Biljouw: p. 91
Arton de Boer: pp. 20, 70, 86-88, 106-107, 150-152, 226-228, 230
Guus Dubbelman: pp. 90, 110-111
Gottfried Junker: 239
Loek van der Klis: back cover, pp. 4, 26-27, 28, 32, 36, 38-39, 42, 60-61, 64, 74, 82, 94, 100, 112, 116, 118, 120, 126-127, 130, 134, 138-139, 140, 141, 146, 148, 153, 155, 168, 174-175, 180, 186-187, 188, 192, 200, 206, 208, 212, 214-215, 216, 220
Adriaan Kok: 98-99
Johannes Niemeijer: front cover, end papers, pp. 8, 12, 14, 18, 24, 34, 40, 46, 50, 54, 56, 58, 68, 78, 84, 92, 96, 104, 108, 124, 132, 136, 144, 154, 158, 160, 162, 166, 172, 176, 178, 182, 194, 196, 204, 218, 222, 234
Peter de Ruig: p. 190 (below)
Harry Verkuylen: pp. 164-165
Theo Jansen: other photos

Translated by John Kirkpatrick
Graphic design by Johannes Niemeijer
Printed by Die Keure, Brugge

© 2007 Theo Jansen and 010 Publishers, Rotterdam
Second edition, 2009
www.010.nl
Courtesy Galerie Akinci, Amsterdam

ISBN 978 90 6450 630 7

Translation of text in illustration on page 9.

Beach Roamers
Why are the dunes as high as they are?
Dunes are high because of the grains of sand that blow onto them, and they are low because of the grains of sand that blow off them. Just about as many grains blow on as off, so dunes stay the same height.
This in contrast with the level of the sea, which keeps on rising and threatens to reduce our national territory to what it was in medieval times. And we all know that the little piece this would leave wouldn't keep too many of us dry.
So the question is how to get more grains of sand onto the dunes.
There should really be animals on the beach that are permanently engaged in loosening sand in large quantities and throwing it into the air so that it can blow onto the dunes.
To make this a reality, I have devised some creatures that could come to influence the ecological balance on the beach, the way beavers do in the Biesbos wetlands. These creatures are assembled from yellow plastic tubing, satay skewers and tape and get their energy from the wind; so they don't need to eat. There are two types:
1) The transverse beach roller.
This 3.6 metre high creature has a head consisting of tentacles that stick into the ground. As the tail invariably catches more wind than the head, it always throws its head into the wind. The propeller at the rear causes a transverse rolling movement. As it rolls, the tentacles stab into the sand at an angle like ski sticks, thrusting the animal forwards. At the same time, the animal throws up sand in a circle.
If it rolls against something, the propeller keeps on turning. A worm wheel coupled to a lever that stretches a spring and so on, causes the propeller blades to suddenly swing ninety degrees on their long axis so that the animal rolls in the opposite direction. This keeps the transverse beach roller permanently on the move.
2) The dune digger.
This creature consists of one-metre lengths of plastic tubing set on the ground at an angle. A windmill at the top, kept in the wind by a long tail, makes the tubes move up and down, causing the animal to crawl. There is also a lateral arm with a foot attached. This foot shifts amounts of sand in a sideways direction and also helps to propel the animal along. Only this propulsion isn't uniform but cyclic; a mechanism causes the dune digger to walk fast and slowly by turns within an hourly cycle. This makes the animal crawl in a spiral, with each step shifting a little sand in the direction of the centre of the spiral. After a few days, a cone-shaped hill will have formed, and after several months, a mountain of sand.

This summer [1990] I shall take the time to make a pair of these animals. In the autumn I shall release them along the coast so that they can profit from the first autumn storms. Perhaps the Dutch coast will look quite different in a year's time.

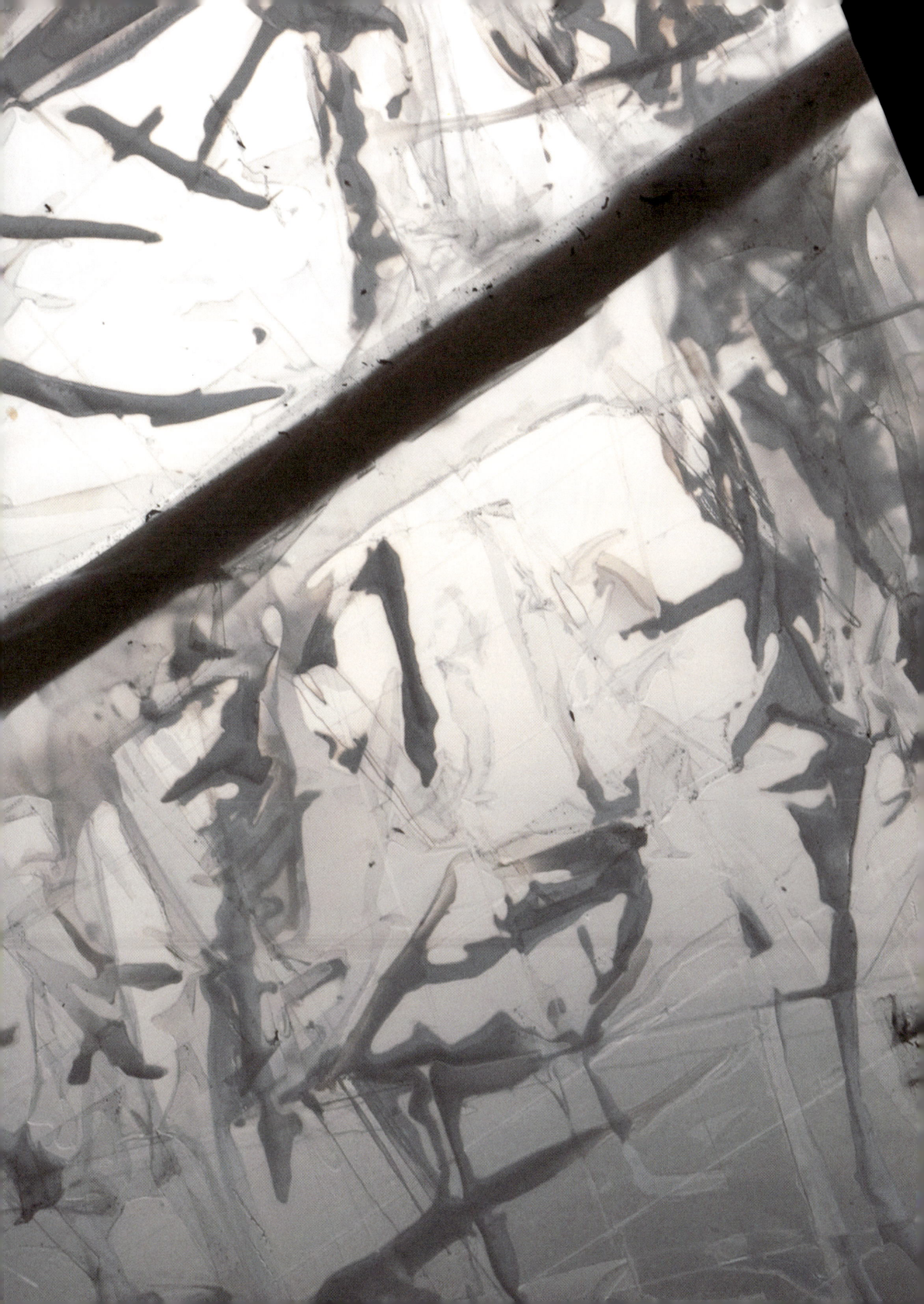

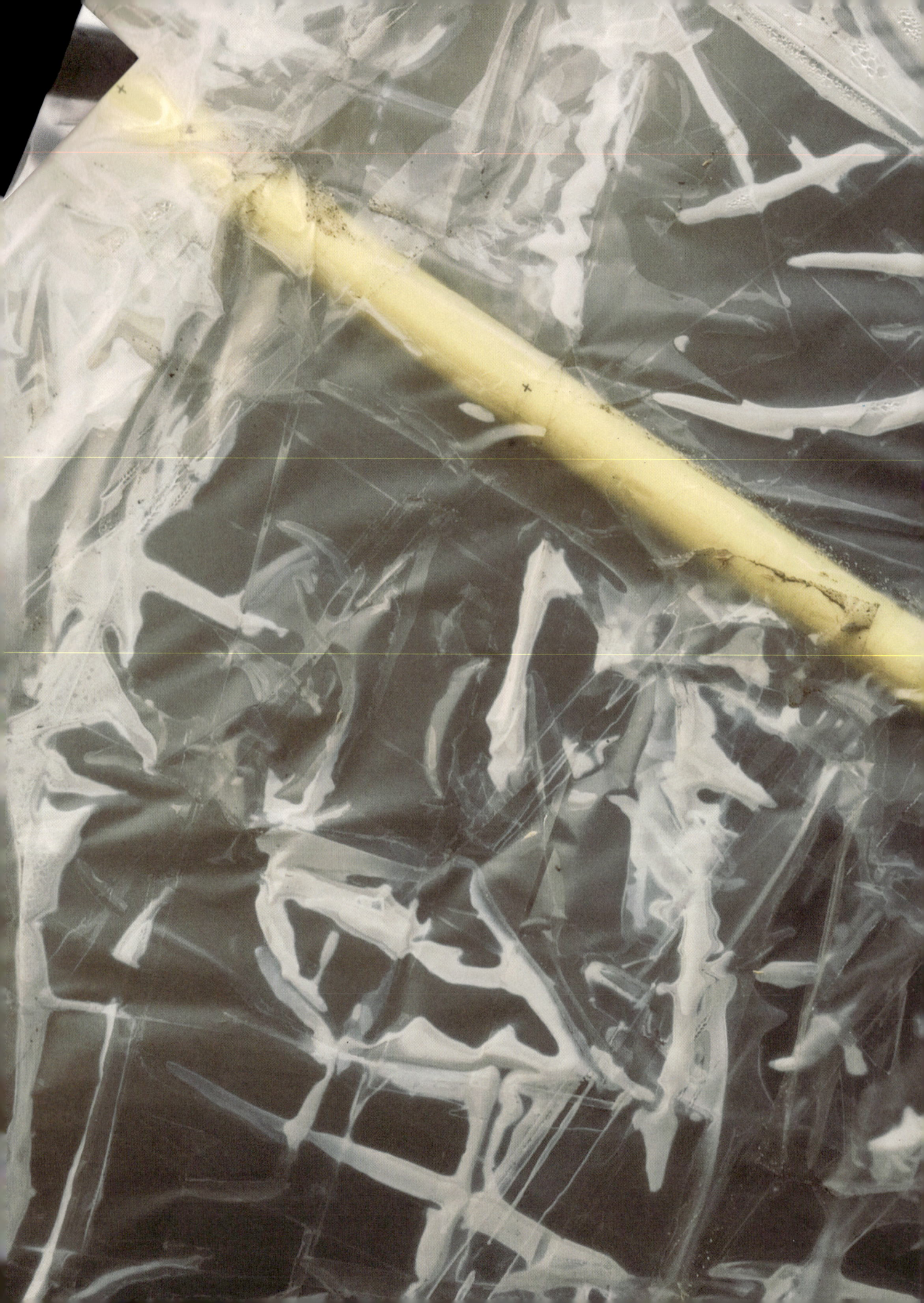